T LYMPHOCYTE SUBPOPULATIONS IN IMMUNOTOXICOLOGY

T LYMPHOCYTE SUBPOPULATIONS IN IMMUNOTOXICOLOGY

Edited by

Ian Kimber
ZENECA Pharmaceuticals, Central Toxicology Laboratory, Macclesfield, Cheshire, UK

and

MaryJane K. Selgrade
US Environmental Protection Agency, National Health and Environmental Effects Research Laboratory, NC, USA

JOHN WILEY & SONS
Chichester • New York • Weinheim • Brisbane • Singapore • Toronto

Baffins Lane, Chichester,
West Sussex PO19 1UD, England
National 01243 779777
International (+44) 1243 779777
e-mail (for orders and customer service enquiries):
cs-books@wiley.co.uk
Visit our Home Page on http://www.wiley.co.uk
or http://www.wiley.com

Other Wiley Editorial Offices

John Wiley & Sons, Inc., 605 Third Avenue,
New York, NY 10158-0012, USA

WILEY-VCH Verlag GmbH, Pappelallee 3,
D-69469 Weinheim, Germany

Jacaranda Wiley Ltd, 33 Park Road, Milton,
Queensland 4064, Australia

John Wiley & Sons (Asia) Pte Ltd, 2 Clementi Loop #02-01,
Jin Xing Distripark, Singapore 129809

John Wiley & Sons (Canada) Ltd, 22 Worcester Road,
Rexdale, Ontario M9W 1L1, Canada

Library of Congress Cataloging-in-Publication Data

T lymphocyte subpopulations in immunotoxicology / edited by Ian Kimber
and MaryJane K. Selgrade
p. cm.
Includes bibliographical references and index.
ISBN 0-471-97194-4 (hbk : alk. paper)
1. Immunotoxicology. 2. T cells. I. Kimber, Ian. II. Selgrade, MaryJane, K.
[DNLM: 1. T-Lymphocytes–immunology. 2. T-Lymphocyte Subsets–immunology.
3. Xenobiotics–immunology. 4. Immunotoxins–immunology. 5. Immunologic
Diseases–chemically induced. QW 568 T626 1998]
RC582. 17. T13 1998
616.07'97–dc21
DNLM/DLC
for Library of Congress 97–44163 CIP

British Library Cataloguing in Publication Data

A catalogue record for this book is available from the British Library
ISBN 0 471 97194 4
Typeset in 10/12pt Times by Keyword Publishing.
Printed and bound in Great Britain by Biddles Ltd, Guildford and King's Lynn.
This book is printed on acid-free paper responsibly manufactured from sustainable forestry, in which at least two trees are planted for each one used for paper production.

Contents

Contributors

Robert W. Baker, Department of Bacteriology, University College London Medical School, Windeyer Building, 46 Cleveland Street, London, W1P 6DB, UK

J.W. Coleman, Department of Pharmacology and Therapeutics, University of Liverpool, Liverpool L69 3BX, UK

Rebecca J. Dearman, Zeneca Central Toxicology Laboratory, Alderley Park, Macclesfield, Cheshire SK10 4TJ, UK

Matthew I. Gilmour, Center for Environmental Medicine and Lung Biology, University of North Carolina, Chapel Hill, NC 27599-7310, USA

Y. Heo, Wadsworth Center, New York State Department of Health, PO Box 509, Albany, NY 12201-0509, USA

Michael P. Holsapple, Health & Environmental Research Laboratories, The Dow Chemical Company, 1803 Building, Midland, MI 48674, USA

Norbert E. Kaminski, Department of Pharmacology & Toxicology, Michigan State University, B440 Life Sciences Building, East Lansing, MI 48824-1317, USA

Deborah J. Kasprowicz, Department of Cell Biology, Neurobiology and Anatomy, Loyola University Medical School, Stritch School of Medicine, 2160 First Avenue, Maywood, IL 60153, USA

David M. Kemeny, Department of Immunology, King's College School of Medicine and Dentistry, Bessemer Road, London, SE5 9PJ, UK

Nancy I. Kerkvliet, Department of Agricultural Chemistry and Environmental Health Sciences Center, Oregon State University, Corvallis, OR 97331, USA

Ian Kimber, Zeneca Central Toxicology Laboratory, Alderley Park, Macclesfield, Cheshire, SK10 4TJ, UK

J.A. Krieger, University of New Mexico Health Sciences Center, Cancer Research and Treatment Center, 900 Camino De Salud, Albuquerque, NM 87131, USA

Lakshmi Krishnan, Institute for Biological Sciences, National Research Council, 100 Sussex Drive, Room 3016, Ottawa, Ontario, K1A 0R6, Canada

David A. Lawrence, Wadsworth Center, New York State Department of Health, PO Box 509, Albany, NY 12201-0509, USA

B. Paige Lawrence, Department of Pharmacological Science, College of Pharmacy, Washington State University, Pullman, WA 99164, USA

Paul A. MacAry, Department of Immunology, King's College School of Medicine and Dentistry, Bessemer Road, London, SE5 9PJ, UK

Tim R. Mosmann, Department of Medical Microbiology & Immunology, University of Alberta, 632 Heritage Medical Research Center, Edmonton, Alberta, T6G 2S2, Canada

Artin Moussavi, Department of Immunology, King's College School of Medicine and Dentistry, Bessemer Road, London, SE5 9PJ, UK

Sahal Al Nakhli, Department of Bacteriology, University College London Medical School, Windeyer Building, 46 Cleveland Street, London, W1P 6DB, UK

Alistair Noble, Department of Immunology, King's College School of Medicine and Dentistry, Bessemer Road, London, SE5 9PJ, UK

David B.G. Oliveira, Division of Renal Medicine, St George's Hospital Medical School, Cranmer Terrace, Tooting, London, SW17 0RE, UK

Stephen B. Pruett, Department of Cellular Biology and Anatomy, Louisiana State University Medical Center, 1501 Kings Highway, Shreveport, LA 71130, USA

D. Susanne Ramer-Quinn, Department of Cell Biology, Neurobiology and Anatomy, Loyola University Medical School, Stritch School of Medicine, 2160 First Avenue, Maywood, IL 60153, USA

Graham A.W. Rook, Department of Bacteriology, University College London Medical School, Windeyer Building, 46 Cleveland Street, London, W1P 6DB, UK

Virginia M. Sanders, Department of Cell Biology, Neurobiology and Anatomy, Loyola University Medical School, Stritch School of Medicine, 2160 First Avenue, Maywood, IL 60153, USA

Mark Schuyler, VA Medical Center, 1501 San Pedro Drive SE, Albuquerque, NM 87108, USA

MaryJane K. Selgrade, Immunotoxicology Branch, National Health and Environmental Effects Research Laboratory, US Environmental Protection Agency, Research Triangle Park, NC 27711, USA

Stephen E. Ullrich, Department of Immunology, PO Box 178, University of Texas, MD Anderson Cancer Center, 1515 Holcombe Boulevard, Houston, TX 77030-4095, USA

Milica Vukmanovic-Stejic, Department of Immunology, King's College School of Medicine and Dentistry, Bessemer Road, London, SE5 9PJ, UK

Beejal Vyas, Department of Immunology, King's College School of Medicine and Dentistry, Bessemer Road, London, SE5 9PJ, UK

Julie Wilder, UNM School of Medicine, 2211 Lomas SE, Albuquerque, NM 87131, USA

Ali Zumla, Academic Infectious Disease Unit, University College London Medical School, Windeyer Building, 46 Cleveland Street, London, W1P 6DB, UK

1 T Lymphocyte Subpopulations in Immunotoxicology: An Introduction

Ian Kimber and MaryJane K. Selgrade

Immunotoxicology is a diverse discipline that encompasses the study of all adverse health effects resulting from the interaction of xenobiotics with the immune system. These interactions fall into two broad categories: 1) immune suppression, in which case the concern is increased risk of infectious or neoplastic disease, or 2) immune stimulation, resulting in allergic or autoimmune disease. This book deals with the role of T lymphocytes in the expression of these immunotoxic effects

During the last 2 decades a number of studies, conducted for the most part in laboratory rodents, has indicated that a variety of chemicals may impair one or more of the components of the immune system, resulting in immune suppression (Burns *et al*., 1996). The mechanisms associated with these effects have received much attention, and there have been attempts, many successful, to characterize in detail the way in which particular chemicals influence innate and adaptive immune function. In many (but not all) instances, T lymphocytes and/or the tissue responsible for their functional maturation (the thymus) have been implicated as the primary or secondary targets for the chemical insult. This comes as no surprise, as it is clear that T cells have a central role in adaptive immune responses.

T lymphocytes also play a pivotal role in the induction and elicitation of chemical and protein allergy and in the pathogenesis of autoimmune disease. Xenobiotics can contribute to allergic disease in several ways. In some cases, xenobiotics, in the form of either chemical haptens or proteins, can act directly as allergens stimulating specific allergic responses. In other cases, xenobiotics may modulate immune responses in a manner that enhances allergic responses to common allergens such as dust mite or animal dander. In some forms of allergic disease, most notably allergic contact dermatitis, a form of delayed-type hypersensitivity, T lymphocytes themselves represent the important effector population that induces directly the adverse reactions; in the case of

---

T Lymphocyte Subpopulations in Immunotoxicology. Edited by Ian Kimber and MaryJane K. Selgrade. Published in 1998 by John Wiley & Sons Ltd. ISBN 0 471 97194 4

contact hypersensitivity, this reaction being characterized by cutaneous inflammation. However, even in those instances where allergic reactions are effected by antibody, or by as yet undefined immunopathological processes, T lymphocytes still play a central role. Most antibody responses are dependent upon, and influenced by, specific T lymphocytes, and products of T lymphocytes (cytokines and other secreted immunoactive molecules) impact directly upon natural immune processes and inflammatory reactions.

An understanding of the impact that toxicants have on T lymphocytes is critical to understanding immunotoxic mechanisms. Immunotoxicants may have direct or indirect effects on T lymphocytes and their precursors that are subtle in nature and which pervert normal control mechanisms. It is apparent also that chemicals vary with respect to the type of reactions they will elicit and the potency of their effects. Awareness of the complexity of interactions between chemicals and the immune system has been coincident with a growing recognition of the complexity of T lymphocytes themselves. Classically, T lymphocytes have been divided into two main populations based on the expression of cell surface markers. These two populations have frequently been referred to as helper T cells ($CD4^+$) and cytotoxic or suppressor T cells ($CD8^+$), although these functions are not restricted exclusively to one or other population. It is now clear that these populations of T cells, once considered to be relatively homogeneous in nature, may display clear functional and phenotypic heterogeneity.

A seminal observation was made in 1986 by Mosmann and colleagues, who described the existence of heterogeneity among cloned murine T helper (Th) lymphocytes; the cells known to provide a necessary second signal for the production of antibody by B lymphocytes in response to most antigens (Mosmann *et al.*, 1986). Two populations, designated Th1 and Th2, were described that differed in their patterns of secreted cytokines. These results were confirmed and extended (Mosmann and Coffman, 1989) and evidence found for a similar, although not identical, heterogeneity among human (Romagnani, 1991) and rat (McKnight *et al.*, 1991) Th cell clones. These differential Th cell populations have also been observed *in vivo* (Bass *et al.*, 1989; Papp *et al.*, 1992). Functional subpopulations of $CD4^+$ Th cells are considered in detail in Chapter 2 of this volume. More recently, the existence of functional subpopulations of $CD8^+$ T lymphocytes has been described also, again characterized by cytokine expression profiles (Croft *et al.*, 1994; Chapter 3, this volume). Much has been learnt of the development and function of $CD4^+$, and to a lesser extent $CD8^+$, subpopulations; although many aspects of their biology have still to be resolved, including the important signals and microenvironmental conditions that drive adaptive immune responses in one or other the direction toward a phenotype of type 1 or type 2 cytokine production. While much remains to be clarified, there can be no doubt that the discovery of functional subsets of Th cells has had a profound influence on the way in which the immune response is viewed. At a simplistic level, it has focused attention on qualitative aspects of immune function, and the type of response induced, as being of equal or even greater importance than quantitative aspects.

With the benefit of hindsight, the evolutionary advantages conferred by an adaptable, as well as an adaptive, immune response are evident. A polarized Th1-type immune response, with the production among other cytokines of interferon γ (IFN-γ), favours cell-mediated immune responses, the differentiation of cytotoxic T lymphocytes, the activation of macrophages and the production of opsonizing and complement fixing antibodies (the latter facilitating opsonization and phagocytosis by neutrophil polymorphonuclear cells). Such responses will in turn confer resistance to, and facilitate resolution of, infection with intracellular pathogens and some extracellular bacteria. In contrast, the hallmark of polarized Th2-type responses is the production of cytokines such as interleukins 4 and 5 (IL-4 and IL-5). In such responses, there is the production of neutralizing IgG antibody, the production of IgE antibody, the activation of mast cells and the recruitment of eosinophils; conditions that provide protection (in general terms) against extracellular pathogens, including multicellular parasites (Abbas *et al.*, 1996). Naturally, this description is an oversimplification and in many instances elements of both Th1 and Th2 responses are required for the most effective elimination of pathogens and resolution of infection. Notwithstanding this, however, Th1 and Th2 cells provide an important insight into the ways in which the immune system can tailor responses to suit the antigenic challenge encountered.

Polarized Th responses also play discrete roles in immune-mediated pathology. As mentioned previously, Th1 responses are associated with delayed-type hypersensitivity reactions, including allergic contact dermatitis, and are also largely responsible for the pathology associated with organ specific autoimmune disease. Th2-like cells are prominent in immediate-type hypersensitivity responses to allergens and are implicated in the pathology associated with systemic autoimmune diseases.

As such, the Th1/Th2 paradigm has been of great value in immunobiology, and the importance of functional dichotomy among T lymphocyte populations has impacted on many areas of immunological uncertainty. Th cell heterogeneity has been implicated in the immunological paradox of pregnancy (where a semi-allogeneic fetus survives and thrives in the mother), in the phenomenon of neonatal tolerance, and in determining the form allergic responses may take (Forsthuber *et al.*, 1996; Romagnani, 1997). The fact that Th1 and Th2 cells are mutually antagonistic provides a plausible mechanism for the observation that some chemicals (e.g. certain metals) can elicit immunosuppressive, autoimmune, and allergenic effects (Kusaka, 1996; Lawrence and McCabe, 1995).

Useful as it is, like all paradigms, the Th1/Th2 cell dichotomy must be treated with caution. The exclusive existence of two distinct classes of differentiated Th cells (and of Tc cells) displaying characteristic cytokine secretion patterns does not prevail in most circumstances, with probably a minority, and sometimes only a small minority, of individual cells displaying the coordinate expression of cytokines that are classically associated with mature Th1 and Th2 phenotypes (Kelso, 1995). It is clear, however, that frequently the overall tenor of immune responses, particularly following prolonged antigenic stimulation, is predominantly of Th1 or Th2 type. If warnings of over-simplification (Allen and Maizels, 1997; Borish and Rosenwasser, 1997) are heeded, then further investigation of functional heterogeneity among T lymphocyte populations will continue to bear fruit.

The chapters contained within this book reveal that a consideration of the interaction of xenobiotics with the immune system in the context of functional diversity among T cells, and the existence of qualitatively polarized immune responses, has already borne impressive fruit in the analysis of immunotoxicity, allergy and autoimmunity. Contained within these pages are considerations of the basic immunobiology of $CD4^+$ Th cell (Chapter 2) and $CD8^+$ Tc cell (Chapter 3) subpopulations. Other chapters explore the role of T lymphocyte subsets in immunotoxicity caused by dioxin (Chapter 4), drugs of abuse (Chapter 5), heavy metals (Chapter 6) and ultraviolet radiation (Chapter 7). There follows a review of the use of Th cell clones in the assessment of immunotoxicity *in vitro* (Chapter 8). The interactions of the immune system with other physiological processes; the sympathetic nervous system (Chapter 9) and the endocrine system (Chapter 10) are discussed in the context of T lymphocyte subpopulations. The impact of T lymphocyte subsets on the characteristics of immune responses to chemical allergens (Chapter 11), on the pathogenesis of human allergic disease (Chapter 12) and on autoimmune responses to chemicals and drugs (Chapter 14) is considered. Finally, the modulation of T lymphocyte subset responses by air pollutants is reviewed (Chapter 13).

Disclaimer

The chapter has been reviewed by the National Health and Environmental Effects Research Laboratory, US Environmental Protection Agency and approved for publication. Approval does not signify that the contents necessarily reflect the views and policies of the Agency, nor does mention of trade names or commercial products constitute endorsement or recommendation for use.

References

Abbas, A.K., Murphy, K.M. and Sher, A. (1996) Functional diversity of helper T lymphocytes. *Nature*, **383**, 787–93.

Allen, J.E. and Maizels, R.M. (1997) Th1-Th2 : reliable paradigm or dangerous dogma. *Immunol. Today*, **18**, 387–92.

Bass, H., Mosmann, T. and Strober. S. (1989) Evidence for mouse Th1 and Th2-like helper T cells *in vivo*: selective reduction of Th1-like cells after total lymphoid irradiation. *J. Exp. Med.*, **170**, 1495–1511.

Borish, L. and Rossenwasser, L. (1997) T_{H1}/T_{H2} lymphocytes : doubt some more. *J. Allergy Clin. Immunol.*, **99**, 161–4.

Burns, L.A., Meade, B.J. and Munson, A.E. (1996) Toxic responses of the immune system, in *Casarett & Doull's Toxicology The Basic Science of Poisons*, 5th edn (eds C.D. Klaassen, M.O. Amdur and J. Doull), McGraw-Hill, New York, pp. 355–402.

Croft, M., Carter, L., Swain, S.L. and Dutton, R.W. (1994) Generation of polarized antigen-specific CD8 effector populations : reciprocal action of interleukin (IL)-4 and IL-12 in promoting type 2 versus type 1 cytokine profiles. *J. Exp. Med.*, **180**, 1715–28.

Forsthuber, T., Yip, H.C. and Lehman, P.V. (1996) Induction of T_{H1} and T_{H2} immunity in neonatal mice. *Science*, **271**, 1728–30.

Kelso, A. (1995) Th1 and Th2 subsets : paradigms lost? *Immunol. Today*, **16**, 374–9.

Kusaka, Y. (1996) Cobalt and nickel induced hard metal asthma, in *Toxicology of Metals* (eds L.W. Chang, L. Magos and T. Suzki), CRC Lewis, Boca Raton, pp. 461–8.

Lawrence, D.A. and McCabe, M.J. (1995) Immune modulation by toxic metals, in *Organ-Specific Metal Toxicology* (eds M. Waalkes, R. Goyer and C. Klaassen), Academic, New York, pp. 305–37.

McKnight, A.J., Barclay, A.J. and Mason, D.W. (1991) Molecular cloning of rat interleukin 4 cDNA and analysis of the cytokine repertoire of subsets of $CD4^+$ T cells. *Eur. J. Immunol.*, **21**, 1187–94.

Mosmann, T.R. and Coffman, R.L. (1989) Heterogeneity of cytokine secretion patterns and functions of helper T cells. *Adv. Immunol.*, **46**, 111–45.

Mosmann, T.R., Cherwinski, H., Bond, M.W., Giedlin, M.A. and Coffman, R.L. (1986) Two types of murine helper T cell clone. 1. Definition according to profiles of lymphokine activities and secreted proteins. *J. Immunol.*, **136**, 2348–57.

Papp, I., Wieder, K.J., Sablinski, T., O'Connell, J., Milford, E.L., Strom, T.B. and Kupiec-Weglinski, J.W. (1992) Evidence for functional heterogeneity of rat $CD4^+$ T cells *in vivo*: differential expression of IL-2 and IL-4 mRNA in recipients of cardiac allografts. *J. Immunol.*, **148**, 1308–14.

Romagnani, S. (1991) Human T_{H1} and T_{H2} subsets: doubt no more. *Immunol. Today*, **12**, 256–7.

Romagnani, S. (1997) The Th1/Th2 paradigm. *Immunol. Today*, **18**, 263–6.

2 Functional Subpopulations of CD4$^+$ T Lymphocytes

Lakshmi Krishnan and Tim R. Mosmann

Effector functions of the immune system are mediated by differentiated lymphocyte populations. B lymphocytes producing antibodies were first distinguished from T lymphocytes, which provide help for B cells (Mitchell and Miller, 1968). Surface marker expression allowed the separation of T lymphocytes into CD4$^+$ (helper) and CD8$^+$ (cytotoxic) cell populations. Activated T cells mediate their function primarily through the production of secretory molecules, cytokines. The heterogeneity of CD4$^+$ T cells based on cytokine secretion patterns was first described more than a decade ago (Mosmann *et al.*, 1986). This chapter will describe the functional diversity of the CD4$^+$ helper subsets; their differentiation pathways; factors influencing their differentiation; and the increasing knowledge of the importance of these subsets in immune regulation.

Functional subsets of CD4$^+$ T helper cells

In the early 1980s, the availability of purified proteins, monoclonal antibodies, and recombinant cDNA clones allowed characterization of a multitude of T cell cytokines. This aided the recognition of heterogeneous populations of T cells. Two distinct cytokine secretion phenotypes were first recognized among differentiated mouse T helper cell clones (Mosmann *et al.*, 1986). Th1 cells secrete interleukin 2 (IL-2), interferon γ (IFN-γ) and lymphotoxin (LT), whereas Th2 cells secrete IL-4, IL-5, IL-6, IL-9, IL-10 and IL-13 (Cherwinski *et al.*, 1987; Mosmann and Sad, 1996). Human T cells also differentiate into similar cytokine-secreting phenotypes, although the synthesis of IL-10 is not as tightly restricted to a single subset as with mouse T cells (Del Prete *et al.*, 1993). Some cytokines such as IL-3, tumour necrosis factor-alpha

T Lymphocyte Subpopulations in Immunotoxicology. Edited by Ian Kimber and MaryJane K. Selgrade.

(TNF-α), granulocyte–macrophage colony stimulating factor (GM-CSF) and chemokines are secreted by both Th1 and Th2 cells (Cherwinski *et al.*, 1987).

Immune responses are often dichotomous, as cytokines secreted by Th1 and Th2 cells mutually inhibit the growth, proliferation and differentiation of the reciprocal cell type. IFN-γ produced by Th1 cells selectively inhibits the proliferation of Th2 cells (Fernandez-Botran *et al.*, 1988; Gajewski and Fitch, 1988), whereas IL-10 produced by Th2 cells suppresses cytokine synthesis of Th1 cells (Fiorentino *et al.*, 1989; Moore *et al.*, 1990). IL-10 inhibits macrophage APC/accessory function by down regulating expression of IL-12 (D'Andrea *et al.*, 1993) and costimulatory molecule B7 (Ding *et al.*, 1993). Reduction in IL-12 production inhibits the induction of Th1 development from naive T cells, whereas lack of co-operative interaction between costimulatory molecule CD28/B7 and IL-12 down regulates IFN-γ production by committed Th1 cells. IL-4 may also be involved in cross-regulating Th1 responses, as IL-4 reduces IL-2 production by $CD4^+$ and $CD8^+$ T cells (Sad *et al.*, 1995; Tanaka *et al.*, 1996).

Recently it has been suggested that Th1 and Th2 may not be the only cytokine phenotypes of $CD4^+$ T cells. Th0 cells expressing both the Th1 as well as Th2 cytokine patterns have been described (Firestein *et al.*, 1989; Street *et al.*, 1990). Cells producing high amounts of TGF-β named Th3 were isolated from the gut of mice orally tolerized to the experimental autoimmune encephalomyelitis (EAE)-inducing antigen, myelin basic protein (Chen *et al.*, 1994). Among long-term T cell clones, several other phenotypes have also been noted. Long-term Th1 clones often lose their ability to produce IL-2, while the synthesis of other cytokine seems to be more stable (Mosmann and Coffman, 1989). It is still unclear whether some of the alternative phenotypes are aberrations occurring in tissue culture or transient developmental phenotypes along a differentiation pathway. At the single cell level, among normal short-term T cell populations, IL-4 and IL-5 are often co-expressed, whereas IL-4 and IFN-γ are reciprocally expressed (Assenmacher *et al.*, 1994; Elson *et al.*, 1995). Interestingly, the Th1 cytokine IFN-γ and the Th2 cytokine IL-10 are co-expressed in some cells. Regardless of the variation in T-cell cytokine synthesis, it is clear that many T-cell clones and *in vivo* responses show a dramatic dichotomy towards either Th1 or Th2 and these remain the primary functional divisions of $CD4^+$ T helper cells.

Functions of Th1 and Th2 cells

The dichotomous cytokine profile of Th1 and Th2 cells also confers distinct functional capabilities on these cell types. Th1 cells are primarily involved in cell-mediated inflammatory reactions, whereas Th2 cells aid the humoral and allergic arms of the immune response.

HUMORAL IMMUNITY

Th2 cells are excellent helpers of B cells both in antigen-specific (Stevens *et al.*, 1988) and polyclonal systems (Coffman *et al.*, 1988). On encounter with a naive resting B cell, a Th2 clone induces activation, growth and antibody production by the B cell (Boom *et al.*, 1988). While activation of the naive B cell requires cell contact with the Th cell, optimum differentiation and proliferation occurs in the presence of IL-4 and IL-5 produced by the Th2 cell (Coffman *et al.*, 1988). Addition of neutralizing anti-IL-5 antibody to the coculture of B and Th2 cells substantially inhibits immunoglobulin (Ig) production (Boom *et al.*, 1988; Rasmussen *et al.*, 1988). The addition of anti-IL-4 has a similar effect, although to a lesser and more variable extent (Boom *et al.*, 1988; Coffman *et al.*, 1988; Killar *et al.*, 1987). The combination of both anti-IL-4 and anti-IL-5 totally inhibits antibody production (Coffman *et al.*, 1988). Activated B cells do not require Th-B cell contact, and can proliferate and produce antibodies in response to even Th2 supernatants or IL-5 (Herron *et al.*, 1988; Rasmussen *et al.*, 1988; Mosmann and Coffman, 1989). Thus, IL-4 and IL-5 are the major 'helper' proteins produced by mouse Th2 cells. Synthesis of Ig production by human B cells is stimulated by a wider range of cytokines, including IL-4, IL-2, IL-6, IL-10 and IL-13 (Banchereau and Rousset, 1992).

The helper function of the Th1 cell is more complex. The assessment of Th1 help for B-cells is complicated by two observations; IFN-γ at high concentrations is inhibitory, whereas at low concentrations it promotes B cell responses (Snapper and Paul, 1987); and Th1 clones can be directly cytotoxic for activated B cells (Janeway *et al.*, 1988; Tite and Janeway, 1984). However, Th1 cells retain their ability to activate and promote differentiation of B cells if sufficient IL-2 is present and the effects of IFN-γ are neutralized, or if IL-4 and IL-5 are added to Th1-induced B cell cultures (Coffman *et al.*, 1988). Interestingly, the helper activity of Th1 clones appears to dominate at low antigen concentrations or at low T:B ratios, whereas cytotoxicity activity begins to dominate at high ratios (Del Prete *et al.*, 1991b).

One important difference between the helper activity of Th1 and Th2 cells is their ability to stimulate the production of specific Ig isotypes. Th2 cells, unlike Th1 cells, induce significant IgE production and are associated with allergic responses (Mosmann and Coffman, 1989). IL-4 is the specific switch factor for mouse B cell IgE production, while human B cells switch to IgE in response to either IL-4 or IL-13 (Coffman and Carty, 1986; Del Prete *et al.*, 1988; Punnonen *et al.*, 1993). IgE production stimulated by Th2 cells can be substantially blocked by either IFN-γ or anti-IL-4 (Coffman and Carty, 1986). Consequently, Th1 clones can induce IgE production if IL-4 and anti-IFN-γ are added to the *in vitro* cultures of B and T cells (Coffman *et al.*, 1988). IL-4 produced by Th2 cells also promotes production of IgG1 (mouse) and IgG4 (human), although some mouse Th1 cells induce IgG1 production even in the absence of IL-4 (Coffman *et al.*, 1988; Banchereau and Rousset, 1992). Th1 clones, in contrast, induce substantially more IgG2a than Th2, as IFNγ induces B cell switching towards this isotype (Snapper and Paul, 1987).

CELL-MEDIATED IMMUNITY

Th1 cells are strong inducers of delayed type hypersensitivity (DTH), a cell-mediated inflammatory reaction. In a system where T cells and antigen were directly injected into mouse footpads, Th1 but not Th2 cells induced DTH (Cher and Mosmann, 1987). IFN-γ is commonly expressed at sites of DTH (Tsicopoulos *et al.*, 1992) and is chemotactic for at least some of the cells that migrate into the inflammatory site causing DTH swelling (Issekutz *et al.*, 1988). However, DTH is a complex reaction, and recently it has been observed that $CD8^+$ T cells producing Th2-pattern of cytokines can induce substantial DTH (Li *et al.*, 1997). Thus, while Th1 cells are able to activate this process efficiently, the reason for the inability of Th2 cells to evoke this response remains unclear. If Th2 cells are preactivated before injection into mouse footpads, an inflammatory reaction can be induced (Müller *et al.*, 1993). In a model in which granulomas are induced by proteins coupled to beads, both Th1- and Th2-inducing proteins induce granulomas, with different cytokine patterns and regulation (Chensue *et al.*, 1995), again suggesting that the association of inflammatory reactions only with Th1 responses is an oversimplification.

Th1 cytokines also influence the activation of cytotoxic and inflammatory cell types. IL-2 and IFN-γ can enhance the generation and/or activation of mature cytotoxic T cells (Erard *et al.*, 1985; Simon *et al.*, 1986). However, IL-4 and IL-5 have also been implicated for similar functions (Mosmann and Coffman, 1989; Pfeifer *et al.*, 1987; Takatsu *et al.*, 1987). On the other hand, IFN-γ and TNF can activate macrophage functions (Belosevic *et al.*, 1988), whereas IL-4 and IL-10 inhibit this activation (Bogdan and Nathan, 1993; Mosmann, 1994).

Th2 cells are associated with eosinophilia. IL-5 appears to be the main eosinophil-stimulating Th2 cytokine. IL-5 can promote the growth and differentiation of eosinophils *in vitro* (Sanderson *et al.*, 1986), and administration of exogenous IL-5 *in vivo* induces eosinophil recruitment (Anderson and Coyle, 1994). Transgenic mice overexpressing the IL-5 gene develop eosinophilia of the peripheral blood, bone marrow and tissues (Dent *et al.*, 1990). Furthermore, administration of anti-IL5 antibody can inhibit allergen-induced eosinophilia in sensitized animals (Akutsu *et al.*, 1995).

Overall, Th1 cells are often associated with cytotoxic and inflammatory reactions, whereas Th2 cells promote B cell help, enhance eosinophil production and are associated with antibody responses and allergy. The cross-regulatory nature of Th1 and Th2 cells may in part explain the well documented reciprocality *in vivo* between antibody responses and DTH (Mosmann and Coffman, 1989) (Figure 2.1).

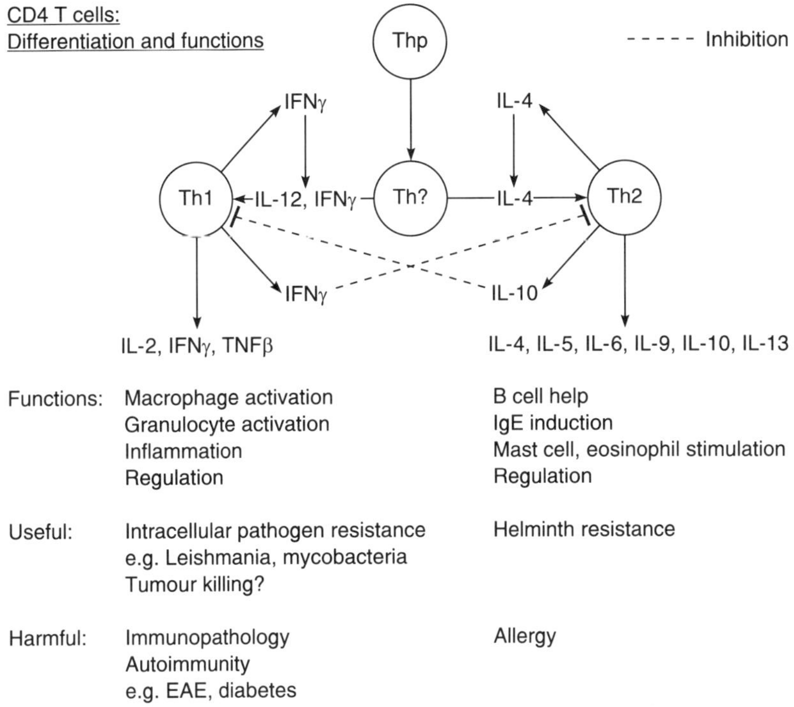

Figure 2.1 CD4 T cell differentiation and functions. This figure shows the positive feedback of Th1 and Th2 cytokines on the differentiation of similar cells, and the negative cross-regulation between the two phenotypes. Note that this diagram shows only differentiation to the two extreme T cell phenotypes, Th1 and Th2.

Differentiation of Th1 and Th2 cells from a common precursor

Th1 and Th2 cells represent phenotypes of differentiated effector T cell populations. On the other hand, naive T cells encountering antigen on an antigen-presenting cell (APC) for the first time produce mainly IL-2 (Swain *et al.*, 1990a). Th1 and Th2 cells can be derived from this bi-potential precursor cell (Thp) that secretes only IL-2 (Sad and Mosmann, 1994). With the addition of both TGF-β and anti-IFN-γ antibody, naive $CD4^+$ T cells can be maintained in a proliferating but uncommitted Thp state. These replicating Thp cells exhibit surface expression of memory markers ($CD44^{hi}$, $CD45RB^{lo}$ and L-selectin^{lo}). Stimulation of these cells with appropriate cytokines allows differentiation into either Th1 or Th2 phenotypes.

The differentiation of Th1 and Th2 cells may proceed through an intermediate Th0-like phenotype expressing both Th1 and Th2 cytokine patterns. This possibility was elegantly highlighted in a transgenic mouse model, in which IL-4 promoter drives the expression of herpes simplex virus 1 thymidine kinase gene (HSV-TK). Gancyclovir, which is toxic to cells expressing HSV-TK, resulted in lineage ablation of both Th1 and Th2 cells, strongly suggesting that IL-4 gene expression occurs during differentiation of both subtypes (Kamogawa *et al.*, 1993). However, it is not clear whether the HSV-TK gene mimics in totality the expression of the endogenous IL-4 gene; or whether m-RNA expression ensures protein production. The possible production of IFN-γ along the Th2 differentiation pathway also needs to be investigated. Studies on cytokine secretion during polarized differentiation of CD4 T cells suggest that secretion of 'inappropriate' cytokines (IL-4 on the Th1 pathway, IFN-γ on the Th2 pathway) occurs at low levels, if at all (P. Akai and T.R. Mosmann, unpublished).

Factors influencing Th1 and Th2 differentiation

Cytokines are the major players that influence the differentiation of naive T cell populations into distinct functional phenotypes. Presence of IL-4 during initial stimulation of naive T cells results in differentiation towards Th2 cells, whereas IFN-γ, IL-12 and TGF-β induce a Th1 phenotype (Seder and Paul, 1994) (Figure 2.1)

The ability of IL-4 to induce a Th2 phenotype was first described in the murine system. When T cells were stimulated with either mitogen or plate bound anti-CD3 in the presence of IL-4, and then restimulated, differentiation to a Th2 but not Th1 phenotype occurred (Swain *et al.*, 1990b). Human peripheral blood T cells stimulated with purified protein derivative (PPD) antigen and IL-4 also produce increasing levels of IL-4 on re-stimulation (Maggi *et al.*, 1992). However, in this case, IFN-γ production was not totally inhibited. One inherent problem with mitogen-induced stimulation of T cells is that it may activate both memory as well as naive T cell populations. To address the factors influencing primary differentiation of T cells, it is essential to focus on naive resting T cell populations. Using an antigen-specific $\alpha\beta$-T-cell receptor transgenic mouse as a more defined source of naive CD4$^+$ T cells, several groups have demonstrated the ability of IL-4 to induce Th2 differentiation. When naive CD4$^+$ T cells derived from these mice are stimulated with antigen and APC in the presence of IL-4, they differentiate into T cells secreting IL-4 but not IFN-γ and IL-2 upon re-stimulation with antigen (Hsieh *et al.*, 1992; Seder *et al.*, 1992; Tanaka *et al.*, 1993). Furthermore, this effect can be accomplished only by IL-4 and cannot so far be reproduced with any other accessory signal/cytokine or APC (O'Garra and Murphy, 1994). The importance of the presence of IL-4 during priming in allowing Th2 differentiation is also highlighted in a number of infection models. For example, *in vivo* neutralization of IL-4 inhibits the aberrant Th2 response against *Leishmania major* in susceptible strains of mice, resulting in containment of infection (Sadick *et al.*, 1990). In mutant Balb/c mice with a disrupted IL-4 gene, two studies showed that *L. major* infection was partially contained, in contrast to the usual progressive infection (Kopf *et al.*, 1996; Noben-Trauth *et al.*, 1996). Although only one of the studies interpreted their findings in this way, both studies showed levelling off of the footpad swelling, as well as reduced parasite loads. CD4^{+} T cells from IL-4-deficient mice also made reduced levels of Th2 cytokines, IL-5, IL-9 and IL-10 after a nematode infection (Kopf *et al.*, 1993). Thus, the presence of IL-4 appears to be critical for directing Th2 differentiation.

A dominant factor influencing the differentiation of Th1 cells appears to be IL-12 (Trinchieri, 1993). IL-12 is produced by multiple cell types, including monocytes, macrophages and B cells, and enhances IFN-γ production by natural killer (NK) and T cells (Trinchieri, 1995). IL-12 can also directly induce the differentiation of Th1 cells from an uncommitted T cell (Seder *et al.*, 1993). However, IL-4 in itself, appears to inhibit the ability of IL-12 to promote Th1 development (Seder *et al.*, 1992), possibly by providing a more powerful stimulus for Th2 differentiation. Interestingly, in some systems IL-12-independent development of Th1 cells has been noted (Gollob and Coffman, 1994; Hsieh *et al.*, 1992; Seder *et al.*, 1992). The role of IFN-γ in inducing Th1 differentiation is still unclear. Neutralization of IFN-γ ablates cells secreting high levels of IFN-γ (O'Garra and Murphy, 1994), suggesting that the IFN-γ augments Th1 cell expansion and may be required for Th1 differentiation. However, in some systems addition of IFN-γ in itself is not sufficient to drive Th1 differentiation and other cytokines may be required (Seder *et al.*, 1993). Studies in a transgenic mouse model suggest that naive T cells but not memory T cells are totally dependent on IFN-γ for Th1 differentiation (O'Garra and Murphy, 1994). TGF-β is another cytokine that favours the development of Th1 reponses (Sad and Mosmann, 1994; Swain *et al.*, 1991). One distinction that needs to be made is that cytokines can influence both primary differentiation as well as effector functions of T cells, and sometimes the effects of the same cytokine on these two processes may be opposing. For example, TGF-β promotes Th1 differentiation, but can also inhibit Th1 effector function (Meade *et al.*, 1992).

In a primary response, antigen-specific T cells are present at a very low frequency, and the initial source of the cytokines that govern T cell differentiation, may include other cell types. IL-12, the key Th1-inducing cytokine, is primarily produced by other cells such as macrophages (Trinchieri, 1995). In *L. major* infection, NK cells are the primary source of the early burst of IFN-γ that promotes a curative Th1 differentiation (Scharton and Scott, 1993). The initial source of IL-4 may include mast cells and basophils (Ben-Sasson *et al.*, 1990; Brown *et al.*, 1987). Recently a subpopulation of CD3^{+}CD4^{+}NK1.1^{+} T cell has been described in mice, that can also produce large amounts of IL-4 within a few hours of polyclonal activation (Yoshimoto and Paul, 1994).

Initially it was speculated that different types of APCs may influence distinct patterns of T cell differentiation. However, it is now known that dendritic cells, macrophages and B cells are all potentially capable of promoting both Th1 and

Th2 differentiation, if the appropriate cytokines are present. However, *in vivo*, APCs, by virtue of their distinct cytokine secreting ability, may selectively influence differentiation towards either Th1 or Th2. For example, IL-12 and IL-10 are both produced by macrophages and IL-10 can inhibit IL-12 production (D'Andrea *et al.*, 1993). Thus, macrophages can either augment or inhibit Th1 differentiation depending on relative amounts of IL-12 and IL-10 produced by these cells. Recently, it has been suggested that costimulatory molecules on APCs may also decisively influence T cell differentiation. Antibodies against B7.1 inhibit development of Th1 responses, whereas anti-B7.2 antibody has the reciprocal effect of suppressing Th2 responses (Kuchroo *et al.*, 1995). However, this effect appears to be complicated, as both B7.1 and B7.2 can induce Th1 and Th2 differentiation and their relative effects are modulated by antigen concentration.

In spite of the well-characterized effects of cytokines on T cell differentiation, it is unclear how the immune system chooses the appropriate response to a given infection. This decision is critical to ensure survival of the host. It can then be envisaged that an intricate network of factors, driven in part by the nature of the invading pathogen, would result in a multitude of cytokine and costimulatory signals that appropriately guide the immune response. Some pathogen products themselves induce appropriate responses. For example, proteinases of multicellular parasites that aid tissue penetration may induce an appropriate Th2 response (Urban *et al.*, 1992). A *Listeria monocytogenes* product induces IL-12 synthesis, thus initiating Th1 differentiation (Hsieh *et al.*, 1993). Antigen dose may be another key regulator of the choice between the Th1 and Th2 differentiation pathways. Early studies showed that both low and high antigen amounts primed for DTH, whereas intermediate antigen doses promoted antibody production (Parish, 1972). In *L. major* infection, a healing phenotype (Th1-like cell-mediated response) can be seen in normally susceptible BALB/c mice, if low infecting doses of the parasite are used (Bretscher *et al.*, 1992). Similarly, a switch to a Th1 response occurs after infection with low numbers of *Trichuris muris* (Bancroft *et al.*, 1994), in this case causing persistence of the parasite in otherwise resistant mice. Thus, *in vivo* experiments indicate a DTH–antibody–DTH progression in correlation to increasing antigen concentration. Recent *in vitro* experiments, however, suggest the opposite; using a wide range of antigen concentrations Th2–Th1–Th2 shifts could be achieved while moving from very low to very high antigen concentrations (Constant *et al.*, 1995; Hosken *et al.*, 1995). The difficulty in interpreting these results may be due to the lack of distinction between effect of antigen dose on primary differentiation and regulation of mature effectors. Differentiated Th1 cells are stimulated by moderate but not high antigen dose, whereas Th2 cells remain responsive to a wide range of antigen concentrations (Fitch *et al.*, 1993).

Stability of differentiated Th1 and Th2 cells

A key question in the regulation of T helper cell development is whether cells that have apparently differentiated into polarized Th1 and Th2 subsets can be influenced to change their cytokine production pattern. Cytokine production by differentiated Th1 and Th2 cells appears to be stable, in that the progeny of one subset have not been observed to convert their cytokine patterns to those of the opposite subset (Mosmann and Coffman, 1989). It has, however, been suggested that Th0-like effector cells can be influenced alternatively by IL-4 and IL-12 to differentiate into Th2 and Th1 cells, respectively (Abehsira-Amar *et al.*, 1992; Manetti *et al.*, 1994). Similarly, memory effectors obtained from unimmunized mice may be regulated in either direction, at the time of primary stimulation (Gollob and Coffman, 1994). In experiments using bulk populations, it is important to distinguish between re-differentiation of committed Th1 or Th2 cells, in contrast to the differentiation and selective outgrowth of uncommitted T cells that may still be present in the population. Although it has not been possible thus far to convert a polarized Th2 clone into a Th1 clone, it is definitely possible to regulate cytokine expression, proliferation and effector functions of committed Th1 and Th2 cells. For example, once a Th1-directed phenotype is induced, antigen-specific MHC-restricted, restimulation by dendritic cells alone is not sufficient to achieve maximal levels of IFN-γ secretion. Additional costimulatory signals, including IL-12, TNF-α, and IL-1 are necessary for upregulation of IFN-γ production (O'Garra and Murphy, 1994). The requirement of IL-1 as a costimulator for growth of Th2 clones has also been demonstrated (Lichtman *et al.*, 1988).

Th1 and Th2 surface markers

Since the discovery of the Th1 and Th2 subsets, there has been tremendous interest in finding appropriate cell surface markers that would allow easy distinction of the two subsets. Initial studies suggested that CD45RB may be expressed preferentially by Th2 clones (Birkeland *et al.*, 1988). It is now clear that this marker is more useful in distinguishing naive versus memory phenotypes rather than subsets according to cytokine profile. Naive T cells are nomally $CD45RB^{hi}$, whereas Th1 and Th2 populations can be either $CD45RB^{hi}$ or $CD45RB^{low}$ (Mosmann and Sad, 1996). Recently, it has been shown that CD30 is expressed preferentially on human activated Th2 cells (Del Prete *et al.*, 1995) and may be a marker for the Th2 phenotype. In another study, it was noted that $CD4^+CD7^+$ T cells mainly produce type 1 cytokines, whereas $CD4^+CD7^-$ cells mainly produce type 2 cytokines (Legac *et al.*, 1992). However, it is unclear whether such differences will prove to be adequate phenotypic markers for Th1 and Th2 cells.

Another important question is what is the phenotype of a long-term memory T cell. Adoptive transfer of transgenic T cells into mice lacking endogenous T cells resulted in the retention of the Th1 and Th2 phenotype of committed populations for at least 20 weeks (Swain, 1995). However, normally *in vivo*, continuous competition between ongoing T cells responses occurs, and this may shorten the lifespan of committed effectors. Activated T cells from mice kept in a 'clean' environment to allow minimum ongoing responses, secrete substantial levels of IL-2 but very little of the other cytokines (Street *et al.*, 1990). This suggests that long-term memory cells may revert to, or retain an IL-2-secreting phenotype similar to naive T cells.

Contribution by other cell types to the Th1/Th2 cytokine pattern

Although, $CD4^+$ T cells are the major sources of the Th1 and Th2 cytokine patterns, many other immune as well as non-immune cell types produce cytokines. Subsets of $\gamma\delta^+$ T cells and $CD8^+$ T cells also segregate into Th1 and Th2 cytokine secreting phenotypes (Chapter 3; Croft *et al.*, 1994; Ferrick *et al.*, 1995; Sad *et al.*, 1995). NK cells produce IFN-γ and TNF (Dunn and North, 1991) and contribute to a Th1-like pattern. Th2 cytokines may be contributed by mast cells, basophils and $CD3^+CD4^+NK1.1^+$ T cells, all of which are capable of producing IL-4 (Mosmann and Sad, 1996). IL-10 is produced by macrophages, keratinocytes and placental cells (Mosmann, 1994). Several cytokines of both Th1 and Th2 pattern are produced in the vicinity of the feto–placental unit during pregnancy, and many of the cell types involved may be non-immune in nature (Lin *et al.*, 1993; Wegmann *et al.*, 1993). Thus, while immune responses *in vivo* often show a polarized cytokine pattern towards Th1 or Th2, T cells may not be the only cells contributing to this shift. The terminology/type 1 versus type 2 pattern of immune response may be more appropriate to describe several *in vivo* situations to include contribution by other cell types to the cytokine pattern.

Th1 and Th2 responses in immune regulation

The Th1 and Th2 cytokine patterns have now been implicated in a number of immune responses, including infection, autoimmunity, allergy and transplantation (summarized in Mosmann and Sad, 1996). *In vivo*, several examples of clear-cut dichotomy between Th1 and Th2 immunity have been documented. However, intermediate stages of immune responses that do not conform to the extremes of Th1/Th2 are also common. Immunity to some infections require sequential activation of both Th1 and Th2 responses. Still others require simultaneous activation of both Th1 and Th2 at separate sites.

Th1/Th2 IN INFECTION

Host immune responses against infection often exhibit a dichotomy towards Th1 or Th2. Several factors contribute to the polarized nature of anti-parasitic immune response. Chronicity of infection leads to persistent antigenic stimulation evoking extreme immunoregulatory states. The diverse *in vivo* habitat of pathogens leads to distinct forms of antigen presentation and T cell activation, consequently evoking alternate effector arms of the immune response. Genetic polymorphism in the nature of parasite antigen processing by the host also influences T cell immunity. The nature of the immune response mounted by the host is pivotal in determining the outcome of parasitic infection. While an appropriate response enables clearance of infection, an inappropriate response not only compromises host defence but can lead to immunopathology. In some cases, pathogens themselves may subvert the immune response towards conditions promoting their replication.

Th1 responses are important for control of a number of intracellular protozoan and bacterial and viral infections (Sher and Coffman, 1992). The curative effect of Th1 responses is exemplified in the mouse *L. major* infection model. Infection of mice with this protozoan parasite can lead to either a self-healing minimal cutaneous lesion or disseminated fatal disease, depending on the genetic background of the animal. Protective Th1 cells develop in resistant C57BL/6 mice, whereas Th2 cells are stimulated in susceptible BALB/c mice (Scott *et al.*, 1988; Heinzel *et al.*, 1989). The disease pattern in both resistant and susceptible strains of mice can be switched by altering the host immune status at the onset of infection. Thus, anti-IFN-γ antibody treatment exacerbates infection in C57BL/6 mice, whereas anti-IL-4 treatment of BALB/c mice ameliorates the disease (Belosevic *et al.*, 1989; Sadick *et al.*, 1990). The regulation of specific effector functions by the cytokines characteristic of the Th1/Th2 subset determines the fate of *L. major* inside the macrophages. IFN-γ and TNF-α enhance activation of macrophages and consequently the killing of *L. major* (Belosevic *et al.*, 1988). In contrast IL-3 and IL-4 can inhibit this activation (Belosevic *et al.*, 1988), and IL-10 can suppress IFN-γ production by Th1 cells (Fiorentino *et al.*, 1989).

Similar curative effects of Th1 responses are also seen in humans. In *L. donovani* infection, lack of IFN-γ production by peripheral blood mononuclear cells during the initial stages of infection predicts progression of disease, whereas individuals producing large amounts of IFN-γ remain asymptomatic (Reiner and Locksley, 1993). Similarly, in leprosy patients, IL-4 producing cells are seen in lepromatous leprosy, whereas IFN-γ producing cells dominate in the tuberculoid form (Salgame *et al.*, 1991). However, three intermediate stages (borderline tuberculoid, borderline, borderline lepromatous) that do not conform to the extremes of Th1/Th2, have also been described (Modlin *et al.*, 1988). These may, nevertheless, be part of a slowly developing immune response.

In contrast, helminthic parasites predominately induce Th2 responses both in mice and humans (Locksley, 1994). The hallmarks of helminth infection, eosinophilia, intestinal mastocytosis and IgE production, are clearly Th2 cytokine controlled; IL-4 enhances IgE production (Coffman and Carty, 1986),

IL-5 promotes eosinophilia (Dent *et al.*, 1990), and elevated IL-3, IL-4 and IL-10 result in mastocytosis (Finkelman *et al.*, 1991). Are these responses manifestations of chronic infections that evade effective host defence, or are they obligatory protective responses that limit parasitization? Murine models of helminthic infections have provided the answers to this question. *Trichuris muris* is a whipworm that colonizes the colon and caecum of mice. In susceptible B10.BR mice, the worm establishes a chronic infection, whereas resistant BALB/k mice completely expel the worm within 21 days after infection (Else and Grencis, 1991). These disparate outcomes correlate to the generation of detrimental Th1 and protective Th2 responses in the susceptible and resistant mice strains, respectively (Else and Grencis, 1991; Else *et al.*, 1994). In the case of *Heligmosomoides polygyrus*, another enteric helminth, although chronic infection establishes after primary challenge, the ensuing Th2 response protects against rechallenge. Thus, anti-IL-4 antibody treatment ablates protective immunity to secondary infection (Urban *et al.*, 1991). *Nippostrongylus brasiliensis* also promotes a strong Th2 response in mice. As in the *T. muris* system, IFN-γ significantly delays worm expulsion (Urban *et al.*, 1993). IL-12 has similar effects and this can be correlated with the capacity of IL-12 to enhance endogenous IFN-γ production (Finkelman *et al.*, 1994). These three models of infection strongly support the view that anti-helminthic Th2 responses are curative, and that IgE and eosinophilia may be unavoidable consequences of protective immunity. In humans too, helminthic infections induce a Th2 response; T cell clones specific for *Toxocara canis* can be clearly defined as Th2 (Del Prete *et al.*, 1991a). At the same time, there is some evidence for a protective role of Th1 responses in certain other helminthic infections. For example, in experimental *Trichinella spiralis*, the rate of intestinal expulsion is enhanced by a strong Th1 response (Pond *et al.*, 1989; Sher and Coffman, 1992). In murine *Schistosoma mansoni* infection, protective immunity against egg-induced pathology can be induced by vaccination; this correlates with an elevated Th1 response (Wynn *et al.*, 1994).

Infection with certain pathogens is associated with stimulation of different subsets at different stages of disease. For example, infection of mice with *Plasmodium chabaudi* leads to sequential activation of Th1 and Th2 cells; during the early acute phase of infection, Th1 cells are activated, whereas at later times Th2 cells provide help for effective antibody production (Langhorne *et al.*, 1989; von der Weid and Langhorne, 1993). In addition, induction of strong Th2 response early in infection is associated with high parasite loads and susceptibility (Stevenson and Tam, 1993), whereas the increased Th1 response that occurs in IL-10-deficient mice also results in death, but in this case probably due to immunopathology rather than an increase in parasite growth (Linke *et al.*, 1996).

Infection by mouse hepatitis virus (MHV) also shows the importance of Th1 and Th2 responses at different stages. When mice are infected with a virulent strain of MHV, the established infection can be cured by mice that mount a

Th1 response, but not by mice that induce a Th2 response. However, if the mice are preimmunized with an attenuated strain of MHV, this induces protective immunity that is due to antibodies (Pope *et al.*, 1996).

In schistosomiasis too, both Th1 and Th2 responses are activated sequentially. Infective larvae penetrate the skin, and develop into sexually mature egg-laying worms. The adult worms migrate and lodge in portal and mesenteric veins of the host, while the eggs released are often swept into the microvasculature of the liver. The immunopathological host reaction against the egg manifests itself as a liver granuloma (Sher and Coffman, 1992). In a murine model of infection with *S. mansoni*, worms stimulate Th1 responses initially, which are subsequently downregulated by egg-induced Th2 responses (Pearce *et al.*, 1991). IL-10 produced by the host in response to egg-antigen plays a key role in downregulating the protective Th1 response, leading to granuloma formation (Stadecker and Flores Villanueva, 1994). While the pathological reactions in schistosomiasis may be attributed to the aberrant Th2 response, the Th2-induced IgE and eosinophils are also involved in acquired resistance to reinfection, and IgE promotes antibody-dependent cell-mediated cytotoxicity (Sher and Coffman, 1992). A further complicating issue is that while single vaccinations induce a strong protective Th1 immunity, multiple antigen exposures stimulate Th2 responses (Pemberton and Wilson, 1995). Thus, a simple dichotomy between Th1 and Th2 cannot account for the multiple facets of immunoregulation in schistosomiasis, and both responses may be equally important.

As in mice, human disease states can exhibit Th1 and Th2 responses at different stages during progression. Infection with *P. falciparum* appears to require both Th1 and Th2 cells for clearance. An uncontrolled Th1 response and consequent elevated TNF levels can lead to the immunopathology seen in cerebral malaria, whereas TNF and IFN-γ have also been implicated in immunity to blood-stage parasites (Troye-Blomberg *et al.*, 1994).

Compartmentalization of the immune response is seen in certain infections (Scott and Kaufmann, 1991). In murine *Trichinella* infection, both Th1 and Th2 responses occur but in different lymphoid organs. In murine cytomegalovirus infection, CD8$^+$ T cells are important for protection in all organs except the salivary glands, where activation of CD4$^+$ T cells controls infection.

Th1/Th2 IN AUTOIMMUNITY

Th1 cells appear to contribute to pathogenesis of several organ-specific autoimmune diseases, while Th2 cells prevent them (Liblau *et al.*, 1995). Animal models of autoimmunity have been helpful in unravelling the role of CD4$^+$ T cell subsets in progression and prevention of autoimmunity.

Experimental autoimmune encephalomyelitis (EAE) is an inflammatory autoimmune disease of the central nervous system (CNS), and serves as an animal model for multiple sclerosis (MS). The disease is characterized by acute onset of paralysis, and perivascular infiltration of the CNS by mononuclear cells. EAE can be induced in a number of animal species by immunization with myelin basic protein (MBP) or proteolipid protein (PLP). Evidence suggests that Th1 cells are involved in initiating the disease process: inflammation in EAE lesions is similar to DTH; a strong positive correlation between development of EAE and DTH to myelin antigens has been documented; and expression of IL-2, IFN-γ and TNF but not IL-4 is seen in the CNS tissues during the acute phase of the disease (Liblau *et al.*, 1995). In contrast, recovering animals showed decreased expression of IFN-γ and concomitantly elevated expression of IL-4, IL-10 and TGF-β (Kennedy *et al.*, 1992; Khoury *et al.*, 1992). More recently, it has been shown that MBP-specific Th1 clones are sufficient to induce EAE, whereas Th2 clones can suppress EAE (van der Veen and Stohlman, 1993). In humans, T cell clones from MS patients specific for MBP are primarily Th1 in nature (Liblau *et al.*, 1995).

Insulin-dependent diabetes mellitus (IDDM) is another autoimmune disease in which the pathogenic role of Th1 cells has been highlighted. Non-obese diabetic (NOD) mice, spontaneously develop diabetes, and exhibit lymphocytic infiltration of the islets of Langerhans. Infiltrates include $CD4^+$ T, $CD8^+$ T and B cells (Liblau *et al.*, 1995). None the less, there have been several reports of $CD4^+$ T cells, by themselves, instigating diabetes (Katz *et al.*, 1993; Reich *et al.*, 1989). Analysis of cytokine profile *in situ* has proved difficult in NOD mice, and the results are conflicting. However, anti-IFN-γ antibodies and systemic administration of IL-4 confer protection against the development of diabetes in NOD mice (Campbell *et al.*, 1991; Rapoport *et al.*, 1993). Similarly, T cell clones that accelerate the onset of diabetes in NOD mice produce mainly Th1 cytokines. Thus, the concept has evolved that a Th1 response contributes to the pathogenicity of the disease, whereas a Th2 response would be protective. However, studies in which Th1 and Th2 cells expressing an identical diabetogenic T cell receptor were tested for their ability to evoke diabetes, suggest that while Th1 cells actively promote diabetes, Th2 cells neither provoke disease, nor provide substantial protection (Katz *et al.*, 1995).

Th1/Th2 IN ALLERGY

The interaction between environmental allergens and the immune system critically modulates the development of allergic responses in humans. Typically, uptake and presentation of allergens by APC to T cells activates Th subsets, which then induce B cells to produce allergen-specific antibodies. Atopic allergic disorders are characterized by elevated allergen-specific IgE responses and eosinophilia. As mentioned earlier, Th2 cells and cytokines can contribute to immunoglobulin isotype switch towards IgE (Coffman *et al.*, 1986), and IL-5 promotes differentiation (Sanderson *et al.*, 1985), vascular adhesion (Walsh *et al.*, 1990) and survival of eosinophils (Lopez *et al.*, 1988).

Thus, substantial research has focused on the role of Th2 cells in the pathological responses in allergic inflammation.

An early indication that Th2 cells are present in atopic disease was obtained when T cell clones specific for house dust mite (HDM) or grass-pollen rye I were generated from peripheral blood lymphocytes of allergic subjects. Most allergen-specific clones produced high levels of IL-4 and IL-5, and little or no IFN-γ and IL-2 after antigen-specific stimulation. Most clones could also induce IgE production in B cells even from non-allergic individuals (Parronchi *et al.*, 1991; Wierenga *et al.*, 1990). Similarly, lymphocytes obtained by bronchoalveolar lavage or nasal biopsies after allergen provocation in allergic individuals have increased expression of Th2 cytokines (Anderson and Coyle, 1994; Durham *et al.*, 1992; Robinson *et al.*, 1992). IL-5 expression is enhanced in bronchial biopsy tissue of patients with atopic asthma (Chapter 12 and Hamid *et al.* (1991). Allergen-specific T cells are also involved in local allergic inflammatory reactions. Skin lesions of atopic dermatitis patients show a high percentage of IL-4 producing cells (Leung, 1995). While Th2 cells may be an important source of cytokines in atopic disorders, other cell types may also be involved. For example, mast cells secrete IL-4, and mast cell metaplasia in the bronchial mucosa is a pathological feature of chronic asthma (Anderson and Coyle, 1994).

Atopic allergy appears to be a result of an aberrant state of predisposition for Th2 cell-directed responses involving high levels of IL-4 and IL-5. What directs this predominance of Th2 responses is not clear. It has been suggested that elevated nitric oxide production during asthmatic episodes may inhibit Th1 responses leading to proliferation of Th2 cells (Barnes and Liew, 1995). Nevertheless, therapeutic strategies aimed at inhibiting Th2 responses, and thereby relieving eosinophilia and pathology of allergies, have gained considerable interest.

Th1/Th2 RESPONSES IN TRANSPLANTATION

It is clear that Th1-like responses are manifested during allograft rejection. IL-2, and IFN-γ expression have been consistently detected in hosts undergoing allograft rejection (Dallman *et al.*, 1991a; O'Connell *et al.*, 1993). In addition, IL-2 can prevent long-term engraftment when given in conjunction with a tolerizing allograft protocol (Dallman *et al.*, 1991b). Interestingly, in hosts receiving tolerizing therapies, a distinct absence of Th1 cytokines, and a concomitant elevated expression of Th2 cytokines in the allograft has been noted (Strom *et al.*, 1996). This has led to the suggestion that immune deviation from Th1 to Th2 can lead to long-term engraftment. However, recent results from gene knockout (KO) mice are confusing. IL-2 or IFN-γ KO mice are able efficiently to reject islet and cardiac allografts (Steiger *et al.*, 1995; Strom *et al.*, 1996). The redundancy in the immune system may explain some of these discrepancies. For example, IL-2 KO mice generate activated CTLs and produce IFN-γ, both of which contribute to graft rejection (Steiger *et al.*, 1995). Alternatively, whether Th2 cytokines promote tolerance is also not clear, as infusion of IL-10 or IL-4 does not by itself promote tolerance in IFN-γ or IL-2 KO mice (Strom *et al.*, 1996). There is a suggestion that a Th2 response, while permissive for long-term engraftment, may lead to a chronic rejection process. However, the effects of some tolerizing protocols are completely obviated by application of neutralizing anti-IL4 antibody (Strom *et al.*, 1996).

Overall, owing to the cross-regulatory effects of cytokines, ongoing Th2 expression can delay rejection by inhibiting Th1 responses which strongly promote rejection. However, whether Th2 cytokines, especially IL-4, have any direct immunosuppressive effects that promote tolerance is not clear. Furthermore, several other cell types, such as CTLs, and cytokines, such as TGF-β, also play key roles in the graft rejection process.

Th1 and Th2 cytokine interactions between pregnancy and infection

Th1/Th2 responses occuring in different compartments of the host immune system also lead to cross-regulation between different systems, as demonstrated by our recent studies on the interactions between pregnancy and concomitant infections.

Several components of the cell-mediated immune response, including NK cells and Th1 cytokines, IL-2, IFN-γ and TNF-α, can lead to fetal demise (Chaouat *et al.*, 1990), whereas IL-10 can protect against spontaneous abortions (Chaouat *et al.*, 1995). While cytokines of both Th1 and Th2 pattern are produced in the vicinity of the placenta (Delassus *et al.*, 1994; Lin *et al.*, 1993; Yelavarthi *et al.*, 1991), some mechanism probably ensures dominance of Th2 responses and inhibition of harmful Th1 immunity. Clinical and experimental evidence favours this concept, as pregnancy inhibits cell-mediated inflammatory responses while maintaining and/or enhancing antibody responses (Da Silva and Spector, 1992; Varner, 1991). Consequently, this leads to a compromise in the ability of the mother to resist infections by intracellular pathogens. *L. major* infection in pregnant mice is exacerbated and this correlates to a reduced production of Th1 cytokines (Krishnan *et al.*, 1996a). However, even this muted Th1 response tends to affect protective mechanisms in pregnancy, as implantation failure and resorptions are increased in infected mice (Krishnan *et al.*, 1996b). The increased rate of pregnancy failure in infected mice correlates to an increased production of Th1 cytokines by the placenta.

This bidirectional cross-talk between pregnancy and infection may be only one of the several interactions that may occur between different compartments of the host immune system. It is still unclear how this type of immune cross-talk is accomplished. Circulating cytokines may cross-regulate responses as they traverse different compartments of the host immune system. This concept, however, questions the traditional view of cytokines as short-range effector molecules. Alternatively, immune cell types may be influenced as they traverse different regions of the host, and may alter cytokine networks at another site. In the case of pregnancy and host-immunity interactions, hormones may aid in this process. For example, progesterone has been shown to induce Th2 responses (Piccinni *et al.*, 1995).

Conclusions

The classification of CD4$^+$ T cells responses into Th1/Th2 has aided our understanding of immune regulation. Th1 and Th2 remain the primary functional divisions of the T helper subset, although the full extent of T cell diversity may be more complex. Several factors influence the differentiation of naive cells, including cytokines, APCs and antigen dose. It has also become increasingly clear that *in vivo* responses involve an interplay of diverse cell types. While an appropriate immune response can lead to host resistance to infection, inappropriate responses often lead to immunopathology. Successful immunity often involves a complex matrix of interlocking factors. This makes sense, as this would make the host more resilient to pathogenic interference.

References

Abehsira-Amar, O., Gibert, M., Joliy, M., Theze, J. and Jankovic, D.L. (1992) IL-4 plays a dominant role in the differential development of Th0 into Th1 and Th2 cells. *J. Immunol.*, **148**, 3820–9.

Akutsu, I., Kojima, T., Kariyone, A., Fukuda, T., Makino, S., and Takatsu, K. (1995) Antibody against interleukin-5 prevents antigen-induced eosinophil infiltration and bronchial hyperreactivity in the guinea pig airways. *Immunol. Lett.*, **45**, 109–16.

Anderson, G.P. and Coyle, A.J. (1994) Th2 and 'Th2-like' cells in allergy and asthma: pharmacological perspectives. *Trends Pharmacol. Sci.*, **15**, 324–32.

Assenmacher, M., Schmitz, J. and Radbruch, A. (1994) Flow cytometric determination of cytokines in activated murine T helper lymphocytes: expression of interleukin-10 in interferon-gamma and in interleukin-4-expressing cells. *Eur. J. Immunol.*, **24**, 1097–101.

Banchereau, J. and Rousset, F. (1992) Human B lymphocytes: phenotype, proliferation, and differentiation. *Adv. Immunol.*, **52**, 125–262.

Bancroft, A.J., Else, K.J. and Grencis, R.K. (1994) Low-level infection with *Trichuris muris* significantly affects the polarization of the CD4 response. *Eur. J. Immunol.*, **24**, 3113–18.

Barnes, P.J. and Liew, F.Y. (1995) Nitric oxide and asthmatic inflammation. *Immunol. Today*, **16**, 128–30.

Belosevic, M., Davis, C.E., Meltzer, M.S. and Nacy, C.A. (1988) Regulation of activated macrophage antimicrobial activities. Identification of lymphokines that cooperate with IFN-gamma for induction of resistance to infection. *J. Immunol.*, **141**, 890–6.

Belosevic, M., Finbloom, D.S., Van Der Meide, P.H., Slayter, M.V. and Nacy, C.A. (1989) Administration of monoclonal anti-IFN-gamma antibodies *in vivo* abrogates natural resistance of C3H/HeN mice to infection with *Leishmania major*. *J. Immunol.*, **143**, 266–74.

Ben-Sasson, S.Z., Le Gros, G., Conrad, D.H., Finkelman, F.D. and Paul, W.E. (1990) Cross-linking Fc receptors stimulate splenic non-B, non-T cells to secrete interleukin 4 and other lymphokines. *Proc. Natl. Acad. Sci. USA*, **87**, 1421–5.

Birkeland, M.L., Metlay, J., Sanders, V.M., Fernandez-Botran, R., Vitetta, E.S., Steinman, R.M. and Pure, E. (1988) Epitopes on CD45R [T200] molecules define differentiation antigens on murine B and T lymphocytes. *J. Mol. Cell. Immunol.*, **4**, 71–85.

Bogdan, C. and Nathan, C. (1993) Modulation of macrophage function by transforming growth factor beta, interleukin-4, and interleukin-10. *Ann. N.Y. Acad. Sci.*, **685**, 713–39.

Boom, W.H., Liano, D. and Abbas, A.K. (1988) Heterogeneity of helper/inducer T lymphocytes. II. Effects of interleukin 4- and interleukin 2-producing T cell clones on resting B lymphocytes. *J. Exp. Med.*, **167**, 1350–63.

Bretscher, P.A., Wei, G., Menon, J.N. and Bielefeldt Ohmann, H. (1992) Establishment of stable, cell-mediated immunity that makes 'susceptible' mice resistant to *Leishmania major*. *Science*, **257**, 539–42.

Brown, M.A., Pierce, J.H., Watson, C.J., Falco, J., Ihle, J.N. and Paul, W.E. (1987) B cell stimulatory factor-1/interleukin-4 mRNA is expressed by normal and transformed mast cells. *Cell*, **50**, 809–18.

Campbell, I.L., Kay, T.W., Oxbrow, L. and Harrison, L.C. (1991) Essential role for interferon-gamma and interleukin-6 in autoimmune insulin-dependent diabetes in NOD/Wehi mice. *J. Clin. Invest.*, **87**, 739–42.

Chaouat, G., Menu, E., Clark, D.A., Dy, M., Minkowski, M. and Wegmann, T.G. (1990) Control of fetal survival in CBA x DBA/2 mice by lymphokine therapy. *J. Reprod. Fertil.*, **89**, 447–58.

Chaouat, G., Meliani, A.A., Martal, J., Raghupathy, R., Elliott, J.F., Mosmann, T.R., and Wegmann, T.G. (1995) IL-10 prevents naturally occurring fetal loss in the CBA x DBA/2 mating combination, and local defect in IL-10 production in this abortion-prone combination is corrected by *in vivo* injection of IFN-tau. *J. Immunol.*, **154**, 4261–8.

Chen, Y., Kuchroo, V.K., Inobe, J., Hafler, D.A. and Weiner, H.L. (1994) Regulatory T cell clones induced by oral tolerance: suppression of autoimmune encephalomyelitis. *Science*, **265**, 1237–40.

Chensue, S.W., Warmington, K.S., Ruth, J.H., Lincoln, P. and Kunkel, S.L. (1995) Cytokine function during mycobacterial and schistosomal antigen-induced pulmonary granuloma formation. Local and regional participation of IFN-gamma, IL-10, and TNF. *J. Immunol.*, **15**, 5969–76.

Cher, D.J. and Mosmann, T.R. (1987) Two types of murine helper T cell clone. II. Delayed-type hypersensitivity is mediated by Th1 clones. *J. Immunol.*, **138**, 3688–94.

Cherwinski, H.M., Schumacher, J.H., Brown, K.D. and Mosmann, T.R. (1987) Two types of mouse helper T cell clone. III. Further differences in lymphokine synthesis between Th1 and Th2 clones revealed by RNA hybridization, functionally monospecific bioassays, and monoclonal antibodies. *J. Exp. Med.*, **166**, 1229–44.

Coffman, R.L. and Carty, J. (1986) A T cell activity that enhances polyclonal IgE production and its inhibition by interferon-gamma. *J. Immunol* **136**:949–954.

Coffman, R.L., Ohara, J., Bond, M.W., Carty, J., Zlotnik, A. and Paul, W.E. (1986) B cell stimulatory factor 1 enhances the IgE response of lipopolysaccharide-activated B cells. *J. Immunol.*, **136**, 4538–41.

Coffman, R.L., Seymour, B.W., Lebman, D.A., Hiraki, D.D., Christiansen, J.A., Shrader, B., Cherwinski, H.M., Savelkoul, H.F., Finkelman, F.D., Bond, M.W. and Mosmann, T.R. (1988) The role of helper T cell products in mouse B cell differentiation and isotype regulation. *Immunol. Rev.*, **102**, 5–28.

Constant, S., Pfeiffer, C., Woodard, A., Pasqualini, T. and Bottomly, K. (1995) Extent of T cell receptor ligation can determine the functional differentiation of naive CD4(+) T cells. *J. Exp. Med.*, **182**, 1591–6.

Croft, M., Carter, L., Swain, S.L. and Dutton, R.W. (1994) Generation of polarized antigen-specific CD8 effector populations: reciprocal action of interleukin IL-4 and IL-12 in promoting type 2 versus type 1 cytokine profiles. *J. Exp. Med.* **180**, 1715–28.

D'Andrea, A., Aste-Amezaga, M., Valiante, N.M., Ma, X., Kubin, M. and Trinchieri, G. (1993) Interleukin 10 (IL-10) inhibits human lymphocyte interferon gamma-production by suppressing natural killer cell stimulatory factor/IL-12 synthesis in accessory cells. *J. Exp. Med.*, **178**, 1041–8.

Da Silva, J.A. and Spector, T.D. (1992) The role of pregnancy in the course and aetiology of rheumatoid arthritis. *Clin. Rheumatol.*, **11**, 189–94.

Dallman, M.J., Larsen, C.P. and Morris, P.J. (1991a) Cytokine gene transcription in vascularised organ grafts: analysis using semiquantitative polymerase chain reaction. *J. Exp. Med.*, **174**, 493–6.

Dallman, M.J., Shiho, O., Page, T.H., Wood, K.J. and Morris, P.J. (1991b) Peripheral tolerance to alloantigen results from altered regulation of the interleukin 2 pathway. *J. Exp. Med.* **173**, 79–87.

Del Prete, G.F., Maggi, E., Parronchi, P., Chretien, I., Tiri, A., Macchia, D., Ricci, M., Banchereau, J., de Vries, J.E. and Romagnani, S. (1988) IL-4 is an essential factor for the IgE synthesis induced *in vitro* by human T cell clones and their supernatants. *J. Immunol.*, **140**, 4193–8.

Del Prete, G.F., De Carli, M., Mastromauro, C., Biagiotti, R., Macchia, D., Falagiani, P., Ricci, M. and Romagnani, S. (1991a) Purified protein derivative of *Mycobacterium tuberculosis* and excretory–secretory antigen(s) of *Toxocara canis* expand *in vitro* human T cells with stable and opposite (type 1 T helper or type 2 T helper) profile of cytokine production. *J. Clin. Invest.*, **88**, 346–50.

Del Prete, G.F., De Carli, M., Ricci, M. and Romagnani, S. (1991b) Helper activity for immunoglobulin synthesis of T helper type 1 (Th1) and Th2 human T cell clones: the help of Th1 clones is limited by their cytolytic capacity. *J. Exp. Med.*, **174**, 809–13.

Del Prete, G., De Carli, M., Almerigogna, F., Giudizi, M.G., Biagiotti, R. and Romagnani, S. (1993) Human IL-10 is produced by both type 1 helper (Th1) and type 2 helper (Th2) T cell clones and inhibits their antigen-specific proliferation and cytokine production. *J. Immunol.*, **150**, 353–60.

Del Prete, G., De Carli, M., Almerigogna, F., Daniel, C.K., D'Elios, M.M., Zancuoghi, G., Vinante, F., Pizzolo, G. and Romagnani, S. (1995) Preferential expression of CD30 by human CD4$^+$ T cells producing Th2-type cytokines. *FASEB J.*, **9**, 81–6.

Delassus, S., Coutinho, G.C., Saucier, C., Darche, S. and Kourilsky, P. (1994) Differential cytokine expression in maternal blood and placenta during murine gestation. *J. Immunol.*, **152**, 2411–20.

Dent, L.A., Strath, M., Mellor, A.L. and Sanderson, C.J. (1990) Eosinophilia in transgenic mice expressing interleukin 5. *J. Exp. Med.*, **172**, 1425–31.

Ding, L., Linsley, P.S., Huang, L.-Y., Germain, R.N. and Shevach, E.M. (1993) IL-10 inhibits macrophage costimulatory activity by selectively inhibiting the up-regulation of B7 expression. *J. Immunol.*, **151**, 1224–34.

Dunn, P.L. and North, R.J. (1991) Early gamma interferon production by natural killer cells is important in defense against murine listeriosis. *Infect. Immun.*, **59**, 2892–900.

Durham, S.R., Ying, S., Varney, V.A., Jacobson, M.R., Sudderick, R.M., Mackay, I.S., Kay, A.B. and Hamid, Q.A. (1992) Cytokine messenger RNA expression for IL-3, IL-4, IL-5, and granulocyte/macrophage-colony-stimulating factor in the nasal mucosa after local allergen provocation: relationship to tissue eosinophilia. *J. Immunol.*, **148**, 2390–4.

Else, K.J. and Grencis, R.K. (1991) Cellular immune responses to the murine nematode parasite *Trichuris muris*. I. Differential cytokine production during acute or chronic infection. *Immunology*, **72**, 508–13.

Else, K.J., Finkelman, F.D., Maliszewski, C.R. and Grencis, R.K. (1994) Cytokine-mediated regulation of chronic intestinal helminth infection. *J. Exp. Med.* **179**, 347–51.

Elson, L.H., Nutman, T.B., Metcalfe, D.D. and Prussin, C. (1995) Flow cytometric analysis for cytokine production identifies T helper 1, T helper 2, and T helper 0 cells within the human $CD4^+$CD27-lymphocyte subpopulation. *J. Immunol.*, **154**, 4294–301.

Erard, F., Corthesy, P., Nabholz, M., Lowenthal, J.W., Zaech, P., Plaetinck, G. and MacDonald, H.R. (1985) Interleukin 2 is both necessary and sufficient for the growth and differentiation of lectin-stimulated cytolytic T lymphocyte precursors. *J. Immunol.*, **134**, 1644–52.

Fernandez-Botran, R., Sanders, V.M., Mosmann, T.R. and Vitetta, E.S. (1988) Lymphokine-mediated regulation of the proliferative response of clones of T helper 1 and T helper 2 cells. *J. Exp. Med.*, **168**, 543–58.

Ferrick, D.A., Schrenzel, M.D., Mulvania, T., Hsieh, B., Ferlin, W.G. and Lepper, H. (1995) Differential production of interferon-gamma and interleukin-4 in response to Th1- and Th2-stimulating pathogens by gamma delta T cells *in vivo*. *Nature*, **373**, 255–7.

Finkelman, F.D., Pearce, E.J., Urban, J.F., Jr. and Sher, A. (1991) Regulation and biological function of helminth-induced cytokine responses. *Immunol. Today*, **12**, A62–A66.

Finkelman, F.D., Madden, K.B., Cheever, A.W., Katona, I.M., Morris, S.C., Gately, M.K., Hubbard, B.R., Gause, W.C. and Urban, J.F. Jr (1994) Effects of interleukin 12 on immune responses and host protection in mice infected with intestinal nematode parasites. *J. Exp. Med.*, **179**, 1563–72.

Fiorentino, D.F., Bond, M.W. and Mosmann, T.R. (1989) Two types of mouse T helper cell. IV. Th2 clones secrete a factor that inhibits cytokine production by Th1 clones. *J. Exp. Med.*, **170**, 2081–95.

Firestein, G.S., Roeder, W.D., Laxer, J.A., Townsend, K.S., Weaver, C.T., Hom, J.T., Linton, J., Torbett, B.E. and Glasebrook, A.L. (1989) A new murine $CD4^+$ T cell subset with an unrestricted cytokine profile. *J. Immunol.*, **143**, 518–25.

Fitch, F.W., McKisic, M.D., Lancki, D.W. and Gajewski, T.F. (1993) Differential regulation of murine T lymphocyte subsets. *Annu. Rev. Immunol.*, **11**, 29–48.

Gajewski, T.F. and Fitch, F.W. (1988) Anti-proliferative effect of IFN-gamma in immune regulation. I. IFN-gamma inhibits the proliferation of Th2 but not Th1 murine helper T lymphocyte clones. *J. Immunol.*, **140**, 4245–52.

Gollob, K.J. and Coffman, R.L. (1994) A minority subpopulation of $CD4^+$ T cells directs the development of naive $CD4^+$ T cells into IL-4-secreting cells. *J. Immunol.*, **152**, 5180–8.

Hamid, Q., Azzawi, M., Ying, S., Moqbel, R., Wardlaw, A.J., Corrigan, C.J., Bradley, B., Durham, S.R., Collins, J.V., Jeffery, P.K. *et al.* (1991) Expression of mRNA for interleukin-5 in mucosal bronchial biopsies from asthma. *J. Clin. Invest.*, **87**, 1541–6.

Heinzel, F.P., Sadick, M.D., Holaday, B.J., Coffman, R.L. and Locksley, R.M. (1989) Reciprocal expression of interferon gamma or interleukin 4 during the resolution or progression of murine leishmaniasis. Evidence for expansion of distinct helper T cell subsets. *J. Exp. Med.*, **169**, 59–72.

Herron, L.R., Coffman, R.L., Bond, M.W. and Kotzin, B.L. (1988) Increased autoantibody production by NZB/NZW B cells in response to IL-5. *J. Immunol.*, **141**, 842–8.

Hosken, N.A., Shibuya, K., Heath, A.W., Murphy, K.M. and O'Garra, A. (1995) The effect of antigen dose on CD4$^+$ T helper cell phenotype development in a T cell receptor-alpha beta-transgenic model. *J. Exp. Med.*, **182**, 1579–84.

Hsieh, C.-S., Heimberger. A.B., Gold, J.S., O'Garra, A. and Murphy, K.M. (1992) Differential regulation of T helper phenotype development by interleukins 4 and 10 in an alpha beta T-cell-receptor transgenic system. *Proc. Natl. Acad. Sci. USA*, **89**, 6065–9.

Hsieh, C.-S., Macatonia, S.E., Tripp, C.S., Wolf, S.F., O'Garra, A. and Murphy, K.M. (1993) Development of Th1 CD4$^+$ T cells through IL-12 produced by Listeria-induced macrophages. *Science*, **260**, 547–9.

Issekutz, T.B., Stoltz, J.M. and v.d.Meide, P. (1988) Lymphocyte recruitment in delayed-type hypersensitivity. The role of IFN-gamma. *J. Immunol.*, **140**, 2989–93.

Janeway, C.A. Jr, Carding, S., Jones, B., Murray, J., Portoles, P., Rasmussen, R., Rojo, J., Saizawa, K., West, J. and Bottomly, K. (1988) CD4$^+$ T cells: specificity and function. *Immunol. Rev.*, **101**, 39–80.

Kamogawa, Y., Minasi, L.A., Carding, S.R., Bottomly, K. and Flavell, R.A. (1993) The relationship of IL-4- and IFN gamma-producing T cells studied by lineage ablation of IL-4-producing cells. *Cell*, **75**, 985–95.

Katz, J.D., Wang, B., Haskins, K., Benoist, C. and Mathis, D. (1993) Following a diabetogenic T cell from genesis through pathogenesis. *Cell*, **74**, 1089–100.

Katz, J.D., Benoist, C. and Mathis, D. (1995) T helper cell subsets in insulin-dependent diabetes. *Science*, **268**, 1185–8.

Kennedy, M.K., Torrance, D.S., Picha, K.S. and Mohler, K.M. (1992) Analysis of cytokine mRNA expression in the central nervous system of mice with experimental autoimmune encephalomyelitis reveals that IL-10 mRNA expression correlates with recovery. *J. Immunol.*, **149**, 2496–505.

Khoury, S.J., Hancock, W.W. and Weiner, H.L. (1992) Oral tolerance to myelin basic protein and natural recovery from experimental autoimmune encephalomyelitis are associated with downregulation of inflammatory cytokines and differential upregulation of transforming growth factor beta, interleukin 4, and prostaglandin E expression in the brain. *J. Exp. Med.*, **176**, 1355–64.

Killar, L., MacDonald, G., West, J., Woods, A. and Bottomly, K. (1987) Cloned, Ia-restricted T cells that do not produce interleukin 4(IL 4)/B cell stimulatory factor 1(BSF-1) fail to help antigen-specific B cells. *J. Immunol.*, **138**, 1674–9.

Kopf, M., Le Gros, G., Bachmann, M., Lamers, M.C. Bluethmann, H. and Köhler, G. (1993) Disruption of the murine IL-4 gene blocks Th2 cytokine responses. *Nature*, **362**, 245–8.

Kopf, M., Brombacher, F., Kohler, G., Kienzle, G., Widmann K.-H., Lefrang, K., Humborg, C., Ledermann, B. and Solbach, W. (1996) IL-4-deficient Balb/c mice resist infection with *Leishmania major*. *J. Exp. Med.*, **184**, 1127–36.

Krishnan, L., Guilbert, L.J., Russell, A.S., Wegmann, T.G., Mosmann, T.R. and Belosevic, M. (1996a) Pregnancy impairs resistance of C57BL/6 mice to *Leishmania major* infection and causes decreased antigen-specific IFN-γ responses and increased production of Th2 cytokines. *J. Immunol.*, **156**, 644–52.

Krishnan, L., Guilbert, L.J., Wegmann, T.G., Belosevic, M. and Mosmann, T.R. (1996b) The Th1 response against *Leishmania major* in pregnant C57Bl/6 mice increases implantation failure and fetal resorptions: correlation with increased IFN-γ and TNF and reduced IL-10 production by placental cells. *J. Immunol.*, **156**, 653–62.

Kuchroo, V.K., Das, M.P., Brown, J.A., Ranger, A.M., Zamvil, S.S., Sobel, R.A., Weiner, H.L., Nabavi, N. and Glimcher, L.H. (1995) B7-1 and B7-2 costimulatory molecules activate differentially the Th1/Th2 developmental pathways: application to autoimmune disease therapy. *Cell*, **80**, 707–18.

Langhorne, J., Gillard, S., Simon, B., Slade, S. and Eichmann, K. (1989) Frequencies of $CD4^+$ T cells reactive with *Plasmodium chabaudi chabaudi*: distinct response kinetics for cells with Th1 and Th2 characteristics during infection. *Int. Immunol.*, **1**, 416–24.

Legac, E., Autran, B., Merle-Beral, H., Katlama, C. and Debre, P. (1992) CD4+CD7-$CD57^+$ T cells: a new T-lymphocyte subset expanded during human immunodeficiency virus infection. *Blood*, **79**, 1746–53.

Leung, D.Y. (1995) Atopic dermatitis: the skin as a window into the pathogenesis of chronic allergic diseases. *J. Aller. Clin. Immunol.*, **96**, 302–18; quiz 31.

Li, L., Sad, S., Kagi, D., Hengartner, H. and Mosmann, T.R. (1997) CD8 Tc1 and Tc2 cells secrete distinct cytokine patterns *in vitro* and *in vivo*, but induce similar inflammatory reactions. *J. Immunol.*, **158**, 4152–61.

Liblau, R.S., Singer, S.M. and McDevitt, H.O. (1995) Th1 and Th2 $CD4^+$ T cells in the pathogenesis of organ-specific autoimmune diseases. *Immunol. Today*, **16**, 34–8.

Lichtman, A.H., Chin, J., Schmidt, J.A. and Abbas, A.K. (1988) Role of interleukin 1 in the activation of T lymphocytes. *Proc. Natl. Acad. Sci. USA*, **85**, 9699–703.

Lin, H., Mosmann, T.R., Guilbert, L., Tuntipopipat, S. and Wegmann, T.G. (1993) Synthesis of T helper 2-type cytokines at the maternal–fetal interface. *J. Immunol.*, **151**, 4562–73.

Linke, A., Kuhn, R., Muller, W., Honarvar, N., Li, C. and Langhorne, J. (1996) *Plasmodium chabaudi chabaudi*: differential susceptibility of gene-targeted mice deficient in IL-10 to an erythrocytic-stage infection. *Exp. Parasitol.*, **84**, 253–63.

Locksley, R.M. (1994) Th2 cells: help for helminths. *J. Exp. Med.*, **179**, 1405–7.

Lopez, A.F., Sanderson, C.J., Gamble, J.R., Campbell, H.D., Young, I.G. and Vadas, M.A. (1988) Recombinant human interleukin 5 is a selective activator of human eosinophil function. *J, Exp, Med.*, **167**, 219–24.

Maggi, E., Parronchi, P., Manetti, R., Simonelli, C., Piccinni, M.P., Rugiu, F.S., De Carli, M., Ricci, M. and Romagnani, S. (1992) Reciprocal regulatory effects of IFN-gamma and IL-4 on the *in vitro* development of human Th1 and Th2 clones. *J. Immunol.*, **148**, 2142–7.

Manetti, R., Gerosa, F., Giudizi, M.G., Biagiotti, R., Parronchi, P., Piccinni, M.P., Sampognaro, S., Maggi, E., Romagnani, S., Trinchieri, G. *et al.* (1994) Interleukin 12 induces stable priming for interferon gamma (IFN-gamma) production during differentiation of human T helper (Th) cells and transient IFN-gamma production in established Th2 cell clones. *J. Exp. Med.*, **179**, 1273–83.

Meade, R., Askenase, P.W., Geba, G.P., Neddermann, K., Jacoby, R.O. and Pasternak, R.D. (1992) Transforming growth factor $\beta 1$ inhibits murine immediate and delayed type hypersensitivity. *J. Immunol.*, **149**, 521–8.

Mitchell, G.F. and Miller, J.F. (1968) Cell to cell interaction in the immune response. II. The source of hemolysin-forming cells in irradiated mice given bone marrow and thymus or thoracic duct lymphocytes. *J. Exp. Med.*, **128**, 821–37.

Modlin, R.L., Melancon-Kaplan, J., Young, S.M., Pirmez, C., Kino, H., Convit, J., Rea, T.H. and Bloom, B.R. (1988) Learning from lesions: patterns of tissue inflammation in leprosy. *Proc. Natl. Acad. Sci. USA*, **85**, 1213–17.

Moore, K.W., Vieira, P., Fiorentino, D.F., Trounstine, M.L., Khan, T.A. and Mosmann, T.R. (1990) Homology of cytokine synthesis inhibitory factor (IL-10) to the Epstein–Barr virus gene BCRFI. *Science*, **248**, 1230–4.

Mosmann, T.R. (1994) Properties and functions of Interleukin 10. *Adv. Immunol.*, **56**, 1–26.

Mosmann, T.R. and Coffman, R.L. (1989) Th1 and Th2 cells: Different patterns of lymphokine secretion lead to different functional properties. *Annu. Rev. Immunol.*, **7**, 145–73.

Mosmann, T.R. and Sad, S. (1996) The expanding universe of T cell subsets – Th1, Th2 and more. *Immunol. Today*, **17**, 138–46.

Mosmann, T.R., Cherwinski, H., Bond, M.W., Giedlin, M.A. and Coffman, R.L. (1986) Two types of murine helper T cell clone. I. Definition according to profiles of lymphokine activities and secreted proteins. *J. Immunol.*, **136**, 2348–57.

Müller, K.M., Jaunin, F., Masouyé, I., Saurat, J.-H. and Hauser, C. (1993) Th2 cells mediate IL-4-dependent local tissue inflammation. *J. Immunol.*, **150**, 5576–84.

Noben-Trauth, N., Kropf, P. and Muller, I. (1996) Susceptibility to *Leishmania major* infection in interleukin-4-deficient mice. *Science*, **271**, 987–90.

O'Connell, P.J., Pacheco-Silva, A., Nickerson, P.W., Muggia, R.A., Bastos, M., Kelley, V.R. and Strom, T.B. (1993) Unmodified pancreatic islet allograft rejection results in the preferential expression of certain T cell activation transcripts. *J. Immunol.*, **150**, 1093–104.

O'Garra, A. and Murphy, K. (1994) Role of cytokines in determining T-lymphocyte function. *Curr. Opin. Immunol.*, **6**, 458–66.

Parish, C.R. (1972) The relationship between humoral and cell-mediated immunity. *Transplant. Rev.*, **13**, 35–66.

Parronchi, P., Macchia, D., Piccinni, M.P., Biswas, P., Simonelli, C., Maggi, E., Ricci, M., Ansari, A.A. and Romagnani, S. (1991) Allergen- and bacterial antigen-specific T-cell clones established from atopic donors show a different profile of cytokine production. *Proc. Natl. Acad. Sci. USA*, **88**, 4538-42.

Pearce, E.J., Caspar, P., Grzych, J.M., Lewis, F.A. and Sher, A. (1991) Downregulation of Th1 cytokine production accompanies induction of Th2 responses by a parasitic helminth, *Schistosoma mansoni*. *J. Exp. Med.*, **173**, 159–66.

Pemberton, R.M. and Wilson, R.A. (1995) T-helper type-1-dominated lymph node responses induced in C57BL/6 mice by optimally irradiated cercariae of *Schistosoma mansoni* are down-regulated after challenge infection. *Immunology*, **84**, 310–16.

Pfeifer, J.D., McKenzie, D.T., Swain, S.L. and Dutton, R.W. (1987) B cell stimulatory factor 1 (interleukin 4) is sufficient for the proliferation and differentiation of lectin-stimulated cytolytic T lymphocyte precursors. *J. Exp. Med.*, **166**, 1464–70.

Piccinni, M.P., Giudizi, M.G., Biagiotti, R., Beloni, L., Giannarini, L., Sampognaro, S., Parronchi, P., Manetti, R., Annunziato, F., Livi, C., and Romagnani, S. (1995) Progesterone favors the development of human T helper cells producing Th2-type cytokines and promotes both IL-4 production and membrane CD30 expression in established Th1 cell clones. *J. Immunol.*, **155**, 128–33.

Pond, L., Wassom, D.L. and Hayes, C.E. (1989) Evidence for differential induction of helper T cell subsets during *Trichinella spiralis* infection. *J. Immunol.*, **143**, 4232–7.

Pope, M., Chung, S.W, Mosmann, T.R., Leibowitz, J.L., Gorczynski, R.M. and Levy, G.A. (1996) Resistance of naive mice to murine hepatitis virus strain 3 requires development of a Th1, but not a Th2, response, whereas pre-existing antibody partially protects against primary infection. *J. Immunol.*, **156**, 3342–9.

Punnonen, J., Aversa, G., Cocks, B.G., McKenzie, A.N.J., Menon, S., Zurawski, G., de Waal Malefyt, R. and de Vries, J.E. (1993) Interleukin 13 induces interleukin 4-independent IgG4 and IgE synthesis and CD23 expression by human B cells. *Proc. Natl. Acad. Sci. USA*, **90**, 3730–4.

Rapoport, M.J., Jaramillo, A., Zipris, D., Lazarus, A.H., Serreze, D.V., Leiter, E.H., Cyopick, P., Danska, J.S. and Delovitch, T.L. (1993) Interleukin 4 reverses T cell proliferative unresponsiveness and prevents the onset of diabetes in nonobese diabetic mice. *J. Exp. Med.*, **178**, 87–99.

Rasmussen, R., Takatsu, K., Harada, N., Takahashi, T. and Bottomly, K. (1988) T cell-dependent hapten-specific and polyclonal B cell responses require release of interleukin 5. *J. Immunol.*, **140**, 705–12.

Reich, E.P., Sherwin, R.S., Kanagawa, O. and Janeway, C.A. Jr (1989) An explanation for the protective effect of the MHC class II I-E molecule in murine diabetes. *Nature*, **341**, 326–8.

Reiner, S.L. and Locksley, R.M. (1993) Cytokines in the differentiation of Th1/Th2 $CD4^+$ subsets in leishmaniasis. *J. Cell. Biochem.*, **53**, 323–8.

Robinson, D.S., Hamid, Q., Ying, S., Tsicopoulos, A., Barkans, J., Bentley, A.M., Corrigan, C., Durham, S.R. and Kay, A.B. (1992) Predominant TH2-like bronchoalveolar T-lymphocyte population in atopic asthma. *N. Engl. J. Med.*, **326**, 298–304.

Sad, S. and Mosmann, T.R. (1994) Single IL-2-secreting precursor CD4 T cell can develop into either Th1 or Th2 cytokine secretion phenotype. *J. Immunol.*, **153**, 3514–22.

Sad, S., Marcotte, R. and Mosmann, T.R. (1995) Cytokine-induced differentiation of precursor mouse $CD8^+$ T cells into cytotoxic $CD8^+$ cells secreting Th1 or Th2 cytokines. *Immunity*, **2**, 271–9.

Sadick, M.D., Heinzel, F.P., Holaday, B.J., Pu, R.T., Dawkins, R.S. and Locksley, R.M. (1990) Cure of murine leishmaniasis with anti-interleukin 4 monoclonal antibody. Evidence for a T cell-dependent, interferon gamma-independent mechanism. *J. Exp. Med.*, **171**, 115–27.

Salgame, P., Abrams, J.S., Clayberger, C., Goldstein, H., Convit, J., Modlin, R.L. and Bloom, B.R. (1991) Differing lymphokine profiles of functional subsets of human CD4 and CD8 T cell clones. *Science*, **254**, 279–82.

Sanderson, C.J., Warren, D.J. and Strath, M. (1985) Identification of a lymphokine that stimulates eosinophil differentiation *in vitro*. Its relationship to interleukin 3, and functional properties of eosinophils produced in cultures. *J. Exp. Med.*, **162**, 60–74.

Sanderson, C.J., O'Garra, A., Warren, D.J. and Klaus, G.G. (1986) Eosinophil differentiation factor also has B-cell growth factor activity: proposed name interleukin 4. *Proc. Natl. Acad. Sci. USA*, **83**, 437–40.

Scharton, T.M. and Scott, P. (1993) Natural killer cells are a source of interferon gamma that drives differentiation of $CD4^+$ T cell subsets and induces early resistance to *Leishmania major* in mice. *J. Exp. Med.*, **178**, 567–77.

Scott, P. and Kaufmann, S.H. (1991) The role of T-cell subsets and cytokines in the regulation of infection. *Immunol. Today*, **12**, 346–8.

Scott, P., Natovitz, P., Coffman, R.L., Pearce, E. and Sher, A. (1988) Immunoregulation of cutaneous leishmaniasis. T cell lines that transfer protective immunity or exacerbation belong to different T helper subsets and respond to distinct parasite antigens. *J. Exp. Med.*, **168**, 1675–84.

Seder, R.A. and Paul, W.E. (1994) Acquisition of lymphokine-producing phenotype by $CD4^+$ T cells. *Annu. Rev. Immunol.*, **12**, 635–73.

Seder, R.A., Paul, W.E., Davis, M.M. and Fazekas de St Groth, B. (1992) The presence of interleukin 4 during *in vitro* priming determines the lymphokine-producing potential of $CD4^+$ T cells from T cell receptor transgenic mice. *J. Exp. Med.*, **176**, 1091–8.

Seder, R.A., Gazzinelli, R., Sher, A. and Paul, W.E. (1993) Interleukin 12 acts directly on $CD4^+$ T cells to enhance priming for interferon gamma production and diminishes interleukin 4 inhibition of such priming. *Proc. Natl. Acad. Sci. USA*, **90**, 10188–92.

Sher, A. and Coffman, R.L. (1992) Regulation of immunity to parasites by T cells and T cell-derived cytokines. *Annu. Rev Immunol.*, **10**, 385–409.

Simon, M.M., Hochgeschwender, U., Brugger, U. and Landolfo, S, (1986) Monoclonal antibodies to interferon-gamma inhibit interleukin 2-dependent induction of growth and maturation in lectin/antigen-reactive cytolytic T lymphocyte precursors. *J. Immunol.*, **136**, 2755–62.

Snapper, C.M. and Paul, W.E. (1987) Interferon-gamma and B cell stimulatory factor-1 reciprocally regulate Ig isotype production. *Science*, **236**, 944–7.

Stadecker, M.J. and Flores Villanueva, P.O. (1994) Accessory cell signals regulate Th-cell responses: from basic immunology to a model of helminthic disease. *Immunol. Today*, **15**, 571–4.

Steiger, J., Nickerson, P.W., Steurer, W., Moscovitch-Lopatin, M. and Strom, T.B. (1995) IL-2 knockout recipient mice reject islet cell allografts. *J. Immunol.*, **155**, 489–98.

Stevens, T.L., Bossie, A., Sanders, V.M., Fernandez-Botran, R., Coffman, R.L., Mosmann, T.R. and Vitetta, E.S. (1988) Regulation of antibody isotype secretion by subsets of antigen-specific helper T cells. *Nature*, **334**, 255–8.

Stevenson, M.M. and Tam, M.F. (1993) Differential induction of helper T cell subsets during blood-stage *Plasmodium chabaudi* AS infection in resistant and susceptible mice. *Clin. Exp. Immunol.* **92**, 77–83.

Street, N.E., Schumacher, J.H., Fong, T.A., Bass, H., Fiorentino, D.F., Leverah, J.A. and Mosmann, T.R. (1990) Heterogeneity of mouse helper T cells: evidence from bulk cultures and limiting dilution cloning for precursors of Th1 and Th2 cells. *J. Immunol.*, **144**, 1629–39.

Strom, T.B., Roy-Chaudhury, P., Manfro, R., Zheng, X.X., Nickerson, P.W., Wood, K. and Bushell, A. (1996) The Th1/Th2 paradigm and the allograft response. *Curr. Opin. Immunol.*, **8**, 688–93.

Swain, S.L. (1995) Generation and *in vivo* persistence of polarized Th1 and Th2 memory cells. *Immunity*, **1**, 543–52.

Swain, S.L., Weinberg, A.D. and English, M. (1990a) CD4+ T cell subsets: lymphokine secretion of memory cells and of effector cells which develop from precursors *in vitro*. *J. Immunol.*, **144**, 1788–99.

Swain, S.L., Weinberg, A.D., English, M. and Huston, G. (1990b) IL-4 directs the development of Th2-like helper effectors. *J. Immunol.*, **145**, 3796–806.

Swain, S.L., Huston, G., Tonkonogy, S. and Weinberg, A. (1991) Transforming growth factor-beta and IL-4 cause helper T cell precursors to develop into distinct effector helper cells that differ in lymphokine secretion pattern and cell surface phenotype. *J. Immunol.*, **147**, 2991–3000.

Takatsu, K., Kikuchi, Y., Takahashi, T., Honjo, T., Matsumoto, M., Harada, N., Yamaguchi, N. and Tominaga, A. (1987) Interleukin 5, a T-cell-derived B-cell differentiation factor also induces cytotoxic T lymphocytes. *Proc. Natl. Acad. Sci. USA*, **84**, 4234–8.

Tanaka, T., Hu-Li, J., Seder, R.A., Fazekas de St Groth, B. and Paul, W.E. (1993) Interleukin 4 suppresses interleukin 2 and interferon gamma production by naive T cells stimulated by accessory cell-dependent receptor engagement. *Proc. Natl. Acad. Sci. USA*, **90**, 5914–18.

Tite, J.P. and Janeway, C.A. Jr (1984) Cloned helper T cells can kill B lymphoma cells in the presence of specific antigen: Ia restriction and cognate vs. noncognate interactions in cytolysis. *Eur. J. Immunol.*, **14**, 878–86.

Trinchieri, G. (1993) Interleukin-12 and its role in the generation of Th1 cells. *Immunol. Today*, **14**, 335–7.

Trinchieri, G. (1995) Interleukin-12: a proinflammatory cytokine with immunoregulatory functions that bridge innate resistance and antigen-specific adaptive immunity. *Annu. Rev. Immunol.*, **13**, 251–76.

Troye-Blomberg, M., Berzins, K. and Perlmann, P. (1994) T-cell control of immunity to the asexual blood stages of the malaria parasite. *Crit. Rev. Immunol.*, **14**, 131–55.

Tsicopoulos, A., Hamid, Q., Varney, V., Ying, S., Moqbel, R., Durham, S.R. and Kay, A.B. (1992) Preferential messenger RNA expression of Th1-type cells (IFN-gamma+, IL-2+) in classical delayed-type (tuberculin) hypersensitivity reactions in human skin. *J. Immunol.*, **148**, 2058–61.

Urban, J.F. Jr, Katona, I.M., Paul, W.E. and Finkelman, F.D. (1991) Interleukin 4 is important in protective immunity to a gastrointestinal nematode infection in mice. *Proc. Natl. Acad. Sci. USA*, **88**, 5513–17.

Urban, J.F. Jr, Madden, K.B., Svetic, A., Cheever, A., Trotta, P.P., Gause, W.C., Katona, I.M. and Finkelman, F.D. (1992) The importance of Th2 cytokines in protective immunity to nematodes. *Immunol. Rev.* **127**, 205–20.

Urban, J.F. Jr, Madden, K.B., Cheever, A.W., Trotta, P.P., Katona, I.M. and Finkelman, F.D. (1993) IFN inhibits inflammatory responses and protective immunity in mice infected with the nematode parasite, *Nippostrongylus brasiliensis*. *J. Immunol.*, **151**, 7086–94.

van der Veen, R.C. and Stohlman, S.A. (1993) Encephalitogenic Th1 cells are inhibited by Th2 cells with related peptide specificity: relative roles of interleukin (IL)-4 and IL-10. *J. Neuroimmunol.*, **48**, 213–20.

Varner, M.W. (1991) Autoimmune disorders and pregnancy. *Semin. Perinatol.*, **15**, 238–50.

von der Weid, T. and Langhorne, J. (1993) The roles of cytokines produced in the immune response to the erythrocytic stages of mouse malarias. *Immunobiology*, **189**, 397–418.

Walsh, G.M., Hartnell, A., Wardlaw, A.J., Kurihara, K., Sanderson, C.J. and Kay, A.B. (1990) IL-5 enhances the *in vitro* adhesion of human eosinophils, but not neutrophils, in a leucocyte integrin (CD11/18)-dependent manner. *Immunology*, **71**, 258–65.

Wegmann, T.G., Lin, H., Guilbert, L.J. and Mosmann, T.R. (1993) Bidirectional cytokine interactions in the maternal–fetal relationship: is successful pregnancy a Th2 phenomenon? *Immunol. Today*, **14**, 353–6.

Wierenga, E.A., Snoek, M., de Groot, C., Chretien, I., Bos, J.D., Jansen, H.M. and Kapsenberg, M.L. (1990) Evidence for compartmentalization of functional subsets of $CD2^+$ T lymphocytes in atopic patients. *J. Immunol.*, **144**, 4651–6.

Wynn, T.A., Eltoum, I., Oswald, I.P., Cheever, A.W. and Sher, A. (1994) Endogenous interleukin 12 (IL-12) regulates granuloma formation induced by eggs of *Schistosoma mansoni* and exogenous IL-12 both inhibits and prophylactically immunizes against egg pathology. *J. Exp. Med.* **179**, 1551–61.

Yelavarthi, K.K., Chen, H.L., Yang, Y.P., Cowley, B.D.J., Fishback, J.L. and Hunt, J.S. (1991) Tumor necrosis factor-alpha mRNA and protein in rat uterine and placental cells. *J. Immunol.*, **146**, 3840–8.

Yoshimoto, T. and Paul, W.E. (1994) $CD4^{pos}$, $NK1.1^{pos}$ T cells promptly produce interleukin 4 in response to *in vivo* challenge with anti-CD3. *J. Exp. Med.* **179**, 1285–95.

3 Functional Subpopulations of CD8$^+$ T Lymphocytes

David M. Kemeny, Milica Vukmanovic-Stejic, Beejal Vyas, Paul A. MacAry, Artin Moussavi and Alistair Noble

Introduction

As reviewed in the previous chapter, the discovery by Mosmann *et al.* (1986) that long-term murine CD4$^+$ T cell clones could be segregated according to their patterns of cytokine secretion initiated extensive investigation into the functional characteristics of these cells. The terms Th1 and Th2 were, in fact, first used by Tada *et al.* in 1978 to distinguish helper T cells that were Lyt-1$^+$,2$^-$,3$^-$ and provided cognate (Th1) and polyclonal (Th2) B cell help, respectively (Tada *et al.*, 1978). CD4$^+$ T cell subsets comparable to murine Th1 and Th2 have been described in the rat (Noble *et al.*, 1993b) and human (Maggi *et al.*, 1988). Other CD4$^+$ T cell subsets include Th0 cells, which exhibit an unrestricted cytokine profile (Firestein *et al.*, 1989), and Th3 cells, which secrete IL-4, IL-10 and high levels of TGF-β (Chen *et al.*, 1995; Miller *et al.*, 1992). Similar studies on CD8$^+$ T cells have now revealed the existence of functionally distinct subpopulations that can be compartmentalized along broadly similar lines to CD4$^+$ T cells, with IFN-γ secretion representing Th1/Tc1 cells, IL-4 secretion representing Th2/Tc2 cells, and an unrestricted cytokine profile representing Th0/Tc0 cells. There is now good evidence that demonstrates the existence of CD8$^+$ T cell subsets in both rodents and humans. In this chapter, what is known about CD8$^+$ T cell subsets and their functions will be discussed.

CD8$^+$ T cell subsets

Historically, CD8$^+$ T cells have been regarded as a homogeneous population of CD4 dependent, cytotoxic cells producing a limited number of cytokines, namely, IFN-γ and TNF-α (Fong and Mosmann, 1990). Their function was

T Lymphocyte Subpopulations in Immunotoxicology. Edited by Ian Kimber and MaryJane K. Selgrade.
 ISBN 0 471 97194 4

thought to be restricted to defence against viral infections and intracellular pathogens. More recently, a number of groups have reported much more heterogeneous cytokine production by $CD8^+$ T cells, indicating a potential for a wider range of functions.

One of the first observations was that some $CD8^+$ cytotoxic T lymphocytes (CTL) were CD4 independent and were capable of producing IL-2 (Glasebrook *et al.*, 1983; Mauer *et al.*, 1982). Subsequently a number of groups have shown that both human and mouse $CD8^+$ T cells are able to produce IL-4 when stimulated with anti-CD3 or mitogens (Mingari *et al.*, 1984; Paliard *et al.*, 1988a,b), especially when primed *in vitro* together with IL-2 and IL-4 (Seder *et al.*, 1992). Le Gros and colleagues reported that purified $CD8^+$ T cells vigorously stimulated in the presence of high concentrations of IL-4 became non-cytolytic, $CD8^-CD4^-$ and made type 2 cytokines (IL-4, IL-5 and IL-10) (Erard *et al.*, 1993). These cells secreted large amounts of IL-4 and were later shown to be capable of switching B cells to make IgE (Le Gros and Erard, 1994). It was also observed that freshly isolated human peripheral blood $CD8^+$ T cells have the potential to make IL-4 at similar (Li *et al.*, 1995), and sometimes greater (Stanciu *et al.*, 1996), levels than those produced by comparable CD4 cells. This is also true in the rat, where it was observed that freshly purified $CD8^+$ T cells stimulated with PMA and ionomycin express more IL-4 mRNA than comparable CD4 T cells (Noble *et al.*, 1995).

$CD8^+$ T cells secreting type 2 cytokines have also been isolated in a variety of diseases, indicating that such cells were more than an *in vitro* phenomenon. Salgame *et al.* (1991a) reported that some of the *M. leprae*-specific $CD8^+$ clones generated from the patients with lepromatous leprosy made IL-4, IL-5 and IFN-γ but not IL-6 or IL-10. These clones were termed type 2 clones or suppressor clones. However, they were atypical, as they were MHC class II restricted (Salgame *et al.*, 1991b). Th2-like $CD8^+$ T cells secreting IL-4 and/or IL-5 and no IFN-γ have also been found in HIV infected individuals (Maggi *et al.*, 1994). At the same time it was reported that neonatal $CD8^+$ T cells could make IL-4 and IL-5 if cultured in the presence of large amounts of IL-4 and IL-2 (Byun *et al.*, 1994).

The studies mentioned above clearly demonstrate that human, murine and rat $CD8^+$ T cells have the potential to produce a much wider array of cytokines than was initially thought. Moreover, $CD8^+$ T cells appear to differentiate in a polarized fashion and can be divided into subsets analogous to those described for CD4 T cells. These subsets are alternatively termed type 1, T1 or Tc1; or type 2, T2 or Tc2. In the mouse, $CD8^+$ T cell subsets comparable to Th1 and Th2 have been generated from naive precursors following stimulation by antigen-presenting cells in the presence of the polarizing cytokines (Croft *et al.*, 1994). This system generates two populations: Tc1 secreting IL-2 and IFN-γ and Tc2 secreting IL-4 and IL-5 and variably reduced amounts of IFN-γ. These results were confirmed by Sad and colleagues using cytokine primed alloantigen-stimulated $CD8^+$ T cells which generated two distinct populations:

Tc1 secreting IL-2 and IFN-γ and Tc2 secreting IL-4, IL-5, IL-6 and IL-10 and no or very little IFN-γ (Sad *et al.*, 1995). Cronin and colleagues were also able to generate Tc1 (IFN-γ and IL-2, no IL-4 or IL-5) and Tc2 (IL-4 and IL-5 no IFN-γ or IL-2) CD8$^+$ T cell clones using a regimen of repeated stimulation without priming with exogenous cytokines (Cronin *et al.*, 1995). Such clear-cut patterns of cytokine production are less readily observed in rat (Noble *et al.*, 1995) or in humans (Vukmanovic-Stejic *et al.*, unpublished observations). The synthesis of IL-2, IL-6, IL-10 and IL-13 does not fit the CD4$^+$ T cell Th1/Th2 paradigm as closely as in the mouse. In addition, the full range of cytokines has not been determined in many studies and the terms Tc1 and Tc2 or T1 and T2 are often used to describe cells based on IFN-γ/IL-4 production. Studies in the rat have shown the existence of at least three different patterns of cytokine production depending on the combination of cytokines used during priming (Noble *et al.*, 1995). None of the polyclonally stimulated populations or antigen specific clones generated in the rat showed a typical Tc2 profile as IFN-γ production was always present. Similar patterns were observed in the CD8$^+$ clones generated from human peripheral blood using mitogen stimulation: these clones could be divided into Tc1 and Tc0 based on IFN-γ and IL-4 production, but when a panel of cytokines including IL-2, -5, -6 and -10 was examined they were more randomly expressed although there was some association between IL-4 and IL-5 (Vukmanovic-Stejic *et al.*, unpublished observations). However, in addition to the human Tc2 clones from HIV patients (see above), two groups have obtained Tc2 clones from the gingiva (Wassenaar *et al.*, 1996) and peritoneum (Birkhofer *et al.*, 1996) of healthy individuals. The cytokine profiles of CD8$^+$ T cell subsets from mice, rats and humans are summarized in Table 3.1.

Table 3.1 Putative cytokine profiles of CD8$^+$ T cell subsets. – = no detectable cytokine production, + = 'low level' production, ++ = 'high level' production. References: 1. Sad *et al.* (1995); 2. Croft *et al.* (1994); 3. Noble, MacAry and Kemeny (1995); 4. MacAry, Holmes and Kemeny (1997); 5. Li *et al.* (1995); 6. Vukmanovic-Stejic, Vyas and Kemeny (unpublished observations); 7. Maggi *et al.* (1994); 8. Wassenaar *et al.* (1996); 9. Birkhofer, Rehbock and Fricke (1996); 10. Wassenaar *et al.* (1996).

Species/subset	IL-2	IFN-γ	IL-4	IL-5	IL-6	IL-10
Mouse Tc1[1,2]	+	++	–	–	–	–
Rat Tc1[3,4]	++	++	–	–	+	+
Human Tc1[5,6]	+	++	–	–	+	+
Rat Tc0[3,4]	+	++	++	+	–	+
Human Tc0[5,6]	+	++	+	++	+	+
Mouse Tc2[1,2]	+	–	+	++	+	+
Human Tc2[7,8,9,10]	+	–	+	+	+	+

$\gamma\delta$ TcR$^+$ CD8 T cell subsets

Some years ago a novel T cell subset was identified that expresses a TcR composed of CD3 associated, disulphide linked, TcR-γ and TcR-δ gene products (Allison *et al.*, 1988; Weiss *et al.*, 1986). In contrast to TcR-$\alpha\beta$ cells, newly activated $\gamma\delta$ T cells selectively home to specific tissues including the skin, lung, reproductive tract and intestine as intra-epithelial lymphocytes (IEL) (Itohara and Farr, 1990). In mice, the most abundant populations of IEL are CD8$^+$ single positive cells, and the development of these cells is comparable to $\alpha\beta$ T cells in that there is a degree of thymic dependency (Tatsumi *et al.*, 1993). However, recent studies in β_2-microglobulin and peptide transporter (TAP) knockout mice have suggested that $\gamma\delta$ T cells are restricted to non-classical MHC molecules such as CD1 and thymus leukaemia antigen (Sydora *et al.*, 1996). It has been proposed that CD8$^+$ IEL play an immunological gatekeeper role given their proximity to mucosal sites, that they are capable of producing IFN-γ and IL-5 (Taguchi *et al.*, 1991), and that they may be essential for the restoration of immunological homeostasis in the lung and gut following inflammatory disease (Hoang *et al.*, 1991; McMenamin *et al.*, 1995). However, precise evidence for this is lacking. It is likely, but as yet unproven, that $\gamma\delta$ CD8 T cells also differentiate into the Tc1 and Tc2 phenotypes, suggesting they could display a range of functions.

Effect of cytokines on CD8$^+$ T cells

As is the case for CD4$^+$ T cells, cytokines are the major factors that can polarize CD8$^+$ T cells into the Tc1/Tc2 phenotypes. Two studies indicated that IL-4 was a strong inducing factor for IL-4 secretion by CD8$^+$ T cells (Erard *et al.*, 1993; Seder *et al.*, 1992). The effect of different cytokines on the differentiation of naive CD8 precursors has subsequently been investigated further. In the mouse, results show that as for CD4 T cells, IL-12 and IFN-γ promote Type 1 and that IL-4, especially in the presence of anti-IFN-γ, promotes type 2 cells (Croft *et al.*; 1994, Sad *et al.*, 1995). In the rat (Noble *et al.*, 1995) IL-4 was found to inhibit the formation of IL-2 producing CD8$^+$ cells and promote IL-4 and IL-5 expression (Figure 3.1). Furthermore, IL-6 expression by CD8$^+$ T cells was reduced after priming with IL-4. In contrast to other reports, rat CD8$^+$ T cells cultured with IL-4 and anti-IFN-γ produced higher levels of IL-10 and IFN-γ, suggesting the existence of three distinct CD8$^+$ subsets (Noble *et al.*, 1995). Naive (neonatal) human CD8$^+$ T cells produced IL-4 and IL-5 after exposure to large amounts of IL-4 and IL-2, while IL-12 enhanced the capacity of primed cells to produce IFN-γ (Byun *et al.*, 1994). IL-12 priming for IFN-γ production was also observed by Trinchieri and colleagues, in addition to priming for high IL-10 synthesis (Gerosa *et al.*, 1996). Experiments performed on CD8$^+$ T cell clones derived from human peripheral blood are in agreement with these observations. IL-4, alone or

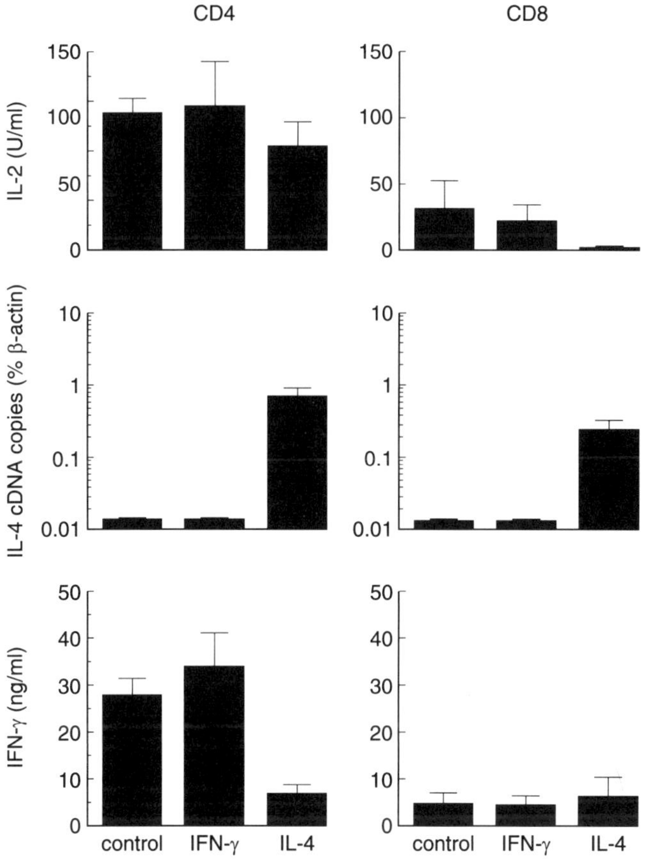

Figure 3.1 Production of IL-2 and IFN-γ, and expression of IL-4 mRNA, in rat splenic $CD4^+$ (left panels), and $CD8^+$ (right panels) T cells. Cells were stimulated in the presence of IFN-γ or IL-4 for 4–6 days, washed and restimulated with PMA and ionomycin before measurement of cytokine synthesis by ELISA or mRNA expression by quantitative PCR.

with IL-2 increases the proportion of clones producing IL-4 and IL-5, while IL-12 had a suppressive effect on the proportion of IL-4 and IL-5 producing clones, but enhanced both the levels of production of IFN-γ and IL-10 and the numbers of IL-10-positive clones.

Once formed, $CD8^+$ effector T cells are still susceptible to the effects of specific cytokines, both in terms of proliferation and function. IL-2 and IL-7 are reported to induce proliferation of Dengue virus-specific human $CD8^+$ CTL clones while a combination of IL-7 and IFN-γ augments their cytotoxic capacity (Livingston *et al.*, 1995). TGF-β and TNF-α have also been reported to stimulate cytotoxic $CD8^+$ T cells (Ioannides *et al.*, 1991; Lee and Rich, 1993).

IL-4 on the other hand, appears to limit $CD8^+$ T cell cytotoxicity (Ikemoto *et al.*, 1995), and in some cases a complete loss of cytotoxic capacity has been reported for IL-4 treated $CD8^+$ cells (Erard *et al.*, 1993). However, other studies indicate that Tc2-like cells retain cytotoxic capacity (Croft *et al.*, 1994; Sad *et al.*, 1995), and that IL-4 can induce CTL (Trenn *et al.*, 1988). IL-12 greatly enhances the proliferation of human Tc1, but not Tc0, clones (Sieling *et al.*, 1994). IL-12 is also reported to increase CTL activity of CD8 effectors (Mehrotra *et al.*, 1993).

Although once established, effector cells maintain a relatively stable cytokine pattern and rarely if ever switch to the opposite phenotype, some groups have reported modulation of cytokine production under the influence of certain cytokines. Tc1 clones stimulated in the presence of IL-4 maintained their level of IFN-γ production and did not acquire the ability to make IL-4 and IL-5. Tc2 clones cultured with IFN-γ maintained the production of IL-4 and IL-5 but some clones gained the ability to make IFN-γ (Sad and Mosmann, 1995). IL-4 inhibits the ability of differentiated Tc1 clones to synthesize IL-2 and induces an anergic state (Sad and Mosmann, 1995). IL-12 can also enhance IFN-γ production from established type 1 clones and transiently induces low levels of IFN-γ production from established type 2 clones (Manetti *et al.*, 1994a).

Expression of cell surface markers on $CD8^+$ T cell subsets

The data currently available on the specific cell surface phenotypes of different $CD8^+$ T cell subsets are somewhat limited. One potential marker, CD28, a costimulatory molecule for T cell activation, has been shown to be differentially expressed on $CD8^+$ T cell subsets isolated from chronic adult periodontitis tissue (Wassenaar *et al.*, 1996). Type 1 $CD8^+$ T cells (IFN-γ producing) were found to be $CD28^+$ and cytolytic whereas type 2 (IL-4, IL-5 producing) were $CD28^-$ and non-cytolytic, thus suggesting that type-1 $CD8^+$ T cells can be identified by their expression of CD28. CD30, a member of the TNF/nerve growth factor receptor superfamily, has been implicated as a marker for Tc0 and Tc2 $CD8^+$ T cells in both healthy and diseased subjects (Agrawal *et al.*, 1996; Manetti *et al.*, 1994b). High CD30 expression was observed on Tc2 cells, whereas intermediate expression was observed in Tc0 $CD8^+$ T cells, thus allowing further differentiation between CD8 subsets.

A small population of $CD8^+$ T cells capable of providing B cell help has been reported. These cells have been shown to produce type-2 cytokines and express CD40L. To distinguish these 'helper' $CD8^+$ T cells from cytotoxic $CD8^+$ T cells producing type-1 cytokines, other than by CD40L expression, the surface marker CDw60 has been proposed (Rieber and Rank *et al.*, 1994; Birkhofer *et al.*, 1996). Cytotoxic activity was found to reside in the $CDw60^-CD8^+$ population, whereas the helper activity was found in the $CDw60^+CD8^+$ population. It is likely that as the functions of distinct CD8$^+$ T cell subsets are clarified, they will be reflected in the cell surface molecules they express.

Effect of CD4$^+$ T cells on CD8$^+$ T cells

For many years immunologists have been aware that CD8$^+$ T cells may require CD4$^+$ T cell help in the form of IL-2, although in some circumstances CD8$^+$ T cells can function autonomously (Mason and Simmonds, 1988). In immunocompromised patients infected with cytomegalovirus CMV, disease severity is related to a deficiency of CD8$^+$ CTL. Survival of adoptively transferred autologous CMV-specific CTL clones in such patients was found to be dependent on the presence of CMV-specific CD4$^+$ T cells (Walter *et al.*, 1995). This can be obviated by transfection of the CD8$^+$ T cells with other growth promoting mechanisms such as the IL-1 receptor (Nagoya *et al.*, 1994).

Other CD8$^+$ T cell functions may also require CD4$^+$ T cells. For example, CD8$^+$ T cell homing to the pancreas in the non-obese diabetic mouse is reportedly CD4$^+$ T cell-dependent (Thivolet *et al.*, 1991), as is their subsequent differentiation into killer cells which act against islet beta cells (Yagi *et al.*, 1992). CD8 dependence on CD4 may be related to the strength of the stimulus (Heath *et al.*, 1993) and only be evident with weak stimulation. Moreover, the development of polarized Tc1 and Tc2 CD8$^+$ T cell responses is no doubt influenced by CD4 T cells, since they generally produce higher levels of cytokines such as IL-4 than comparable CD8$^+$ T cells, and this is supported by data showing the existence of Tc2 cells in the presence of a vigorous Th2 response (Coyle *et al.*, 1995).

CD8$^+$ T cell function

Historically, CD4$^+$ T cells have been regarded as helper cells that assist B cell antibody production, and CD8$^+$ T cells as cytotoxic cells that kill virally infected cells. CD8$^+$ T cells are normally MHC class-I restricted while CD4$^+$ T cells are generally MHC class-II restricted. This view of the division of labour and MHC restriction of these two T cell subsets is, in many ways, still true today. There are, however, a number of exceptions and CD4$^+$ T cells are now known to be capable of killing (Kuge *et al.*, 1995; Rahelu *et al.*, 1993), while CD8$^+$ T cells may provide help for antibody responses (Erard *et al.*, 1993; Holmes *et al.*, 1996; Takahashi *et al.*, 1991). The discovery of the Th1 and Th2 subsets of CD4 T cells had a huge impact on our understanding of immunology because the different cytokine profiles were closely associated with specific functions; i.e. Th1 cells were inflammatory T cells, while Th2 cells were helpers for antibody production. Such a correlation between cytokine profile and function has yet to be defined clearly for CD8$^+$ T cells.

In the past $CD8^+$ T cells have been divided into cytotoxic T cells and suppressor T cells, which switch off established immune responses or maintain tolerance to particular antigens. However, the existence of suppressor cells as a distinct subset of $CD8^+$ T cells has been questioned (Moller, 1988) and the newly described $CD8^+$ cell subsets do not fit neatly into these categories. The question therefore remains as to whether heterogeneity of cytokine production in $CD8^+$ T cells is important for their function. A hypothetical scheme for $CD8^+$ T cell differentiation and functional diversity is illustrated in Figure 3.2. Since some $CD8^+$ T cells can produce potent immunoregulatory cytokines such as IL-4 (Li *et al.*, 1995), IL-10 (Salgame *et al.*, 1991a) and TGF-β (Weiner *et al.*, 1994), they may play distinct roles in both the induction and effector phases of the immune response which are considered below.

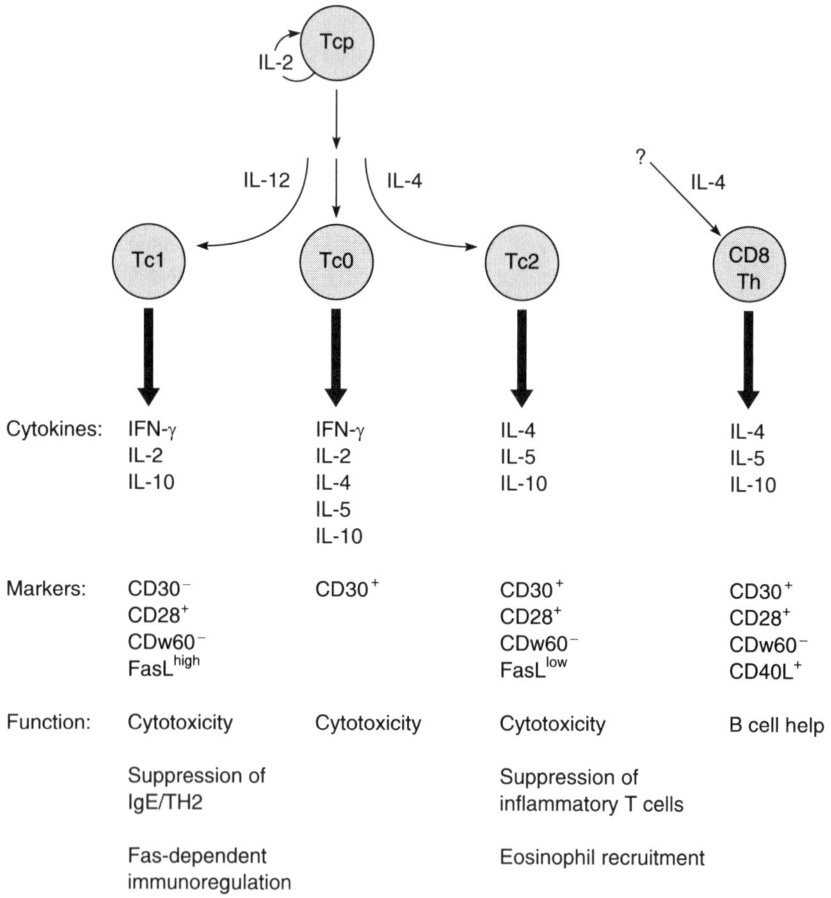

Figure 3.2 Hypothetical scheme for the differentiation and activities of $CD8^+$ T cell subsets. FasL: Fas(CD95) ligand; CD40L: CD40 ligand; Tcp: naive CD8 precursor.

CYTOTOXICITY

The prime function of CD8$^+$ T cells is to kill virally infected target cells and tumour cells. They are generally more cytotoxic than CD4$^+$ T cells. Unlike NK cells, CD8$^+$ T cells can undergo clonal selection. Although the mouse Tc1 and Tc2 subsets are both cytotoxic (Croft *et al.*, 1994) the mechanisms they use for killing, and hence their effectiveness in eliminating a particular target, may differ. Tc1-like CD8$^+$ T cells from perforin-knockout mice kill more efficiently than Tc2-like CD8s using the Fas pathway (Carter and Dutton, 1995) and CD8$^+$ T cells from perforin knockout mice more readily secrete cytokines (Sad *et al.*, 1997). Thus there appears to be a segregation of Fas-ligand-mediated cytotoxicity and other, cytokine-dependent CD8$^+$ T cell functions.

Non-cytotoxic CD8$^+$ T cells have been described (Erard *et al.*, 1993; Rieber and Rank, 1994), and mice undergoing an acute Th2 anti-helminth response showed reduced CD8-dependent cytotoxicity towards an unrelated virus (Actor *et al.*, 1993). In addition, alloreactive Tc2 populations showed a reduced graft-versus-leukaemia effect *in vivo* and failed to induce graft-versus-host disease, unlike Tc1 cells (Fowler *et al.*, 1997). The dependence of CTL development on CD4 T cell-derived IL-2 may also differ between CD8 subsets that secrete different levels of IL-2 (Noble *et al.*, 1995). Autonomous CTL (Mason and Simmonds, 1988) which produce sufficient IL-2 to drive their own expansion would be required in the response to intracellular Ag that failed to enter the MHC class II pathway, while those CD8$^+$ T cells that fail to produce IL-2 would clearly be dependent on CD4 T cells.

B CELL HELP

Some cases of non-cytotoxic CD8$^+$ T cells that could provide help for Ig synthesis have been reported, for example in HIV infection (Maggi *et al.*, 1994), and in CD8$^+$ T cell clones derived from skin window chamber cells from delayed type hypersensitivity (DTH) reactions elicited by either PPD or Candida (MacPhee *et al.*, 1993). CD8$^+$ T cells vigorously stimulated *in vitro* in the presence of high concentrations of IL-4 became CD4$^-$CD8$^-$, secreted large amounts of IL-4 and were capable of switching B cells to make IgE (Erard *et al.*, 1993; Le Gros and Erard, 1994). This is confirmed by the finding that some CD8$^+$ clones express CD40L, a critical signal for antibody production (Cronin *et al.*, 1995) and can provide B cell help *in vitro* for switching to IgE (Fujihashi *et al.*, 1993; Maggi *et al.*, 1994; Paganelli *et al.*, 1995) and for IgG synthesis (Takahashi *et al.*, 1991). It is therefore likely that CD8$^+$ T cells can contribute to antibody responses and that, as for CD4 cells, this is associated with reduced cell-mediated immunity.

RECRUITMENT AND ACTIVATION OF INFLAMMATORY CELLS

Effective immunity to pathogens depends on effective collaboration between cells of the innate and the adaptive immune system which is mediated by cytokines. Given the spectrum of cytokines now known to be produced by $CD8^+$ T cells, it seems likely that they could be involved in recruiting and stimulating macrophages, NK cells, neutrophils, eosinophils or mast cells/basophils in order to eliminate infection, in addition to their direct cytotoxicity to intracellular microbes. The ability of $CD8^+$ T cells to produce IFN-γ, even after a primary stimulation (Noble *et al.*, 1995), indicates that they have the potential to activate macrophages. By stimulating MHC class II expression on a variety of cells types (Pober *et al.*, 1983), IFN-γ signals the arrival of a new pathogen to the immune system. Thus, by virtue of their ability to produce IFN-γ rapidly and copiously, $CD8^+$ T cells may act as an early warning system that helps initiate inflammation. IL-5-producing $CD8^+$ T cells may help recruit eosinophils to the lung in allergic asthma (Coyle *et al.*, 1995; Till *et al.*, 1995). In other peripheral tissues where $\gamma\delta$, intra-epithelial Tc2-like $CD8^+$ T cells have been found (Taguchi *et al.*, 1991), it is possible that IL-4 production could stimulate mast cell growth and activation (Arai *et al.*, 1990). $CD8^+$ T cells have been shown capable of eliciting DTH reactions to viral antigens (Kundig *et al.*, 1992).

T CELL-MEDIATED SUPPRESSION

The confusing array of T suppressor cells, factors and mechanisms of action that have been described make it difficult to associate specific cytokine profiles with phenomena of immune suppression. Various pathways have been described by which suppressor T cells become activated. These include recognition of peptides from foreign antigen, the TCR of activated CD4 cells, or non-antigen-specific factors in association with classical (McMenamin and Holt, 1993) and non-classical (Noble *et al.*, 1998b) MHC class I or MHC class II (Salgame *et al.*, 1991b). The effector mechanisms by which suppressor cells modulate the immune response are also unclear. Suppression may be mediated by the production of particular cytokines such as IL-4, TGF-β or IFN-γ, or by killing specific effector target cells (Bloom *et al.*, 1992). Perhaps the clearest association of a CD8 cytokine profile with suppression is that seen in human leprosy, where $CD8^+$ T cells that secrete IL-4 and IL-10 block the inflammatory Th1 response (Salgame *et al.*, 1991a). $CD8^+$ cell derived TGF-β has been implicated in oral tolerance (Weiner *et al.*, 1994).

As well as preventing excessive responses to foreign antigen, $CD8^+$ suppressor cells may help to maintain self-tolerance. $CD8^+$ T cells express high levels of Fas-ligand mRNA and are particularly efficient at Fas-dependent killing. Recent data suggest that $CD8^+$ T cells can kill autologous, activated CD4 T cells by inducing Fas-mediated apoptosis (Noble *et al.*, 1998a). Activation-induced cell death of self-reactive T cells (mediated by Fas or a lack of 'danger' signals) is thought to be important in the prevention of autoimmune responses (Matzinger, 1994), as evidenced by the development of autoimmunity in Fas or Fas-ligand deficient mice (Nagata and Suda, 1995). Taken together, these data suggest that $CD8^+$ T cells expressing high levels of Fas-ligand or producing other apoptosis-promoting molecules such as TNF-α (Zheng *et al.*, 1995), might suppress anti-self responses or prevent hypersensitivity to foreign antigen.

Infectious disease

Cytotoxic $CD8^+$ T cells are primarily involved in fighting infection caused by intracellular viruses and bacteria. For example, HIV-specific $CD8^+$ T cells are protective in HIV-1 Scid-hu mice (van Kuyk *et al.*, 1994) and $CD8^+$ cell numbers are positively associated with long-term survival in HIV-infected patients (Zanussi *et al.*, 1996). $CD8^+$ T cells are also involved in protection against *Cryptococcus neoformans* (Huffnagle *et al.*, 1994) where depletion of $CD8^+$ T cells reduces survival of infected mice (Mody *et al.*, 1994). Recently $CD8^+$ T cells with new cytokine profiles have been described in infectious diseases. Loss of HIV-specific $CD8^+$ T cells is associated with disease progression and a shift in cytokine production within the infected host from a Th1 to a Th2 pattern (Barker *et al.*, 1995) and Tc2-like $CD8^+$ T cells have been described in HIV-infected patients (Romagnani *et al.*, 1994). IL-4 and IL-5 secreting $CD8^+$ T cells are also present in the lung (Ying, 1997), with especially high levels of IL-5 production in the lungs of asthmatics (Till, 1995). In the immune response to inhaled viruses, virus specific Tc2-like $CD8^+$ T cells are associated with eosinophilia (Coyle *et al.*, 1995), and such cells could exacerbate asthma during viral infection. IL-5 producing $CD8^+$ T cells are also found in allergic eosinophilic gastroenteritis (Jaffe *et al.*, 1994), and Tc2 cells, secreting IL-4 and IL-10, have been described in the skin of patients with cutaneous leishmaniasis (Uyemura *et al.*, 1993). Increased IL-4 production by $CD8^+$ T cells is reported in patients with B-chronic lymphocytic leukaemia (Mu *et al.*, 1997).

$CD8^+$ T cells in allergy

CHEMICAL ALLERGY

Sensitization with different chemical allergens induces polarized Th1/Th2 immune responses following chronic exposure to 2,4-dinitrochlorobenzene (DNCB) and trimellitic anhydride (TMA), respectively (Dearman *et al.*, 1995; Chapter 11, this volume). The cytokines detected (IL-4, IFN-γ and IL-10) were mainly the product of $CD4^+$ T cells (Dearman *et al.*, 1996), although most of the IFN-γ detected following sensitization with TMA was derived from $CD8^+$ T cells. Drug-specific $CD8^+$ T cells have been documented in individuals allergic to the drugs sulphonamides, phenytoin, and carbamazepine (Mauri-Hellweg *et al.*, 1995) and in the lesions of patients with penicillin allergy (Hertl *et al.*, 1993). In toluene diisocyanate (TDI) asthma, IL-5 secreting $CD8^+$ T cells have been cloned from lung (Del Prete *et al.*, 1993; Maestrelli *et al.*, 1994). Hapten-specific $CD8^+$ T cells have also been found in the lesions of patients with contact sensitivity reactions to nickel (Bour *et al.*, 1994).

Xu and colleagues report that sensitization with 2,4-dinitrofluorobenzene (DNFB) or oxazolone, induces lymph node $CD8^+$ T cells that produce IFN-γ but not IL-4 or IL-10, and $CD4^+$ T cells producing IL-4 and IL-10 and no or little detectable IFN-γ (Xu *et al.*, 1996). In their animals, they found that depletion of $CD8^+$ T cells reduced the contact sensitivity response to DNFB and oxazolone, while depletion of $CD4^+$ T cells resulted in increased and prolonged contact sensitivity, suggesting that Th2-like $CD4^+$ T cells suppressed a $CD8^+$ T cell-mediated contact sensitivity response. By contrast, $CD8^+$ Tc2-like cells generated by sub-sensitizing doses of picryl chloride or oxazolone were found to give rise to hapten-specific tolerance (Steinbrink *et al.*, 1996).

Ptak and Askenase (1992) have described a regulatory population of $CD8^+$ $\gamma\delta$ T cells that assisted chemical contact sensitivity effector $\alpha\beta$ T cells. These cells were found to be non-antigen-specific, non-MHC-restricted and $CD3^+$ (Askenase *et al.*, 1995). It appears likely that distinct subsets of $CD8^+$ T cells that can function as effector cells or regulatory cells in drug allergy and contact sensitivity will be described; but until these different families of T cells are cloned, it is unlikely that their mode of action and the different immunoregulatory pathways will be determined.

IgE REGULATION BY $CD8^+$ T CELLS

The ability of $CD8^+$ T cells to regulate the production of IgE is a well-recognized example of T cell suppression which may be vital to prevent allergic disease. Using an early CD8 marker, the Lyt-2 antigen (Hirst *et al.*, 1975), it became clear that the cells that inhibited IgE responses were $CD8^+$ T cells (Marcelletti and Katz, 1984a,b; Okumura *et al.*, 1977). Rat (Sedgwick and Holt, 1984, 1985) and mouse (McMenamin *et al.*, 1994; Renz *et al.*, 1994) IgE inhibitory $CD8^+$ T cells have been generated following either inhalation of antigen or parenteral immunization (Diaz-Sanchez and Kemeny, 1990, 1991; Diaz-Sanchez *et al.*, 1993a,b; Thorpe *et al.*, 1989). MHC class-I restricted mouse $CD8^+$ T cells specific for exogenous protein (bovine casein) that suppress IgE have been described (Hisatsune *et al.*, 1995) and more recently rat MHC class-I restricted ovalbumin-specific clones have been generated which also suppress IgE (MacAry *et al.*, 1998). MHC class-II restricted and $\gamma\delta$ TcR^+ $CD8^+$ T cells that suppress IgE have also been described (Chen *et al.*, 1992). Depletion of $CD8^+$ T cells in immunized rats enhances antigen-specific IL-4 and suppress IFN-γ synthesis (Holmes *et al.*, 1997), suggesting that $CD8^+$ T cells switch the $CD4^+$ T cell response from a Th2 to a Th1 phenotype.

Secretion of IFN-γ by $CD8^+$ T cells is the most obvious explanation for their ability to inhibit IgE production. IFN-γ production during the formative stages of a $CD4^+$ T cell response could inhibit the growth of Th2 cells directly, whilst favouring the expansion of Th1 cells (McMenamin and Holt, 1993; McMenamin *et al.*, 1994, 1995); alternatively, they could stimulate macrophages to produce nitric oxides as part of a non-specific suppressive process (Albina *et al.*, 1991). $CD8^+$ T cell-derived IFN-γ could act directly on B-cells to inhibit IgE synthesis (Snapper and Paul, 1987) and cytolytic $CD8^+$ T cells could target Ag-specific B cells or antigen presenting cells. However, the ability of OVA-specific $CD8^+$ T cell clones to inhibit IgE is unrelated to the levels of IFN-γ (Holmes *et al.*, 1998) they produce or to their *in vitro* cytolytic activity for OVA pulsed targets (MacAry *et al.*, 1998). Thus IgE suppressor cells may be distinct from the Tc1, IFN-γ-producing subset and may produce other cytokines that suppress IgE responses. It seems likely that the mechanisms underlying immunoregulatory activities of $CD8^+$ T cells are complex and that several factors including specific cytokines, limited cytotoxicity, cell motility and survival could combine to mediate the effects of these cells *in vivo*.

Conclusions

During the past few years, there has been a tremendous expansion in our understanding of the different subpopulations of $CD8^+$ T cells. We now know that, as for $CD4^+$ T cells, they can be divided into subsets according to the cytokines they secrete. These subsets may differ in their cytotoxic capacities and in some cases provide help for B cell antibody production or be involved in induction of inflammatory responses. In addition, there are a host of cross-regulatory networks between different $CD4^+$ and $CD8^+$ subsets that control the magnitude and duration of immune responses. The observation that some antigens that are normally presented by MHC class II and seen by $CD4^+$ T cells, can be presented by MHC class I and stimulate $CD8^+$ T cells increases the possibilities for such interactions and this might be targeted by DNA vaccines (Hsu *et al.*, 1996a,b). During the next few years, we can expect that our understanding of $CD8^+$ T cell subsets and their role in immunity will increase. The challenge is to understand how these cells function *in vivo*.

References

Actor, J.K., Shirai, M., Kullberg, M.C., Buller, A., Sher, A. and Berzofsky, J.A. (1993) Helminth infection results in decreased virus-specific CD8 cytotoxic T-cell and Th1 cytokine responses as well as delayed virus clearance. *Proc. Natl. Acad. Sci. USA*, **90**, 948–52.

Agrawal, B., Reddish, M. and Longenecker, B.M. (1996) CD30 expression on human CD8 T cells isolated from peripheral blood lymphocytes of normal donors. *J. Immunol.*, **157**, 3229–34.

Albina, J. E., Abate, J. A. and Henry, W. Jr (1991) Nitric oxide production is required for murine resident peritoneal macrophages to suppress mitogen-stimulated T cell proliferation. Role of IFN-γ in the induction of the nitric oxide-synthesizing pathway. *J. Immunol.*, **147**, 144–8.

Allison, J.P., Havran, W.L., Asarnow, D. and Tigelaar, R.A. (1988). $\gamma\delta$ antigen receptors of Thy-1^+ dendritic epidermal cells: implications for thymic differentiation. *Immunol. Res.*, **7**, 292–301.

Arai, K., Lee, F., Miyajima A., Miyatake S., Arai N. and Yokota T. (1990) Cytokines: coordinators of immune and inflammatory responses. *Annu. Rev. Immunol.*, **59**, 783–836.

Askenase, P. W., Szczepanik, M., Ptak, M., Paliwal, V. and Ptak, W. (1995) $\gamma\delta$ T cells in normal spleen assist immunized alpha beta T cells in the adoptive cell transfer of contact sensitivity. Effect of *Bordetella pertussis*, cyclophosphamide, and antibodies to determinants on suppressor cells. *J. Immunol.*, **154**, 3644–53.

Barker, E., Mackewicz, C. E. and Levy, J. A. (1995) Effects of TH1 and TH2 cytokines on $CD8^+$ cell response against human immunodeficiency virus: implications for long-term survival. *Proc. Natl. Acad. Sci. USA*, **92**, 11135–9.

Birkhofer, A., Rehbock, J. and Fricke, H. (1996) T lymphocytes from the normal human peritoneum contain high frequencies of Th2-type $CD8^+$ T cells. *Eur. J. Immunol.*, **26**, 957–60.

Bloom, B. R., Modlin, R. L. and Salgame, P. (1992) Stigma variations: observations on suppressor T cells and leprosy. *Annu. Rev. Immunol.*, **10**, 453–88.

Bour, H., Nicolas, J. F., Garrigue, J. L., Demidem, A. and Schmitt, D. (1994) Establishment of nickel-specific T cell lines from patients with allergic contact dermatitis: comparison of different protocols. *Clin. Immunol. Immunopathol.*, **73**, 142–5.

Byun, D. G., Demeure, C. E., Yang, L. P., Shu, U., Ishihara, H., Vezzio, N., Gately, M. K. and Delespesse, G. (1994) *In vitro* maturation of neonatal human CD8 T lymphocytes into IL-4- and IL-5-producing cells. *J. Immunol.*, **153**, 4862–71.

Carter, L. L. and Dutton, R. W. (1995) Relative perforin- and Fas-mediated lysis in T1 and T2 CD8 effector populations. *J. Immunol.*, **155**, 1028–31.

Chen, Y., Takata, M., Maiti, P. K., Rector, E. S. and Sehon, A. H. (1992) Characterization of suppressor T cell clones derived from a mouse tolerized with conjugates of ovalbumin and monomethoxypolyethylene glycol. *Cell Immunol.*, **142**, 16–27.

Chen, Y., Inobe, J. and Weiner, H. L. (1995) Induction of oral tolerance to myelin basic protein in CD8-depleted mice: both $CD4^+$ and $CD8^+$ cells mediate active suppression. *J. Immunol.*, **155**, 910–16.

Coyle, A. J., Erard, F., Bertrand, C., Walti, S., Pircher, H. and Le Gros, G. (1995) Virus-specific $CD8^+$ cells can switch to interleukin 5 production and induce airway eosinophilia. *J. Exp. Med.*, **181**, 1229–33.

Croft, M., Carter, L., Swain, S. L. and Dutton, R. W. (1994) Generation of polarized antigen-specific CD8 effector populations: reciprocal action of interleukin (IL)-4 and IL-12 in promoting type 2 versus type 1 cytokine profiles. *J. Exp. Med.*, **180**, 1715–28.

Cronin, D.C., Stack, R. and Fitch, F. W. (1995) IL-4-producing $CD8^+$ T cell clones can provide B cell help. *J. Immunol.*, **154**, 3118–27.

Dearman, R. J., Basketter, D. A. and Kimber, I. (1995) Differential cytokine production following chronic exposure of mice to chemical respiratory and contact allergens. *Immunology*, **86**, 545–50.

Dearman, R. J., Moussavi, A., Kemeny, D. M. and Kimber, I. (1996) Contribution of $CD4^+$ and $CD8^+$ T lymphocyte subsets to the cytokine secretion patterns induced in mice during sensitization to contact and respiratory chemical allergens. *Immunology*, **89**, 502–10.

Del Prete, G. F., De Carli, M., D'Elios, M. M., Maestrelli, P., Ricci, M., Fabbri, L. and Romagnani, S. (1993) Allergen exposure induces the activation of allergen-specific Th2 cells in the airway mucosa of patients with allergic respiratory disorders. *Eur. J. Immunol.*, **23**, 1445–9.

Diaz-Sanchez, D. and Kemeny, D. M. (1990) The sensitivity of rat $CD8^+$ and $CD4^+$ T cells to ricin *in vivo* and *in vitro* and their relationship to IgE regulation. *Immunology*, **69**, 71–7.

Diaz-Sanchez, D. and Kemeny, D.M. (1991) Generation of a long-lived IgE response in high and low responder strains of rat by co-administration of ricin and antigen. *Immunology*, **72**, 297–303.

Diaz-Sanchez, D., Lee, T. H. and Kemeny, D. M. (1993a) Ricin enhances IgE responses by inhibiting a subpopulation of early-activated $CD8^+$ T cells. *Immunology*, **78**, 226–36.

Diaz-Sanchez, D., Noble, A., Staynov, D. Z., Lee, T. H. and Kemeny, D. M. (1993b) Elimination of IgE regulatory $CD8^+$ T cells *in vivo* differentially modulates the ability of splenocytes to produce IL-4 and IFNγ but not IL-2. *Immunology*, **78**, 513–19.

Erard, F., Wild, M. T., Garcia-Sanz, J. A. and Le Gros, G. (1993) Switch of CD8 T cells to noncytolytic CD8-CD4- cells that make TH2 cytokines and help B cells. *Science*, **260**, 1802–5.

Firestein, G. S., Roeder, W. D., Laxer, J. A., Townsend, K. S., Weaver, C. T., Hom, J. T., Linton, J., Torbett, B. E. and Glasebrook, A. L. (1989) A new murine $CD4^{(+)}$ T cell subset with an unrestricted cytokine profile. *J. Immunol.*, **143**, 518–25.

Fong, T. A. and Mosmann, T. R. (1990) Alloreactive murine $CD8^+$ T cell clones secrete the Th1 pattern of cytokines. *J. Immunol.*, **144**, 1744–52.

Fowler, D. H., Breglio, J., Nagel, G., Echaus, M. A. and Gress, R. E. (1997) Allospecific CD8 Tc1 and Tc2 populations in graft-versus-leukemia effect and graft-versus-host disease. *J. Immunol.*, **157**, 4811–21.

Fujihashi, K., Yamamoto, M., McGhee, J. R. and Kiyono, H. (1993) alpha beta T cell receptor-positive intraepithelial lymphocytes with $CD4^+$, $CD8^-$ and $CD4^+$, $CD8^+$ phenotypes from orally immunized mice provide Th2-like function for B cell responses. *J. Immunol.*, **151**, 6681–91.

Gerosa, F., Paganin, C., Peritt, D., Paiola, F., Scupoli, M. T., Aste-Amezaga, M., Frank, I. and Trinchieri, G. (1996) Interleukin-12 primes human CD4 and CD8 T cell clones for high production of both interferon-gamma and interleukin-10. *J. Exp. Med.*, **183**, 6, 2559–69.

Glasebrook, A. L., Kelso, A. and MacDonald, H. R. (1982) Cytolytic T lymphocyte clones that proliferate autonomously to specific alloantigenic stimulation. II. Relationship of the Lyt-2 molecular complex to cytolytic activity, proliferation, and lymphokine secretion. *J. Immunol.*, **130**, 1545–51.

Heath, W. R., Kjer-Nielsen, L. and Hoffmann, M. W. (1993) Avidity for antigen can influence the helper dependence of $CD8^+$ T lymphocytes. *J. Immunol.*, **151**, 5993–6001.

Hertl, M., Geisel, J., Boecker, C. and Merk, H. F. (1993) Selective generation of $CD8^+$ T-cell clones from the peripheral blood of patients with cutaneous reactions to beta-lactam antibiotics. *Br. J. Dermatol.*, **128**, 619–26.

Hisatsune, T., Nishijima, K., Kohyama, M., Kato, H. and Kaminogawa, S. (1995) $CD8^+$ T cells specific to the exogenous antigen: Mode of antigen recognition and possible implication in immunosuppression. *J. Immunol.*, **154**, 88–96.

Hirst, J.A., Beverley, P.C., Kisielow, P., Hoffmann, M.K. and Oettgen H.F. (1975) Ly antigens: markers of T cell function on mouse spleen cells. *J. Immunol.*, **115**, 1555–7.

Hoang, P., Dalton, H.R. and Jewell, D.P. (1991): Human colonic intra-epithelial lymphocytes are suppressor cells. *Clin. Exp. Immunol.*, **85**, 498–503.

Holmes, B. J., Diaz-Sanchez, D., Lawrence, R. A., Maizels, R. M. and Kemeny, D. M. (1996) The contrasting effects of CD8$^+$ T cells on primary, established and *Nippostrongylus brasiliensis*-induced IgE responses. *Immunology*, **88**, 252–60.

Holmes, B. J., MacAry, P. A., Noble, A. and Kemeny, D. M. (1997) Antigen-specific CD8$^+$ T cells inhibit immunoglobulin E responses and interleukin-4 production by CD4$^+$ T cells. *Eur. J. Immunol.*, **27**, 2657–65.

Hsu, C. H., Chua, K. Y., Tao, M. H., Huang, S. K. and Hsieh, K. H. (1996a) Inhibition of specific IgE response *in vivo* by allergen-gene transfer. *Int. Immunol.*, **8**, 1405–11.

Hsu, C. H., Chua, K. Y., Tao, M. H., Lai, Y. L., Wu, H. D., Huang, S. K. and Hsieh, K. H. (1996b) Immunoprophylaxis of allergen-induced immunoglobulin E synthesis and airway hyperresponsiveness *in vivo* by genetic immunization. *Nat. Med.*, **2**, 540–4.

Huffnagle, G. B., Lipscomb, M. F., Lovchik, J. A. Hoag, K. A. and Street, N. E. (1994) The role of CD4$^+$ and CD8$^+$ T cells in the protective inflammatory response to a pulmonary cryptococcal infection. *J. Leukoc. Biol.*, **55**, 35–42.

Ikemoto, K., Pollard, R. B., Fukumoto, T., Morimatsu, M. and Suzuki, F. (1995) CD8$^+$ type-2 T cells enhance the severity of acute herpes virus infection in mice. *Immunol. Lett.*, **47**, 63–72.

Ioannides, C. G., Rashed, S., Fisk, B., Fan, D., Itoh, K. and Freedman, R. S. (1991) Lymphocytes infiltrating ovarian malignant ascites: modulation of IL-2-induced proliferation by IL-4 and of selective increase in CD8$^+$ T cells by TNF-alpha. *Lymphokine Cytokine Res.*, **10**, 307–15.

Itohara, S. and Farr, A.G. (1990) Homing of a $\gamma\delta$ thymocyte subset with homogeneous T-cell receptors to mucosal epithelia. *Nature*, **343**, 754–8.

Jaffe, J. S., James, S. P., Mullins, G. E., Braun-Elwert, L., Lubensky, I. and Metcalfe, D. D. (1994) Evidence for an abnormal profile of interleukin-4 (IL-4), IL-5, and γ-interferon (γ-IFN) in peripheral blood T cells from patients with allergic eosinophilic gastroenteritis. *J. Clin. Immunol.*, **14**, 299–309.

Kuge, S., Miura, Y., Nakamura, Y., Mitomi, T., Habu, S. and Nishimura, T. (1995) Superantigen-induced human CD4$^+$ helper/killer T cell phenomenon. Selective induction of Th1 helper/killer T cells and application to tumor immunotherapy. *J. Immunol.*, **154**, 1777–85.

Kundig, T. M., Althage, A., Hengartner, H. and Zinkernagel, R. M. (1992) Skin test to assess virus-specific cytotoxic T-cell activity. *Proc. Natl. Acad. Sci. USA*, **89**, 7757–61.

Le Gros, G. and Erard, F. (1994) Non-cytotoxic, IL-4, IL-5, IL-10 producing CD8$^+$ T cells: their activation and effector functions. *Curr. Opin. Immunol.*, **6**, 453–7.

Lee, H. M. and Rich, S. (1993) Differential activation of CD8$^+$ T cells by transforming growth factor-beta 1. *J. Immunol.*, **151**, 668–77.

Li, Y., Richards, D., Noble, A. and Kemeny, D. M. (1995) Cytokine production by highly purified human CD8$^+$ T cells. *Int. Arch. Allergy Immunol.*, **107**, 354–5.

Livingston, P. G., Toomey, S., Kurane, I., Janus, J. and Ennis, F. A. (1995) Modulation of the functions of dengue virus-specific human CD8$^+$ cytotoxic T cell clone by IL-2, IL-7 and IFN-γ. *Immunol. Invest.*, **24**, 619–29.

MacAry, P.A., Holmes, B.J. and Kemeny D.M. (1997) Ovalbumin-specific, MHC class I restricted, $\alpha\beta$ positive, Tc1 and Tc0 CD8$^+$ T cell clones mediate the *in vivo* inhibition of rat IgE. *J. Immunol.*, in press.

MacPhee, M. J., Gordon, J., Christou, N. V., Sanchez-Cantu, L. and Rode, H. H. (1993) Cells recovered from human DTH reactions: phenotypic and functional analysis. *Cell Immunol.*, **151**, 80–96.

Maestrelli, P., Del Prete, G. F., De Carli, M., D'Elios, M. M., Saetta, M., Di Stefano, A., Mapp, C. E., Romagnani, S. and Fabbri, L. M. (1994) CD8 T-cell clones producing interleukin-5 and interferon-γ in bronchial mucosa of patients with asthma induced by toluene diisocyanate. *Scand. J. Work Environ. Health*, **20**, 376–81.

Maggi, E., Del Prete, G., Macchia, D., Parronchi, P., Tiri, A., Chretien, I., Ricci, M. and Romagnani, S. (1988) Profiles of lymphokine activities and helper function for IgE in human T cell clones. *Eur. J. Immunol.*, **18**, 1045–50.

Maggi, E., Giudizi, M. G., Biagiotti, R., Annunziato, F., Manetti, R., Piccinni, M. P., Parronchi, P., Sampognaro, S., Giannarini, L., Zuccati, G. and Romagnani, S. (1994) Th2-like $CD8^+$ T cells showing B cell helper function and reduced cytolytic activity in human immunodeficiency virus type 1 infection. *J. Exp. Med.*, **180**, 489–95.

Manetti, R., Gerosa, F., Giudizi, M. G., Biagiotti, R., Parronchi, P., Piccinni, M.P., Sampognaro, S., Maggi, E., Romagnani, S. and Trinchieri, G. (1994a) Interleukin 12 induces stable priming for interferon gamma (IFN-gamma) production during differentiation of human T helper (Th) cells and transient IFN-gamma production in established Th2 cell clones. *J. Exp. Med.*, **179**, 4, 1273–83.

Manetti, R., Annunziato, F., Biagiotti, R., Giudizi, M. G., Piccinni, M. P., Giannarini, L., Sampognaro, S., Parronchi, P., Vinante, F., Pizzolo, G., Maggi, E. and Romagnani, S. (1994b) CD30 expression by $CD8^+$ T cells producing type 2 helper cytokines. Evidence for large numbers of $CD8^+CD30^+$ T cell clones in human immunodeficiency virus infection. *J. Exp. Med.*, **180**, 2407–11.

Marcelletti, J.F. and Katz, D.H. (1984a) FcR epsilon$^+$ lymphocytes and regulation of the IgE antibody system. I. A new class of molecules, termed IgE-induced regulants (EIR), which modulate FcR epsilon expression by lymphocytes. *J. Immunol.*, **133**, 2821–8.

Marcelletti, J.F. and Katz, D.H. (1984b) FcR epsilon$^+$ lymphocytes and regulation of the IgE antibody system. II. FcR epsilon$^+$ B lymphocytes initiate a cascade of cellular and molecular interactions that control FcR epsilon expression and IgE production. *J. Immunol.*, **133**, 2829–36.

Mason, D. W. and Simmonds, S. J. (1988) The autonomy of $CD8^+$ T cells *in vitro* and *in vivo*. *Immunology*, **65**, 249–57.

Matzinger, P. (1994) Tolerance, danger, and the extended family. *Annu. Rev. Immunol.*, **12**, 991–1045.

Mauer, S. C., Hussey, R. E., Penta, A. C., Fitzgerald, K. A., Stadler, B. M., Schlossman, S. F. and Reinherz, E. L. (1982) Cellular origin of interleukin-2 (IL-2) in man: evidence for stimulus-restricted IL-2 production by $T4^+$ and $T8^+$ T lymphocytes. *J. Immunol.*, **129**, 1076–9.

Mauri-Hellweg, D., Bettens, F., Mauri, D., Brander, C., Hunziker, T. and Pichler, W. J. (1995) Activation of drug-specific $CD4^+$ and $CD8^+$ T cells in individuals allergic to sulfonamides, phenytoin, and carbamazepine. *J. Immunol.*, **155**, 462–72.

McMenamin, C. and Holt, P. G. (1993) The natural immune response to inhaled soluble protein antigens involves major histocompatability complex (MHC) class-I-restricted $CD8^+$ T cell mediated, but MHC class II-restricted $CD4^+$ T cell-dependent immune deviation resulting in selective suppression of immunoglobulin E production. *J. Exp. Med.*, **178**, 889–99.

McMenamin, C., Pimm, C., McKersey, M. and Holt, P. G. (1994) Regulation of IgE responses to inhaled antigen in mice by antigen-specific $\gamma\delta$ T cells. *Science*, **265**, 1869–71.

McMenamin, C., McKersey, M., Kuhnlein, P., Hunig, T. and Holt, P. G. (1995) $\gamma\delta$ T cells down-regulate primary IgE responses in rats to inhaled soluble protein antigens. *J. Immunol.*, **154**, 4390–4.

Mehrotra, P. T., Wu, D., Crim, J. A., Mostowski, H. S. and Siegel, J. P. (1993) Effects of IL-12 on the generation of cytotoxic activity in human $CD8^+$ T lymphocytes. *J. Immunol.*, **151**, 5, 2444–52.

Miller, A., Lider, O., Roberts, A. B., Sporn, M. B. and Weiner, H. L. (1992) Suppressor T cells generated by oral tolerization to myelin basic protein suppress both *in vitro* and *in vivo* immune responses by the release of transforming growth factor beta after antigen-specific triggering. *Proc. Natl. Acad. Sci. USA*, **89**, 421–5.

Mingari, M.C., Moretta, A., Maggi, E., Pantaleo, G., Gerosa, F., Romagnani, S. and Moretta, L. (1984) Frequent coexpression of cytolytic activity and lymphokine production among human T lymphocytes. Production of B cell growth factor and interleukin 2 by $T8^+$ and $T4^+$ cytolytic clones. *Eur. J. Immunol.*, **14**, 1066–9.

Mody, C. H., Chen, G. H., Jackson, C., Curtis, J. L. and Toews, G. B. (1994) *In vivo* depletion of murine CD8 positive T cells impairs survival during infection with a highly virulent strain of *Cryptococcus neoformans. Mycopathologia*, **125**, 7–17.

Moller, G. (1988) Do suppressor T cells exist? *Scand. J. Immunol.*, **27**, 247–50.

Mosmann, T. R., Cherwinski, H., Bond, M. W., Giedlin, M. A. and Coffman, R. L. (1986) Two types of murine helper T cell clones. 1. Definition according to profiles of lymphokine activities and secreted proteins. *J. Immunol.*, **135**, 2348–57.

Mu, X., Kay, N. E., Gosland, M. P. and Jennings, C. D. (1997) Analysis of blood T-cell cytokine expression in B-chronic lymphocytic leukaemia: evidence for increased levels of cytoplasmic IL-4 in resting and activated $CD8^+$ T cells. *Br. J. Haematol.*, **96**, 733–5.

Nagata, S. and Suda, T. (1995) Fas and Fas ligand: lpr and gld mutations. *Immunol. Today*, **16** 39–43.

Nagoya, S., Greenberg, P. D., Yee, C., Weisser, K. E., Sugawara, H., Widmer, M. B., Slack, J., Dower, S. K., Lupton, S. D. and Overell, R. W. (1994) Helper T cell-independent proliferation of $CD8^+$ cytotoxic T lymphocytes transduced with an IL-1 receptor retrovirus. *J. Immunol.*, **153**, 1527–35.

Noble, A., Diaz-Sanchez, D., Staynov, D., Lee, T. and Kemeny, D. M. (1993a) Elimination of IgE regulatory rat $CD8^+$ T cells *in vivo* increases the co-ordinate expression of the Th2 cytokines IL-4, IL-5 and IL-10. *Immunology*, **80**, 326–9.

Noble, A., Staynov, D. and Kemeny, D. M. (1993b) Generation of rat Th2-like cells *in vitro* is IL-4 dependent and is inhibited by IFN-γ. *Immunology*, **79**, 562–7.

Noble, A., MacAry, P. A. and Kemeny, D. M. (1995) IFN-γ and IL-4 regulate the growth and differentiation of $CD8^+$ T cells into subpopulations with distinct cytokine profiles. *J. Immunol.*, **155**, 2928–37.

Noble, A., Pestano, G.A. and Cantor, H. (1998a) Suppression of CD4 T-cell responses by CD8 T-cells: superantigen-activated CD8 cells induce unidirectional Fas-mediated apoptosis of antigen-activated CD4 cells. *J. Immunol.*, **160**, 559–65.

Noble, A., Zhao, Z.S. and Cantor, H. (1998b) Suppression of immune responses by CD8 cells: Qa-1 on activated B cells stimulates CD8 cell suppression of Th2 responses. *J. Immunol.*, **160**, 566–71.

Okumura, K., Takemori, T., Tokuhisa, T. and Tada, T. (1977) Specific enrichment of the suppressor T cell bearing I-J determinants: parallel functional and serological characterizations. *J. Exp. Med.*, **146**, 1234–45.

Paganelli, R., Scala, E., Ansotegui, I. J., Ausiello, C. M., Halapi, E., Fanales-Belasio, E., D'Offizi, G., Mezzaroma, I., Pandolfi, F., Fiorilli, M., Cassone, A. and Aiuti, F. (1995) $CD8^+$ T lymphocytes provide helper activity for IgE synthesis in human immunodeficiency virus-infected patients with hyper-IgE. *J. Exp. Med.*, **181**, 423–8.

Paliard, X., de Waal Malefijt, R., Yssel, H., Blanchard, D., Chretien, I., Abrams, J., de Vries, J. and Spits, H. (1988a) Simultaneous production of IL-2, IL-4, and IFN-γ by activated human $CD4^+$ and $CD8^+$ T cell clones. *J. Immunol.*, **141**, 849–55.

Paliard, X., Malefijt, R. W., de Vries, J. E. and Spits, H. (1988b) Interleukin-4 mediates CD8 induction on human $CD4^+$ T-cell clones. *Nature*, **335**, 642–4.

Pober, J.S., Collins, T., Gimbrone, M.A. Jr, Cotran, R.S., Gitlin, J.D., Fiers, W., Clayberger, C., Krensky, A.M., Burakoff, S.J. and Reiss, C.S. (1983) Lymphocytes recognize human vascular endothelial and dermal fibroblast Ia antigens induced by recombinant immune interferon. *Nature*, **305**, 726–9.

Ptak, W. and Askenase, P. W. (1992) $\gamma\delta$ T cells assist alpha beta T cells in adoptive transfer of contact sensitivity. *J. Immunol.*, **149**, 3503–8.

Rahelu, M., Williams, G. T., Kumararatne, D. S., Eaton, G. C. and Gaston, J. S. (1993) Human $CD4^+$ cytolytic T cells kill antigen-pulsed target T cells by induction of apoptosis. *J. Immunol.*, **150**, 4856–66.

Renz, H., Lack, G., Saloga, J., Schwinzer, R., Bradley, K., Loader, J., Kupfer, A., Larsen, G. L. and Gelfand, E. W. (1994) Inhibition of IgE production and normalization of airways responsiveness by sensitized CD8 T cells in a mouse model of allergen-induced sensitization. *J. Immunol.*, **152**, 351–60.

Rieber, E. P. and Rank, G. (1994) CDw60: a marker for human $CD8^+$ T helper cells. *J. Exp. Med.*, **179**, 1385–90.

Romagnani, S., Maggi, E. and Del Prete, G. (1994) An alternative view of the Th1/Th2 switch hypothesis in HIV infection. *Aids Res. Hum. Retroviruses*, **10**, iii–ix.

Sad, S. and Mosmann, T. R. (1995) Interleukin (IL) 4, in the absence of antigen stimulation, induces an anergy-like state in differentiated $CD8^+$ TC1 cells: loss of IL-2 synthesis and autonomous proliferation but retention of cytotoxicity and synthesis of other cytokines. *J. Exp. Med.*, **182**, 1505–15.

Sad, S., Marcotte, R. and Mosmann, T. R. (1995) Cytokine-induced differentiation of precursor mouse $CD8^+$ T cells into cytotoxic $CD8^+$ T cells secreting Th1 or Th2 cytokines. *Immunity*, **2**, 271–9.

Sad, S., Kagi, G. and Mosmann, T. R. (1997) Perforin and Fas killing by $CD8^+$ T cells limits their cytokine synthesis and proliferation. *J. Exp. Med.*, **184**, 1543–7.

Salgame, P., Abrams, J. S., Clayberger, C., Goldstein, H., Convit, J., Modlin, R. L. and Bloom, B. R. (1991a) Differing lymphokine profiles of functional subsets of human CD4 and CD8 T cell clones. *Science*, **254**, 279–82.

Salgame, P., Convit, J. and Bloom, B. R. (1991b) Immunological suppression by human $CD8^+$ T cells is receptor dependent and HLA-DQ restricted. *Proc. Natl. Acad. Sci. USA*, **88**, 2598–602.

Seder, R. A., Boulay, J. L., Finkelman, F., Barbier, S., Ben-Sasson, S. Z., Le Gros, G. and Paul, W. E. (1992) $CD8^+$ T cells can be primed *in vitro* to produce IL-4. *J. Immunol.*, **148**, 1652–6.

Sedgwick J.D. and Holt P.G. (1984) Suppression of IgE responses in inbred rats by repeated respiratory tract exposure to antigen: responder phenotype influences isotype specificity of induced tolerance. *Eur. J. Immunol.*, **14**, 893–7.

Sedgwick J.D. and Holt P.G. (1985) Induction of IgE-secreting cells and IgE isotype-specific suppressor T cells in the respiratory lymph nodes of rats in response to antigen inhalation. *Cell. Immunol.*, **94**, 182–94.

Sieling, P. A., Wang, X. H., Gately, M. K., Oliveros, J. L., McHugh, T., Barnes, P. F., Wolf, S. F., Golkar, L., Yamamura, M., Yogi, Y., Uyemura, K., Rea, T. H. and Modlin, R. L. (1994) IL-12 regulates T helper type 1 cytokine responses in human infectious disease. *J. Immunol.*, **153**, 3639–47.

Snapper, C. M. and Paul, W. E. (1987) Interferon-γ and B cell stimulatory factor-1 reciprocally regulate Ig isotype production. *Science*, **236**, 944–7.

Stanciu, L. A., Shute, J., Holgate, S. T. and Djukanovic, R. (1996) Production of IL-8 and IL-4 by positively and negatively selected $CD4^+$ and $CD8^+$ human T cells following a four-step cell separation method including magnetic cell sorting (MACS). *J. Immunol. Methods*, **189**, 107–15.

Steinbrink, K., Sorg, C. and Macher, E. (1996) Low zone tolerance to contact allergens in mice: a functional role for $CD8^+$ T helper type 2 cells. *J. Exp. Med.*, **183**, 759–68.

Sydora, B.C., Brossay, L., Hagenbaugh, A., Kronenberg, M. and Cheroutre, H. (1996) TAP-independent selection of CD8 intestinal intraepithelial lymphocytes. *J. Immunol.*, **156**, 4209–16.

Tada, T., Takemori, T., Okumura, K., Nonaka, M. and Tokuhisa, T. (1978) Two distinct types of helper T cells involved in the secondary antibody response: independent and synergistic effects of Ia^- and Ia^+ helper T cells. *J. Exp. Med.*, **147**, 446–58.

Taguchi, T., Aicher, W. K., Fujihashi, K., Yamamoto, M., McGhee, J. R., Bluestone, J. A. and Kiyono, H. (1991) Novel function for intestinal intraepithelial lymphocytes. Murine $CD3^+$, gamma/delta TCR^+ T cells produce IFN-γ and IL-5. *J. Immunol.*, **147**, 3736–44.

Takahashi, T., Gray, J. D. and Horwitz, D. A. (1991) Human $CD8^+$ lymphocytes stimulated in the absence of $CD4^+$ cells enhance IgG production by antibody-secreting B cells. *Clin. Immunol. Immunopathol.*, **58**, 352–65.

Tatsumi, Y., Pena, J.C., Matis, L., Deluca, D. and Bluestone, J.A. (1993) Development of T cell receptor $\gamma\delta$ cells. *J. Immunol.*, **156**, 3030–9.

Thivolet, C., Bendelac, A., Bedossa, P., Bach, J. F. and Carnaud, C. (1991) $CD8^+$ T cell homing to the pancreas in the nonobese diabetic mouse is $CD4^+$ T cell-dependent. *J. Immunol.*, **146**, 85–8.

Thorpe, S.C., Murdoch, R.D. and Kemeny, D.M. (1989) The effect of the castor bean toxin, ricin, on rat IgE and IgG responses. *Immunology*, **68**, 307–11.

Till, S., Li, B., Durham, S., Humbert, M., Assoufi, B., Huston, D., Dickason, R., Jeannin, P., Kay, A. B. and Corrigan, C. J. (1995) Secretion of the eosinophil-active cytokines interleukin-5, granulocyte/macrophage colony-stimulating factor and interleukin-3 by bronchoalveolar lavage $CD4^+$ and $CD8^+$ T cell lines in atopic asthmatics, and atopic and non-atopic controls. *Eur. J. Immunol.*, **25**, 2727–31.

Trenn, G., Takayama, H., Hu-Li, J., Paul, W.E. and Sitorsky, M.V. (1988) B cell stimulatory factor 1 (IL-4) enhances the development of cytotoxic T cells from Lyt-2^+ resting murine T lymphocytes. *J. Immunol.*, **140**, 1110–6.

Uyemura, K., Pirmez, C., Sieling, P. A., Kiene, K., Paes-Oliveira, M. and Modlin, R. L. (1993) $CD4^+$ type 1 and $CD8^+$ type 2 T cell subsets in human leishmaniasis have distinct T cell receptor repertoires. *J. Immunol.*, **151**, 7095–104.

van Kuyk, R., Torbett, B. E., Gulizia, R. J., Leath, S., Mosier, D. E. and Koenig, S. (1994) Cloned human $CD8^+$ cytotoxic T lymphocytes protect human peripheral blood leukocyte-severe combined immunodeficient mice from HIV-1 infection by an HLA-unrestricted mechanism. *J. Immunol.*, **153**, 4826–33.

Walter, E. A., Greenberg, P. D., Gilbert, M. J., Finch, R. J., Watanabe, K. S., Thomas, E. D. and Riddell, S. R. (1995) Reconstitution of cellular immunity against cytomegalovirus in recipients of allogeneic bone marrow by transfer of T-cell clones from the donor. *N. Engl. J. Med.*, **333**, 1038–44.

Wassenaar, A., Reinhardus, C., Abraham-Inpijn, L. and Kievits, F. (1996) Type-1 and type-2 $CD8^+$ T-cell subsets isolated from chronic adult periodontitis tissue differ in surface phenotype and biological functions. *Immunology*, **87**, 113–18.

Weiner, H.L., Friedman, A., Miller, S.J., Khoury, A., Al-Sabbagh, Santos, L., Sayegh, M., Nussenblatt, R.B., Trentham, D.E. and Hafler, D.A. (1994) Oral tolerance: immunologic mechanisms and treatment of animal and human organ-specific autoimmune diseases by oral administration of autoantigens. *Annu. Rev. Immunol.*, **12**, 809–37.

Weiss, A., Newton, M. and Crommie, D. (1986) Expression of T3 in association with a molecule distinct from the T cell antigen receptor heterodimer. *Proc. Natl. Acad. Sci. USA*, **83**, 6998–703.

Xu, H., DiIulio, N. A. and Fairchild, R. L. (1996) T cell populations primed by hapten sensitization in contact sensitivity are distinguished by polarized patterns of cytokine production: interferon-γ-producing (Tc1) effector $CD8^+$ T cells and interleukin (IL) 4/IL-10-producing (Th2) negative regulatory $CD4^+$ T cells. *J. Exp. Med.*, **183** 1001–12.

Yagi, H., Matsumoto, M., Kunimoto, K., Kawaguchi, J., Makino, S. and Harada, M. (1992) Analysis of the roles of $CD4^+$ and $CD8^+$ T cells in autoimmune diabetes of NOD mice using transfer to NOD athymic nude mice. *Eur. J. Immunol.*, **22**, 2387–93.

Ying, S., Humbert, M., Barkans, J., Corrigan, C. J., Pfister, R., Menz, G., Larche, M., Robinson, D. S., Durham, S. R. and Kay, A. B. (1995) Expression of IL-4 and IL-5 mRNA and protein product by $CD4^+$ and $CD8^+$ T cells, eosinophils, and mast cells in bronchial biopsies obtained from atopic and non-atopic (intrinsic) asthmatics. *J. Immunol.*, **158**, 3539–44.

Zanussi, S., Simonelli, C., D'Andrea, M., Caffau, C., Clerici, M., Tirelli, U. and DePaoli, P. (1996) $CD8^+$ lymphocyte phenotype and cytokine production in long-term non-progressor and in progressor patients with HIV-1 infection. *Clin. Exp. Immunol.*, **10**, 220–4.

Zheng, L., Fisher, G., Miller, R. E., Peschon, J., Lynch, D. H. and Lenardo, M. J. (1995) Induction of apoptosis in mature T cells by tumor necrosis factor. *Nature*, **377**, 348–51.

4 T Lymphocyte Subpopulations and TCDD Immunotoxicity

Nancy I. Kerkvliet

Introduction

2,3,7,8-Tetrachlorodibenzo-*p*-dioxin (TCDD) is the prototype and most toxic member of a large group of structurally similar chlorinated aromatic compounds that are widely dispersed in the environment. These chemicals represent a unique group of environmental contaminants in that much of their biological activity is initiated by binding to an intracellular receptor, the aryl hydrocarbon receptor (AhR). The AhR, together with another protein (Ah receptor nuclear translocator, ARNT), functions as a heterodimeric ligand-activated transcription factor to induce a battery of structural genes coding for cytochrome P4501A1 and other enzymes important in xenobiotic metabolism (Nebert *et al.*, 1993). In addition, the transcriptional activation or repression of other genes by the activated AhR is postulated to underlie the tissue and cell-specific toxic effects of TCDD and related chemicals (Hankinson, 1995). Based on the identification of at least one AhR-binding core sequence 'GCGTG' in the promoter region, many genes for growth factors, growth factor receptors and cytokines have the potential to be transcriptionally regulated by TCDD (Lai *et al.*, 1996).

The immune system is a very sensitive target for the toxic effects of TCDD. In mice, immune dysfunction is characterized by suppression of acquired immunity, with both antibody and cell-mediated responses susceptible to suppression under different exposure paradigms (reviewed by Vos and Luster, 1989; Kerkvliet, 1995). Structure–activity studies using polychlorinated dioxin, furan and biphenyl congeners that differ in their affinity for the AhR indicate that the immunosuppressive potency of individual congeners is directly correlated with their affinity for binding to the AhR (Kerkvliet *et al.*, 1990a,b). Allelic variation at the Ah locus also results in strains of mice that differ in the ability of their AhR to bind ligand, which correlates with their sensitivity to

T Lymphocyte Subpopulations in Immunotoxicology. Edited by Ian Kimber and MaryJane K. Selgrade.
 ISBN 0 471 97194 4

TCDD toxicity (Birnbaum *et al.*, 1990; Poland and Glover, 1990; Vecchi *et al.*, 1983). C57Bl/6 mice (Ah^{bb}) represent the most sensitive strain; DBA/2 mice (Ah^{dd}) represent the least sensitive strain and require higher doses of TCDD to induce the same effects seen at lower doses in C57Bl/6 mice. Immunotoxicity studies in congenic C57Bl/6 (Ah^{dd}) mice have verified the important role of the AhR in suppression of both humoral and cell-mediated immune responses by TCDD (Kerkvliet *et al.*, 1990a,b). However, on the basis of protein and/or message analysis, lymphocytes in general, and T cells in particular, express relatively low amounts of the AhR compared with cells such as hepatocytes, epithelial cells, and myeloid cells (Greenlee *et al.*, 1985; Hayashi *et al.*, 1995; Lawrence *et al.*, 1996; Masten and Shiverick, 1997; vanden Heuvel *et al.*, 1994). These findings suggest that some of the immunotoxicity of TCDD may be induced indirectly through non-lymphoid tissues that express higher levels of the AhR.

In contrast to the suppression of acquired immunity, non-specific immune functions mediated by macrophages, neutrophils, and natural killer cells appear to be more resistant to suppression by TCDD exposure and, in some cases, show enhanced responsiveness following TCDD treatment (House *et al.*, 1990; Mantovani *et al.*, 1980; Vos *et al.*, 1978). For example, neutrophilia and increased inflammatory infiltrates in the liver and skin are pathological attributes of TCDD toxicity in several species (Puhvel and Sakamoto, 1988; Vos *et al.*, 1973, 1974; Weissberg and Zinkl, 1973). Furthermore, an increased inflammatory response to antigen challenge has been reported in TCDD-treated mice. This effect was characterized by a 2- to 4-fold increase in the numbers of neutrophils and macrophages locally infiltrating the site of antigen injection (Kerkvliet and Oughton, 1993; Moos *et al.*, 1994). However, the kinetics of the cellular influx was not altered by TCDD. The enhancement of inflammatory responses following TCDD exposure may be associated with an increased production of inflammatory mediators such as tumour necrosis factor-α (TNF-α), transforming growth factor-α and β (TGF-α, TGF-β), interleukin (IL)-1β, and plasminogen activator inhibitor-2 (PLA-2) (Gaido and Maness, 1994; Moos *et al.*, 1994; Sutter *et al.*, 1991). Although a cause-effect relationship between enhanced inflammation and depressed immunity in TCDD-treated mice has been postulated, a definitive link has not yet been confirmed (Kerkvliet and Oughton, 1993; Lawrence *et al.*, 1996; Moos and Kerkvliet, 1994).

Although the immunotoxic effects of TCDD have been studied for over 25 years, it is still not possible to define a TCDD-induced immunotoxicity syndrome. The literature on this subject is replete with conflicting data that were obtained using different animal models or differing experimental conditions. Because the AhR is polymorphic, different species and strains of laboratory animals are likely to vary significantly in their sensitivity to TCDD immuntoxicity, and the AhR status of outbred laboratory animals is relatively undefined. Likewise, for reasons not yet understood, rats appear to be much more resistant to some of the immunosuppressive effects of TCDD as compared to mice and guinea pigs (Smialowicz *et al.*, 1994; Vos *et al.*, 1973). In addition to AhR differences, the likely role of indirect effects in TCDD immunotoxicity confounds comparison between data derived from whole animal studies and data derived from *in vitro* systems. Thus, in order to avoid excessive confusion, the following review of TCDD effects on T cells will focus primarily on positive data (derived mostly from *in vivo* studies in mice) that implicate various T cell subsets as targets of TCDD, rather than on documentation of negative results (derived mostly from studies *in vitro* or in rats) that may suggest no effect on the same T cell subsets.

Influence of TCDD on T cell-mediated immunity

TCDD exposure has been shown to affect many immune responses *in vivo* that are mediated by T cells. In the first report of TCDD immunotoxicity by Vos *et al.* (1973), as little as 40 ng TCDD per kg body weight given once weekly for 8 weeks was shown to suppress the delayed hypersensitivity response (DHR) of guinea pigs to tuberculin. Subsequent studies have shown that the DHR of rabbits to tuberculin was decreased by TCDD at dosages of 1.0 μg/kg for 8 weeks (Sharma *et al.*, 1973), and the DHR of mice to tuberculin (Sharma *et al.*, 1973) or oxazolone (Clark *et al.*, 1981) was significantly impaired at weekly TCDD doses of 1 μg/kg or higher. Likewise, contact sensitization to dinitrofluorobenzene was reduced in Swiss mice fed diets containing 50 ppb TCDD for 6 weeks (Hinsdill *et al.*, 1980). Chronic (Clark *et al.*, 1981) as well as acute (Kerkvliet *et al.*, 1990a) exposure to TCDD has been reported to suppress the cytotoxic T lymphocyte (CTL) response of C57Bl/6 mice responding to allogeneic tumour cells. The acute ED_{50} was 7.5 μg/kg (DeKrey and Kerkvliet, 1995). The clinical relevance of the suppression of cell-mediated immune function by TCDD is evidenced by the increased susceptibility of TCDD-treated mice to bacterial infections such as *Salmonella* sp. (Hinsdill *et al.*, 1980), and *Listeria monocytogenes* (Hinsdill *et al.*, 1980) as well as viral influenza (House *et al.*, 1990; Burleson *et al.*, 1996) and Herpesvirus type II infections (Clark *et al.*, 1983).

Phenotypic analysis of T cell subsets in TCDD-treated mice

Several studies have examined the influence of TCDD exposure on lymphocyte subsets in the spleen or peripheral blood in an attempt to identify the targets of TCDD toxicity. In the majority of studies, little if any significant change in the frequency of T cells or T cell subsets (CD4 versus CD8) has been observed in TCDD-treated mice (Clark *et al.*, 1981; Kerkvliet and Brauner, 1990; Smialowicz *et al.*, 1994). However, more recent studies that have utilized multiparameter analysis of activation markers on specific T cell subsets have revealed some interesting changes that implicate an effect of TCDD exposure on T helper cells. Oughton *et al.* (1995) reported on the influence of long-term (14–15 months) low-dose (0.2 μg/kg per week) TCDD exposure on lymphocyte phenotypes in the spleen of C57Bl/6 mice. During the time-frame of the experiment, several age-related changes in lymphocyte phenotypes were documented in T cells that reflect the normal shift from naive to activated/memory phenotypes in both the $CD4^+$ and $CD8^+$ cell compartments during the ageing process (Ernst *et al.*, 1990; Mobley and Daily, 1992). TCDD exposure did not alter the overall frequency or number of $CD4^+$ T in the spleen but produced a significant decrease in the proportion of $CD4^+$ cells that expressed a memory phenotype ($CD44^{hi}$, $CD45RB^{lo}$). Since the animals were not intentionally immunized with any antigens but were housed under conventional rather than barrier conditions, these results suggest that the TCDD-treated mice experienced an overall reduced response to environmental antigens encountered during the extended experimental period, resulting in the development of fewer memory T cells. This interpretation is consistent with the immunosuppressive effects of TCDD. Similar phenotypic changes were reported by Neubert *et al.* (1993) in studies on marmosets, wherein chronic exposure to TCDD for 8 weeks at a concentration of 1.5 ng/kg decreased the ratio of activated helper-inducer ($CDw29^+$) to naive ($CD45RA^+$) $CD4^+$ cells without altering the overall percentage of $CD4^+$ T cells. Likewise, a similar trend was reported by Tonn *et al.* (1996), wherein the percentage of naive $CD45RA^+$ $CD4^+$ cells was increased in the blood of TCDD-exposed industrial workers from 12.0 ± 6.0 to 16.2 ± 5.5 ($\bar{x} \pm$ SD). Although this change was not significantly different from the age-matched controls, it supports the trend seen in animal studies. These workers ($n = 10$) had been exposed to TCDD for several years prior to analysis (body burden 43–874 pg/g blood fat) and had significantly lower MLR and responsiveness to IL-2, which the authors suggested was reflective of a persistent decrease in T helper cell function. However, these changes were subclinical in that the workers were generally healthy and, with one exception, had no history of increased susceptibility to infections or cancer. None the less, taken together, these results identify the $CD4^+$ naive/memory cell phenotype ratio as a potential biomarker of TCDD immunotoxicity.

T helper cells in TCDD immunotoxicity

A number of studies have shown that antibody production is suppressed in laboratory animals following TCDD exposure. The majority of these studies have utilized sheep erythrocytes (SRBC) as the antigen (Clark *et al.*, 1981; Hinsdill *et al.*, 1980; Holsapple *et al.*, 1986; Kerkvliet and Brauner, 1990; Vecchi *et al.*, 1980), although tetanus toxoid (Hinsdill *et al.*, 1980; Vos *et al.*, 1973), trinitrophenol (TNP)-conjugated *Brucella abortus* (Clark *et al.*, 1981), type III pneumococcal polysaccharide (Vecchi *et al.*, 1980), dinitrophenol (DNP)-Ficoll (Holsapple *et al.*, 1986); lipopolysaccharide (LPS) (Holsapple *et al.*, 1986); TNP-LPS (Kerkvliet *et al.*, 1990a) and allogeneic tumour cells

(Kerkvliet *et al.*, 1996) have also been used in various studies. Although antibody responses to both T helper cell-dependent (TD) and T helper cell-independent (TI) antigens are suppressed by TCDD, higher doses of TCDD are usually required to suppress TI responses (Harper *et al.*, 1995; Holsapple *et al.*, 1986; House *et al.*, 1990; Kerkvliet *et al.*, 1990a). For example, in C57Bl/6 mice the antibody response to SRBC, a TD antigen, has an ED_{50} of only 0.7 μg TCDD/kg, while the ED_{50} for the antibody response to TNP-LPS, a TI antigen, is more than 10-fold higher (7–14 μg/kg; Harper *et al.*, 1995; Kerkvliet *et al.*, 1990a). These results suggest that T helper cells are functionally impaired by low doses of TCDD and this impairment contributes to the suppression of antibody production in TCDD-treated mice.

In more definitive studies by Tomar and Kerkvliet (1992), a hapten-carrier system was used to determine the effects of TCDD on the generation of carrier-specific T helper cells that are required for B cells to produce anti-hapten antibody. Results of these studies showed that the *in vitro* anti-TNP antibody response to TNP-SRBC was significantly lower when the carrier-primed T cells were obtained from TCDD-treated mice rather than vehicle-treated mice, directly demonstrating the ability of TCDD to interfere with T helper cell function.

In addition to classic T helper cell activity for TD antibody responses, T cells have also been shown to augment antibody responses to type 2 TI antigens like DNP-Ficoll (Barth and Singal 1973; Braley-Mullen, 1982; Markham *et al.*, 1977). The mechanism by which these non-specific T helper cells are activated and participate in increasing the antibody response to type 2 TI antigens is still not known (Janeway and Travers, 1996). None the less, a possible influence of dioxin exposure on this subset of T cells was suggested by studies using the AhR-binding dioxin congener, 1,2,3,4,6,7,8-heptachlorodioxin (HpCDD). In these studies, the potency of HpCDD to suppress the antibody response to DNP-Ficoll and TNP-LPS differed significantly, with ED_{50}s of 127 versus 516 μg HpCDD/kg, respectively (Kervliet *et al.*, 1987). Because the difference in sensitivity to suppression between the two TI responses could be related to a suppressive effect of HpCDD on T cells, the role of T cells in the suppression of the TI response to DNP-Ficoll was evaluated using T cell-deficient nude mice (Kerkvliet *et al.*, 1987). Results of these studies showed that nu/nu mice required a 5-fold higher dose of HpCDD to achieve the same degree of immune suppression as compared with their nu/+ littermates. These results indicated that T cells indeed played a significant role in the suppression of the antibody response to DNP-Ficoll by HpCDD.

Th1 and Th2 T cell subsets in TCDD immunotoxicity

Although there is ample evidence that TCDD suppresses T cell activity leading to suppressed IgM and IgG responses, there is little information available in the literature relating TCDD effects to specific T helper subsets (e.g., Th1 and

Th2). For example, only two studies have characterized the isotype of antibody produced following TCDD exposure, which could provide some insight into selective effects of TCDD on Th1- versus Th2-dependent responses. It is generally accepted that IFN-γ production by Th1 cells promotes IgG2a production while IL-4 production by Th2 cells promotes IgG1 as well as IgE production (Chapter 2, this book; Siebenkotten and Radbruch, 1995). Early studies by Hinsdill *et al.* (1980) reported that TCDD suppressed IgE titres in mice in response to tetanus toxin immunization, indicating that Th2-dependent responses are sensitive to suppression by TCDD. More recently, Kerkvliet *et al.* (1996) characterized the suppression of the alloantibody response to P815 mastocytoma cells in TCDD-treated mice. Results of these studies showed that C57Bl/6 mice injected with P815 tumour cells produced both IgG1 and IgG2a isotypes of alloantibody in addition to IgM antibody (Figure 4.1). In mice treated with TCDD (15 μg/kg), the IgM and IgG2a titres were profoundly suppressed whereas the IgG1 response was only slightly affected. When cytokine production was evaluated during the same time period, TCDD appeared to selectively suppress the production of Th1-type cytokines (IL-2, IFN-γ , TNF), while the production of Th-2-type cytokines (IL-4, IL-6) was unaffected. Interestingly, when spleen cells from P815-injected mice were selectively depleted of $CD4^+$ or $CD8^+$ cells prior to cytokine analysis, only cultures containing $CD8^+$ cells produced detectable levels of IL-2 and IFN-γ. These results suggested that $CD8^+$ T helper cells secreting Th1-type cytokines [recently designated Tc-1 cells by Sad *et al.* (1995)] were altered following TCDD exposure. However, even though cytokine production by $CD4^+$ cells was not detected *in vitro* in these studies, *in vivo* depletion of $CD8^+$ cells prior to the injection of P815 tumour cells had only a small effect on the cytotoxic alloantibody titres, whereas $CD4^+$ cell depletion ablated the alloantibody response (Kerkvliet *et al.*, 1996). Thus, it is assumed that $CD4^+$ cells *in vivo* are producing cytokines that are necessary for alloantibody production. Furthermore, since the IgG1 but not the IgG2a response was suppressed by TCDD, it would appear that both Th1 cells and Tc1 cells are targets of TCDD suppression.

The profound suppression of IFN-γ production that occurred following TCDD exposure of P815 tumour-challenged mice without concomitant suppression of IL-4 production (Kerkvliet *et al.*, 1996), suggests that TCDD exposure might create a biased cytokine milieu that favours Th2/Tc2 type responses. However, it is not clear from these studies that the selective effect of TCDD on cytokine production was elicited by TCDD exposure *per se* or by the nature of the P815 tumour antigens that induced the cytokine response. That is to say, if the nature of the antigenic stimulus induced a dominant Th2-type cytokine response, would TCDD suppress Th2 cytokine production as potently as it suppresses Th1-type cytokines? Although this question has not been directly investigated, evidence available currently suggests that the effect of TCDD on cytokine production is significantly influenced by the nature of the antigenic stimulus. This evidence is strongest for TNF-α production. Thus,

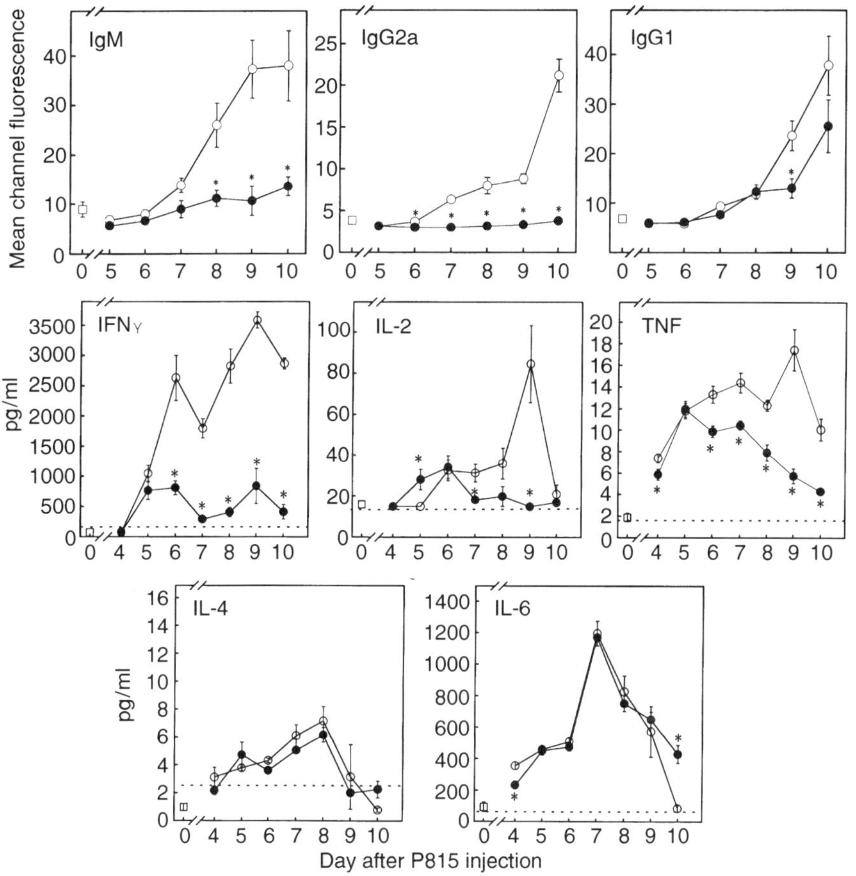

Figure 4.1 Parallel effects of TCDD exposure on the isotypes of anti-P815/H-2[d] alloantibodies and on cytokine production during the allograft response to P815 mastocytoma cells in C57Bl/6 mice. Mice were given a single oral dose of vehicle (○) or TCDD (15 μg/kg) (●) 1 day before the i.p. injection of 1×10^7 P815 tumour cells. Plasma and spleen cells were obtained from five or six mice per treatment on the indicated days following P815 inoculation. Antibody isotypes in plasma were determined by flow cytometric analysis as described in Kerkvliet *et al.* (1996). Data are presented as the mean channel of the fluorescence (MCF) histograms ± SEM. Cytokine production was evaluated in culture supernatants of spleen cells incubated for 6 h with P815 tumour cells as described in Kerkvliet *et al.* (1996). Cytokine levels produced by naive spleen cells are indicated by (□). * indicates $p < 0.05$ between TCDD- and vehicle-treated mice.

TNF-α production in anti-CD3-treated mice was not altered by TCDD exposure (Prell *et al.*, 1995), whereas in P815-injected mice it was significantly decreased (Kerkvliet *et al.*, 1996), and in LPS- or SRBC-injected mice, it was significantly increased (Clark *et al.*, 1991; Moos *et al.*, 1994). Similarly, IFN-γ production was only slightly suppressed by TCDD treatment in anti-CD3 treated mice in comparison with its profound suppression following P815 injection (Prell *et al.*, 1995). These differential effects of TCDD on TNF-α and IFN-γ production may, in part, reflect the type of cell that is making the cytokine (e.g. T cells or macrophages). However, T cell production of IL-2 is also differentially affected by TCDD exposure. IL-2 production was elevated in TCDD-treated mice early after P815 tumour injection, which was followed by later suppression (Kerkvliet *et al.*, 1996), while in anti-CD3 injected mice, IL-2 production was not altered by TCDD exposure (Prell *et al.*, 1995). Similarly, Lundberg *et al.* (1992) reported that antigen-specific T cell production of IL-2 was suppressed by TCDD, whereas IL-2 production in response to anti-CD3 was not affected. Thus, the ability of TCDD to modulate cytokine production appears to be influenced by many factors, including 1) the nature of the antigen, 2) the type of cell that is producing the cytokine, and 3) the state of activation of the cytokine-producing cell.

T suppressor cells in TCDD immunotoxicity

Clark *et al.* (1981) were the first to suggest that TCDD suppressed CTL activity in mice by a mechanism that involved suppressor cells. In their studies, TCDD exposure did not alter the frequency of CTL precursors but appeared to induce a population of suppressor cells in the thymus which prevented the generation of effector CTL *in vitro*. These results were provocative because they occurred in mice at exposure levels of TCDD as low as 10 ng/kg weekly over a 4-week period. Subsequent to this report, several investigators have attempted, without success, to replicate these findings using the same, or variations of, the model system used by Clark and coworkers (Kerkvliet and Baecher-Steppan, 1988; Hanson and Smialowicz, 1995; Holsapple *et al.*, 1991). In addition, Dooley *et al.* (1990) failed to find evidence for the induction of non-specific suppressor T cells in mice following TCDD exposure that could inhibit the *in vitro* antibody response to SRBC or DNP-Ficoll. Thus, at the present time, there is no convincing evidence that TCDD induces immunotoxicity via altered suppressor T cell activity.

Mechanisms of suppression of T cell function by TCDD

From the above discussion, it is clear that T cell responses are suppressed in animals following TCDD exposure. However, despite a relatively large descriptive database for TCDD immunotoxicity, surprisingly little is known about the underlying biochemical and molecular mechanisms that mediate these effects in T cells. Progress in this area has been greatly hampered by the inability to demonstrate direct effects of TCDD on T cells when TCDD is added to *in vitro* cultures or when T cells from TCDD-treated mice are activated *in vitro* (DeKrey and Kerkvliet, 1995; Lawrence *et al.*, 1996; Lundberg *et al.*, 1992; Prell *et al.*, 1995). For example, when TCDD was added directly to cultures of

T cell clones, neither proliferation nor cytokine production (IL-2, IL-4, IFN-γ, IL-3) was significantly altered (Lawrence *et al.*, 1996). Likewise, when TCDD was added to whole splenocyte-derived mixed lymphocyte cultures, neither the proliferative response nor the generation of CTL activity was altered (DeKrey and Kerkvliet, 1995). These results suggest either that the primary cell target is not present in the spleen cell culture or that culture conditions somehow compensate for the toxicity of TCDD. Furthermore, even if T cells are obtained from mice treated with an immunosuppressive dose of TCDD, their responsiveness *in vitro* is generally not altered (Lundberg *et al.*, 1992; Prell *et al.*, 1995), suggesting either that TCDD does not affect resting T cells or, again, that culture conditions compensate for the defect.

There is some evidence to suggest that culture conditions can profoundly influence the ability to detect the immunotoxic effects of TCDD *in vitro*. Morris *et al.* (1991), for example, have reported that the ability of TCDD to suppress the *in vitro* antibody response to SRBC was critically dependent on the type and concentration of serum used in the culture medium. Only three of 23 lots of fetal bovine serum (FBS) were able to support a full dose-dependent suppression of the anti-SRBC response, and, in serum-free cultures, TCDD caused a 15-fold enhancement of the antibody response. In contrast, when cultures were supplemented with autologous mouse serum instead of FBS, dose-dependent suppression of the antibody response was observed (Harper *et al.*, 1995; Morris *et al.*, 1991). These results suggest that many batches of FBS contain compensating factors that overcome TCDD-induced immunotoxicity. Following this report, Lawrence *et al.* (1996) examined the influence of serum conditions on the ability of T cell clones to respond to TCDD. However, like cultures containing FBS, cytokine production by T cells cultured in 2.5% mouse serum was still not influenced by TCDD, and serum-free cultures failed to support normal cytokine production, even in the absence of TCDD. Thus, culture conditions do not appear to explain the lack of direct effects of TCDD on T cells.

More compelling evidence for an indirect effect of TCDD on T cells is derived from studies on the distribution and behaviour of the AhR in lymphoid tissue. For T cells to be direct targets of TCDD effects, they should express the AhR and it should be activated to a DNA-binding form following ligand binding. However, Lawrence *et al.* (1996) found that, on a per cell basis, the amount of AhR protein in T cells was more than 3-fold lower than the level found in hepa cells, a murine hepatocyte cell line. In addition, while treatment of whole spleen cells with TCDD induced binding of the AhR nuclear complex to a consensus dioxin-response element (DRE) (Williams *et al.*, 1996), no DRE binding was detected with nuclear preparations from purified splenic T cells or T cell clones (Lawrence *et al.*, 1996). These results suggest that T cells may lack a factor required for DNA binding of the AhR complex, or they may express a repressor factor that inhibits AhR binding as previously described in a mutant hepa cell line (Watson *et al.*, 1992). The inability to bind DNA did not appear to be related to ARNT expression, which was readily detected in both the cytoplasm and nucleus of T cells (Lawrence *et al.*, 1996).

Antigen presentation as a target for TCDD

If TCDD does not directly alter T cells, antigen presentation to the T cell becomes an interesting potential target for this chemical. Impaired activation of T cells could explain many of the downstream immune events that are suppressed in TCDD-treated mice such as T cell cytokine production and CTL activity. Supporting this hypothesis, the ability of TCDD to suppress the CTL response to P815 tumour cells was shown to decrease as the time of treatment with TCDD was delayed following allosensitization and, if TCDD exposure was delayed until 4 days after allosensitization, there was no effect on the development of CTL activity measured at its peak on day 10 (Figure 4.2; Kerkvliet *et al.*, 1996).

It is now well recognized that the activation of naive T lymphocytes requires at least two signals. The first signal is an antigen-specific signal through the T cell receptor (TCR) that results from the binding of antigen presented on the surface of the APC in association with MHC molecules (Bretscher, 1992). $CD8^+$ T cells bind antigens displayed with class I MHC proteins and $CD4^+$ T cells bind antigens displayed with class II MHC proteins. The CD4 and CD8 molecules interact with non-polymorphic regions of the respective MHC

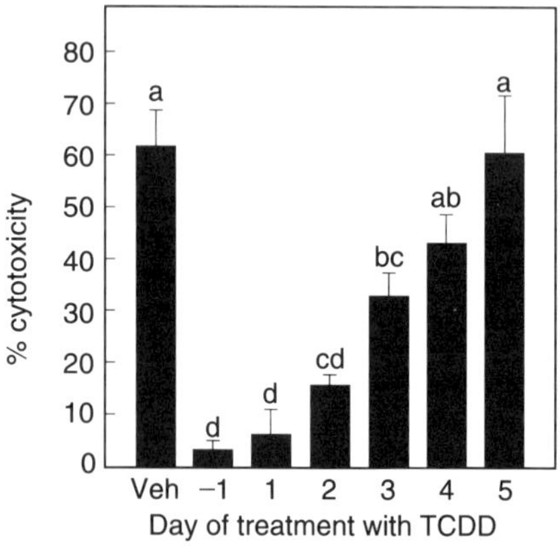

Figure 4.2 Influence of timing of TCDD exposure on the degree of suppression of the CTL response to P815 mastocytoma. C57Bl/6 mice received a single oral dose of TCDD (15 μg/kg) on the indicated days relative to P815 tumour cell inoculation on day 0. All mice were killed on day 10 and CTL activity of spleen cells was determined in a standard ^{51}Cr release assay as described in Kerkvliet *et al.* (1996). Statistically significant differences ($p < 0.05$) between treatment groups are indicated by different letters. If a letter is shared by two treatment groups, they are not significantly different from each other.

proteins and participate in the activation of the T cell. The binding of peptide-MHC to the TCR triggers a series of intracellular biochemical reactions in the T cells, including tyrosine phosphorylation of various proteins and activation of the phosphatidyl-inositol pathway. This pathway is sufficient to cause some IL-2 production and T cell proliferation (Rabinowitz *et al.*, 1996) but is not sufficient to induce the differentiation of T cells. A second, costimulatory signal, provided by other receptor–ligand interactions between the T cell and APC, is necessary for optimal T cell proliferation, cytokine production, and acquisition of effector functions (Dubey *et al.*, 1995; Gimmi *et al.*, 1991). In fact, TCR signalling without costimulation leads to a state of T cell unresponsiveness (anergy) upon antigen restimulation *in vitro* (Chen and Nabavi, 1994; Guerder *et al.*, 1994) or the induction of apoptosis (Mueller *et al.*, 1989; Van Parijs *et al.*, 1996).

Certain immunotoxic effects of TCDD are consistent with the induction of anergy and/or apoptosis of antigen-activated T cells. For example, in studies using anti-CD3 antibody to activate T cells *in vivo*, TCDD exposure led to an unexpected increase in the percentage of T cells that entered the cell cycle and an increased uptake of ^{3}HTdR (Neumann *et al.*, 1993). This increased T cell proliferation was followed by a selective decrease in the number of $CD4^+$ cells and an increase in the frequency of cells expressing an apoptotic phenotype (Prell *et al.*, 1995). These results suggest that TCDD may promote a type of activation-dependent cell death in $CD4^+$ T cells. This effect did not appear to be a due to increased production of TNF-α, a cytokine known to induce apoptosis in peripheral T cells (Zheng *et al.*, 1995). Recently, Rhile *et al.* (1996) reported that C57Bl/6 lpr/lpr mice were less sensitive than wild-type mice to suppression of antigen-specific T cell proliferation following treatment with low doses of TCDD. Since the lpr mutation results in a defect in the expression of Fas antigen (CD95), an important molecule involved in apoptosis of T cells, it is possible that Fas-mediated apoptosis plays a role in TCDD immunotoxicity. Interestingly, Zhang *et al.* (1997) recently reported that Th1 cells are more prone to Fas-mediated apoptosis in comparison with Th2 cells, providing a mechanism to explain preferential suppression of Th1 cytokine production following TCDD exposure (Kerkvliet *et al.*, 1996).

In contrast to apoptosis of $CD4^+$ cells, the induction of energy appears to be more consistent with the immunosuppressive effects of TCDD involving $CD8^+$ cells. For example, in the P815 tumour allograft model, cytokine production by $CD8^+$ cells appears to be initiated normally in TCDD-treated mice, based on normal levels of IL-2, IFN-γ, and TNF produced on days 4 and 5 after P815 sensitization (Figure 4.1). In fact, IL-2 production was significantly elevated in TCDD-treated mice on day 5 compared to vehicle controls. However, on day 6, Tc-1 cells in TCDD-treated mice appear to stop secreting cytokines compared with the sustained increase in cytokine production in vehicle-treated mice. At the same time, CTL activity fails to develop in TCDD- treated mice. These results suggest that Tc-1 cells received some activation signals, began to produce IL-2, but then stopped their progression. Based on the results of recent preliminary studies, it appears that these Tc-1 cells do not undergo apoptosis but rather enter an anergic state, since CTL activity could be 'rescued' in TCDD-treated mice by the injection of high doses of IL-2 on days 7–9 (Figure 4.3; Baecher-Steppan and Kerkvliet, unpublished data). The induction of anergy following TCDD exposure is a finding wholly consistent with impaired antigen presentation.

Several different receptor/counter-receptor pairs of molecules have been identified on the surface of APC and T cells that are important for the optimum activation of naive T cells (Bretscher, 1992). However, by far the most well-characterized costimulatory interaction is transduced by B7-CD28 interaction (Allison, 1994). The ligation of CD28 on T cells by either of its natural ligands expressed on the surface of APC, B7-1 (CD80) or B7-2 (CD86), enhances T cell proliferation as well as the acquisition of T cell effector functions, including lymphokine production, CTL generation and T helper activity. Blocking B7 costimulation *in vivo* has been shown to inhibit several T cell-dependent immune responses, including allograft rejection and antibody production to SRBC. Since TCDD has been shown to suppress these same immune functions, Prell and Kerkvliet (1997) investigated the possible role of B7 expression in TCDD- induced suppression of the P815 tumour allograft response. The results of these studies indicated that B7-2 is the dominant co-

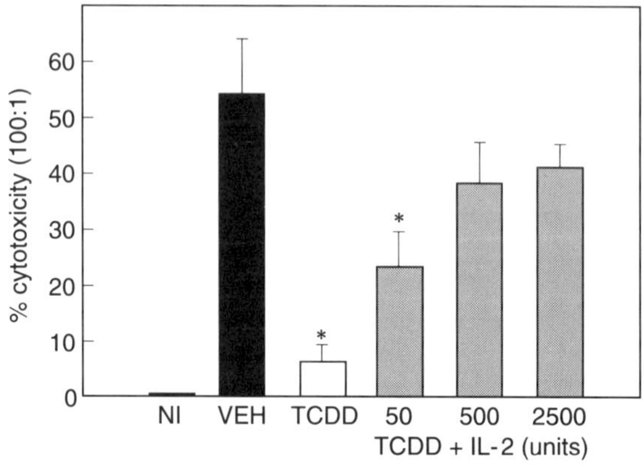

Figure 4.3 Reconstitution of the CTL response in TCDD-treated mice by the injection of IL-2. C57Bl/6 mice were treated TCDD (15 μg/kg) 1 day prior to allosensitization with P815 tumour cells (H-2^d). On days 7, 8 and 9 following P815 allosensitization, mice were injected i.p. with the indicated dose of IL-2. Mice were killed on day 10 for assessment of splenic CTL activity against ^{51}Cr-labelled P815 tumour cells. CTL activity generated in IL-2 injected mice was allospecific based on the inability of the same pools of cells to kill ^{51}Cr-labelled C3H (H-2^k) spleen cell blasts. Data represent the mean CTL activity $\pm$ SEM of six mice per treatment. *$p < 0.05$ compared to vehicle-treated control. NI = non-immune spleen cells.

stimulatory B7 molecule induced on APC during the allograft response to P815 tumour cells, and that the expression of B7-2 on splenic B cells and macrophages was reduced following treatment with an immunosuppressive dose of TCDD. Furthermore, providing exogenous B7-mediated costimulation via the injection of B7-transfected P815 cells was able to restore the CTL but not the alloantibody response in TCDD-treated mice. The injection of B7-transfected tumour cells was also able to bypass the absolute requirement for $CD4^+$ cells in the development of CTL activity to P815 tumour cells, a phenomenon that has also been demonstrated with B7-transfected syngeneic tumours (Gajewski *et al.*, 1995; Harding and Allison, 1993). These results are consistent with a model in which MHC class I^+, class II^- B7-transfected P815 tumour cells can directly activate $CD8^+$ CTL precursors and drive their differentiation into cytokine producing cells and CTL effector cells but are incapable of directly activating $CD4^+$ cells as helper cells for antibody production. Importantly, these results demonstrate that CTL precursors in TCDD-treated mice are functional and able to differentiate into effector cells if they are provided with adequate costimulation via B7, suggesting that defective costimulation of T cells plays an important role in TCDD immunotoxicity.

Summary

The immunotoxicity of TCDD and related AhR ligands has been the subject of many studies over the past 25 years. T cell-mediated immunity, as well as T cell-dependent antibody production, is suppressed following exposure of laboratory animals to TCDD, and susceptibility to bacterial and viral infection is increased in TCDD-treated rodents. TCDD-induced immune suppression is also evidenced by reduced numbers of peripheral T cells that express a memory phenotype following chronic TCDD exposure. Evidence currently available suggests that TCDD does not directly alter T cell function but may mediate immune suppression indirectly via alterations in antigen presentation. The pathways by which TCDD alters antigen presentation are under investigation.

References

Allison, J.P. (1994) CD28/B7 interactions in T-cell activation. *Curr. Opin. Immunol.*, **6**, 414–18.

Barth, R.F. and Singla, O. (1973) Differential effects of concanavalin A on T helper dependent and independent antibody responses. *Cell. Immunol.*, **9**, 96–103.

Birnbaum, L.S., McDonald, M.M., Blair, P.C., Clark, A.M. and Harris, M.S. (1990) Differential toxicity of 2,3,7,8-tetrachlorodibenzo-*p*-dioxin (TCDD) in C57B1/6 mice congenic at the Ah locus. *Fund. Appl. Toxicol.*, **15**, 186–200.

Braley-Mullen, H. (1982) Differential effect of activated T amplifier cells on B cells responding to thymus-independent type 1 and type 2 antigens. *J. Immunol.*, **121**, 484–9.

Bretscher, P. (1992) The two-signal model of lymphocyte activation twenty-one years later. *Immunol. Today*, **13**, 74–6.

Burleson, G.R., Lebrec, H., Yang, Y.G., Ibanes, J.D., Pennington, K.N. and Birnbaum, L.S. (1996) Effect of 2,3,7,8-tetrachlorodibenzo-*p*-dioxin (TCDD) on influenza virus host resistance in mice. *Fund Appl. Toxicol.*, **29**, 40–7.

Chen, C. and Nabavi, N. (1994) In vitro induction of T cell anergy by blocking B7 and early T cell costimulatory molecule ETC-1/B7-2. *Immunity*, **1**, 147–54.

Clark, D.A., Gauldie, J., Szewczuk, M.R. and Sweeney, G. (1981) Enhanced suppressor cell activity as a mechanism of immunosuppression by 2,3,7,8-tetrachlorodibenzo-*p*-dioxin. *Proc. Soc. Exp. Biol. Med.*, **168**, 290–9.

Clark, D.A., Sweeney, G., Safe, S., Hancock, E., Kilburn, D.G. and Gauldie, J. (1983) Cellular and genetic basis for suppression of cytotoxic T cell generation by haloaromatic hydrocarbons. *Immunopharmacology*, **6**, 143–53.

Clark, G.C., Taylor, M.J., Tritscher, A.M. and Lucier, G.W. (1991) Tumor necrosis factor involvement in 2,3,7,8-tetrachlorodibenzo-*p*-dioxin-mediated endotoxin hypersensitivity in C57B1/6J mice congenic at the Ah locus. *Toxicol. Appl. Pharmacol.* **111**, 422–31.

DeKrey, G.K. and Kerkvliet, N.I. (1995) Suppression of cytotoxic T lymphocyte activity by 2,3,7,8-tetrachlorodibenzo-*p*-dioxin occurs *in vivo*, but not *in vitro*, and is independent of corticosterone elevation. *Toxicology*, **97**, 105–12.

Dooley, R.K., Morris, D.L. and Holsapple, M.P. (1990) Elucidation of cellular targets responsible for tetrachlorodibenzo-*p*-dioxin (TCDD)-induced suppression of antibody responses: II. The role of the T-lymphocyte. *Immunopharmacology*, **19**, 47–58.

Dubey, C., Croft, M. and Swain, S.L. (1995) Costimulatory requirements of naive $CD4^+$ T cells. ICAM-1 or B7-1 can costimulate naive CD4 T cells activation but both are required for optimum response *J. Immunol.*, **155**, 45–57.

Ernst, D.N., Hobbs, M.V., Torbett, B.E., Glasebrook, A.L., Rehse, M.A., Bottomly, K., Hayakawa, K., Hardy, R.R. and Weigle, W.O. (1990) Differences in expression profiles of CD45RB, Pgp-1, and 3G11 membrane antigens and in the patterns of lymphokine secretion by splenic CD4+ T cells from young and aged mice. *J. Immunol.*, **145**, 1295–302.

Gaido, K.W. and Maness, S.C. (1994) Regulation of gene expression and acceleration of differentiation in human keratinocytes by 2,3,7,8-tetrachlorodibenzo-*p*-dioxin. *Toxicol. Appl. Pharmacol.*, **127**, 199–208.

Gajewski, T.F., Renauld, J.-C., Van Pel, A. and Boon, T. (1995) Costimulation with B7-1, IL-6, and IL-12 is sufficient for primary generation of murine antitumor cytolytic T lymphocytes *in vitro*. *J. Immunol.*, **154**, 5637–45.

Gimmi, D., Freeman, G.J., Gribbon, J.G., Sugita, K., Freedman, A.S., Morimoto, C. and Nadler, L.M. (1991) B cell surface antigen B7 provides a costimulatory signal that induces T cells to proliferate and secrete interleukin 2. *Proc. Nat. Acad. Sci. USA*, **88**, 6575–80.

Greenlee, W.F., Dold, K.M., Irons, R.D. and Osborne, R. (1985) Evidence for direct action of 2,3,7,8-tetrachlorodibenzo-*p*-dioxin (TCDD) on thymic epithelium. *Toxicol. Appl. Pharmacol.*, **79**, 112–19.

Guerder, S., Meyerhoff, J. and Flavell, R. (1994) The role of the T cell costimulator B7-1 in autoimmunity and the induction and maintenance of tolerance to peripheral antigen. *Immunity*, **1**, 155–64.

Hankinson, O. (1995) The aryl hydrocarbon receptor complex. *Annu. Rev. Pharmocol. Toxicol.*, **35**, 307–40.

Hanson, C.D. and Smialowicz, R.J. (1994) Evaluation of the effect of low-level 2,3,7, 8-tetrachlorodibenzo-*p*-dioxin exposure on cell-mediated immunity. *Toxicology*, **88**, 213–24.

Harding, F.A. and Allison, J.P. (1993) CD28-B7 interactions allow the induction of $CD8^+$ cytotoxic T lymphocytes in the absence of exogenous help. *J. Exp. Med.*, **177**, 1791–7.

Harper, N., Steinberg, M., Thomsen, J. and Safe, S. (1995) Halogenated aromatic hydrocarbon-induced suppression of the plaque-forming cell response in B6C3F1 splenocyte cultures with allogenic mouse serum: Ah receptor structure activity relationships. *Toxicology*, **99**, 199–206.

Hayashi, S., Okabe-Kado, J., Honma, Y. and Kawajiri, K. (1995) Expression of Ah receptor (TCDD receptor) during human monocyte differentiation. *Carcinogenesis*, **16**, 1403–9.

Hindsill, R.D., Couch, D.L. and Speirs, R.S. (1980) Immunosuppression in mice induced by dioxin (TCDD) in feed. *J. Environ. Pathol. Toxicol.*, **4**, 401–25.

Holsapple, M.P., McCay, J.A. and Barnes, D.W. (1986) Immunosuppression without liver induction by subchronic exposure to 2,7-dichlorobenzo-*p*-dioxin in adult female B6C3F1 mice. *Toxicol. Appl. Pharmacol.*, **83**, 445–55.

Holsapple, M.P., Snyder, N.K., Wood, S.C. and Morris, D.L. (1991) A review of 2,3,7,8-tetrachlorodibenzo-*p*-dioxin induced changes in immunocompetence: 1991 update. *Toxicology*, **69**, 219–55.

House, R.V., Lauer, L.D. and Murray, M.J. (1990) Examination of immune parameters and host resistance mechanisms in B6C3F1 mice following adult exposure to 2,3,7,8-tetrachlorodibenzo-*p*-dioxin. *J. Toxicol. Environ. Health*, **31**, 203–15.

Janeway, C.A. and Travers, P. (1996) *Immunobiology. The Immune System in Health and Disease*. Garland Publishing Inc., New York.

Kerkvliet, N.I. (1995) Immunological effects of chlorinated dibenzo-*p*-dioxins. *Environ. Health Perspect.*, **103**, 47–53.

Kerkvliet, N.I. and Baecher-Steppan, L. (1988) Suppression of allograft immunity by 3, 4, 5, 3′, 4′, 5′-hexachlorobiphenyl. II. Effects of exposure on mixed lymphocyte reactivity and induction of suppressor cell activity *in vitro*. *Immunopharmacology*, **16**, 13–23.

Kerkvliet, N.I. and Brauner, J.A. (1987) Mechanisms of 1,2,3,4,6,7,8-heptachlorodibenzo-*p*-dioxin (HpCDD)-induced humoral immune suppression: evidence of primary defect in T-cell regulation. *Toxicol. Appl. Pharmacol.*, **87**, 18–31.

Kerkvliet, N.I. and Oughton, J.A. (1993) Acute inflammatory response to SRBC in mice treated with 2,3,7,8-tetrachlorodibenzo-*p*-dioxin (TCDD): phenotypic and functional analysis of peritoneal exudate cells. *Toxicol. Appl. Pharmacol.* **119**, 248–57.

Kerkvliet, N.I., Baecher-Steppan, L., Shepherd, D.M., Oughton, J.A., Voerderstrasse, B.A. and DeKrey, G.K. (1996) Inhibition of TC-1 cytokine production, effector cytotoxic T lymphocyte development and alloantibody production by 2,3,7,8-tetrachlorodibenzo-*p*-dioxin. *J. Immunol.*, **157**, 2310–19.

Kerkvliet, N.I., Baecher-Steppan, L., Smith, B.B., Youngberg, J.A., Henderson, M.C. and Buhler, D.R. (1990a) Role of the Ah locus in suppression of cytotoxic T lymphocyte activity by halogenated aromatic hydrocarbons (PCBs and TCDD): structure–activity relationships and effects in C57B1/6 mice congenic at the Ah locus. *Fundam. Appl. Toxicol.*, **14**, 532–41.

Kerkvliet, N.I., Steppan, L.B., Brauner, J.A., Deyo, J.A., Henderson, M.C. and Buhler, D.R. (1990b) Influence of the Ah locus on the humoral immunotoxicity of 2,3,7,8-tetrachlorodibenzo-*p*-dioxin (TCDD): evidence for Ah receptor-dependent and Ah receptor-independent mechanisms of immunosuppression. *Toxicol. Appl. Pharmacol.*, **105**, 26–36.

Lai, Z.-W., Pineau, T. and Esser, C. (1996) Identification of dioxin-responsive elements (DREs) in the 5′ regions of putative dioxin-inducible genes. *Chem.-Biol. Interact.*, **100**, 97–112.

Lawrence, B.P., Leid, M. and Kerkvliet, N.I. (1996) Distribution and behavior of the Ah receptor in murine T lymphocytes *Toxicol. Appl. Pharmacol.*, **138**, 275–84.

Lundberg, K., Dencker, L. and Kjell-Olov, G. (1992) 2,3,7,8-Tetrachlorodibenzo-*p*-dioxin (TCDD) inhibits the activation of antigen-specific T-cells in mice. *Int. J. Immunopharmacol.*, **14**, 699–705.

Mantovani, A., Vecchi, A., Luini, W., Sironi, M., Candiani, G.P., Spreafico, F. and Garattini, S. (1980) Effect of 2,3,7,8-tetrachlorodibenzo-*p*-dioxin on macrophage and natural killer cell-mediated cytotoxicity in mice. *Biomedicine*, **32**, 200–4.

Markham, R.B., Stashak, P.W., Prescott, B., Amsbaugh, D.F. and Baker, P.J. (1977) Effect of concanavalin A on lymphocyte interactions involved in the antibody response to type III pneumococcal polysaccharide. II. Ability of suppressor T cells to act on both B cells and amplifier T cells to limit the magnitude of the antibody response. *J. Immunol.* **119**, 1163–8.

Masten, S.A. and Shiverick, K.T. (1996) Characterization of the aryl hydrocarbon receptor complex in human B lymphocytes: evidence for a distinct nuclear DNA-binding form. *Arch. Biochem. Biophys.*, **336**, 297–308.

Mobley, J.L. and Dailey, M.O. (1992) Regulation of adhesion molecule expression by CD8 T cells *in vivo*. I. Differential regulation of $gp90^{MEL-14}$ (LECAM-1), Pgp-1, LFA-1, and VLA-4α during the differentiation of cytotoxic T lymphocytes induced by allografts. *J. Immunol.*, **148**, 2348–56.

Moos, A.B. and Kerkvliet, N.I. (1994) Inhibition of tumor necrosis factor fails to restore 2,3,7,8-tetrachlorodibenzo-*p*-dioxin (TDCC)-induced suppression of the antibody response to sheep red blood cells. *Toxicol. Lett.*, **81**, 175–81.

Moos, A.B., Baecher-Steppan, L. and Kerkvliet, N.I. (1994) Acute inflammatory responses to sheep red blood cells in mice treated with 2,3,7,8-tetrachlorodibenzo-*p*-dioxin: the role of proinflammatory cytokines, IL-1 and TNF. *Toxicol. Appl. Pharmacol.*, **127**, 331–5.

Morris, D.L., Jeong, H.G., Stevens, W.D., Chum, Y.J. and Holsapple, M.P. (1994) Serum modulation of the effects of TCDD on the *in vitro* antibody response and on enzyme induction in primary hepatocytes *Immunopharmacology*, **27**, 93–105.

Mueller, D.L., Jenkins, N.K. and Schwartz, R.H. (1989) Clonal expansion versus functional clonal inactivation: a costimulatory signalling pathway determines the outcome of T cell antigen receptor occupancy. *Annu. Rev. Immunol.*, **7**, 445–66.

Nebert, D.W., Puga, A. and Vasiliou, V. (1993) Role of the Ah receptor and the dioxin-inducible [Ah] gene battery in toxicity, cancer, and signal transduction. *Ann. NY Acad. Sci.*, **685**, 624–40.

Neubert, R., Golor, G., Stahlmann, R., Helge, H. and Neubert, D. (1992) Polyhalogenated dibenzo-*p*-dioxins and dibenzofurans and the immune system. Effects of multiple-dose treatment with 2,3,7,8-tetrachlorodibenzo-*p*-dioxin on peripheral lymphocyte subpopulations of a non-human primate (*callithrix jaochus*). *Arch. Toxicol.*, **66**, 250–9.

Neuman, C. M., Oughton, J. A. and Kerkvliet, N.I. (1993) Anti-CD3-induced T-cell activation – II. Effect of 2,3,7,8-tetrachlorodibenzo-*p*-dioxin. *Int. J. Immunopharmacol.,* **15**, 543–50.

Oughton, J.A., Pereira, C.B., DeKrey, G.K., Collier, J.M., Frank, A.A. and Kerkvliet, N.I. (1995) Phenotypic analysis of spleen, thymus, and peripheral blood cells in aged C57B1/6 mice following long-term exposure to 2,3,7,8-tetrachlorodibenzo-*p*-dioxin. *Fundam. Appl. Toxicol.*, **25**, 60–9.

Poland, A. and Glover, E. (1990) Characterization and strain distribution pattern of the murine Ah receptor specified by the Ah^{d} and Ah^{b-3} alleles. *Mol. Pharmacol.*, **38**, 306–12.

Prell, R.A. and Kerkvliet, N.I. (1997) Involvement of altered B7 expression in dioxin immunotoxicity. B7 transfection restores the CTL but not the alloantibody response to the P815 mastocytoma. *J. Immunol.*, **158**, 2695–703.

Prell, R.A., Oughton, J.A. and Kerkvliet, N.I. (1995) Effect of 2,3,7,8-tetrachlorodibenzo-*p*-dioxin on anti-CD3 induced changes in T cells subsets and cytokine production. *Int. J. Immunopharmacol.*, **17**, 951–61.

Puhvel, S.M. and Sakamoto, M. (1988) Effect of 2,3,7,8-tetrachlorodibenzo-*p*-dioxin on murine skin. *J. Invest. Dermatol.*, **90**, 354–8.

Rabinowitz, J.D., Beeson, C. Wulfing, C., Tate, K., Allen, P.M., Davis, M.M. and McConnell, H.M. (1996) Altered T cell receptor ligands trigger a subset of early T cell signals. *Immunity*, **5**, 125–35.

Rhile, M.J., Nagarkatti, M. and Nagarkatti, P.S. (1996) Role of Fas apoptosis and MHC genes in 2,3,7,8-tetrachlorodibenzo-*p*-dioxin (TCDD)-induced immunotoxicity of T cells. *Toxicology*, **110**, 153–67.

Sad, S., Marcotte, R. and Mosmann, T.R. (1995) Cytokine-induced differentiation of precursor mouse $CD8^+$ T cells into cytotoxic $CD8^+$ T cells secreting Th1 or Th2 cytokines. *Immunity*, **2**, 271–9.

Sharma, R.P., Kociba, R.J. and Gehring, P.J. (1978) Immunotoxicological effects of 2,3,7,8-tetrachlorodibenzo-*p*-dioxin in laboratory animals. *Toxicol. Appl. Pharmacol.*, **45**, 333.

Siebenkotten, G. and Radbruch, A. (1995) Towards a molecular understanding of immunoglobulin class switching. *Immunologist*, **3**, 141–6.

Smialowicz, R.J., Riddle, M.M., Williams, W.C. and Diliberto, J.J. (1994) Effects of 2,3,7,8-tetrachlorodibenzo-*p*-dioxin (TCDD) on humoral immunity and lymphocyte subpopulations: differences between mice and rats. *Toxicol. Appl. Pharmacol.*, **124**, 248–56.

Sutter, T.R., Guzman, K., Dold, K.M. and Greenlee, W.F. (1991) Targets for dioxin: genes for plasminogen activator inhibitor-2 and interleukin-1β. *Science*, **254**, 415–18.

Tomar, R.S. and Kerkvliet, N.I. (1991) Reduced T-helper cell function in mice exposed to 2,3,7,8-tetrachlorodibenzo-*p*-dioxin (TCDD). *Toxicol. Lett.*, **57**, 55–64.

Tonn, T., Esser, C., Schneider, E.M., Steinmann-Steiner-Haldenstatt, W. and Gleichmann, E. (1996) Persistence of decreased T-helper cell function in industrial workers 20 years after exposure to 2,3,7,8-tetrachlorodibenzo-*p*-dioxin. *Environ. Health Perspect.*, **104**, 422–6.

Vanden Heuvel, J.P., Clark, G.C., Kohn, M.C., Tritscher, A.N., Greenlee, W.F., Lucier, G.W. and Bell, D.A. (1994) Dioxin-responsive genes: examination of dose-response relationships using quantitative reverse treanscriptase-polymerase chain reaction. *Cancer Res.*, **54**, 62–8.

Van Parijs, L., Ibraghimov, A. and Abbas, A.K. (1996) The roles of costimulation and Fas in T cell apoptosis and peripheral tolerance. *Immunity*, **4**, 321–8.

Vecchi, A., Mantovanni, A., Sironi, M., Luini, W., Cairo, M. and Garattini, S. (1980) Effect of acute exposure to 2,3,7,8-tetrachlorodibenzo-*p*-dioxin on humoral antibody production in mice. *Chem. Biol. Interact.*, **30**, 337–42.

Vecchi, A., Sironi, M., Canegrati, M.A., Recchis, M. and Garattini, S. (1983) Immunosuppressive effects of 2,3,7,8-tetrachlorodibenzo-*p*-dioxin in strains of mice with different susceptibility to induction of aryl hydrocarbon hydroxylase. *Toxicol. Appl. Pharmacol.*, **68**, 434–81.

Vos, J.G. and Luster, M.I. (1989) Immune alterations, in *Halogenated Biphenyls, Terphenyls, Naphthalenes, Dibenzodioxins and Related Products* (eds. R.D. Kimbrough and S. Jensen), Elsevier Science Publishers B.V. (Biomedical Division), Amsterdam, pp. 295–322.

Vos, J.G., Moore, J.A. and Zinkl, J.G. (1973) Effect of 2,3,7,8-tetrachlorodibenzo-*p*-dioxin on the immune system. *Environ. Health Perspect.*, **5**, 149–62.

Vos, J.G., Moore, J.A. and Zinkl, J.G. (1974) Toxicity of 2,3,7,8-tetrachlorodibenzo-*p*-dioxin (TCDD) in C57Bl/6 mice. *Toxicol. Appl. Pharmacol.*, **29**, 229–41.

Vos, J.G., Kreeftenberg, J.G., Engle, H.W.B., Minderhud, A. and Van Noorle Jansen, L.M. (1978) Studies on 2,3,7,8-tetrachlorodibenzo-*p*-dioxin-induced immune suppression and decreased resistance to infection: endotoxin hypersensitivity, serum zinc concentrations and effect of thymosin treatment. *Toxicology*, **9**, 75–86.

Watson, A.J., Weir-Brown, K.I., Bannister, R.M., Chu, F.-F., Reisz-Porszasz, S., Fujii-Kuriyama, Y., Sogawa, K. and Hankinson, O. (1992) Mechanism of action of a repressor of dioxin-dependent induction of Cyp1a1 gene transcription. *Mol. Cell. Biol.*, **12**, 2115–23.

Weissberg, J.B. and Zinkl, J.G. (1973) Effects of 2,3,7,8-tetrachlorodibenzo-*p*-dioxin upon hemostasis and hematologic function in the rat. *Environ. Health Perspect.*, **5**, 119–23.

Williams, C.E., Crawford, R.B., Holsapple, M.P. and Kaminski, N.E. (1996) Identification of functional aryl hydrocarbon receptor and aryl hydrocarbon receptor nuclear translocator in murine splenocytes *Biochem. Pharmacol.*, **52**, 771–80.

Zhang, X., Brunner, T., Carter, L., Dutton, R.W., Rogers, P., Bradley, L., Sato, T., Reed, J.C., Green, D. and Swain, S.L. (1997) Unequal death in T helper cell (Th)1 and Th2 effectors: Th1, but not Th2, effectors undergo rapid Fas/FasL-mediated apoptosis. *J. Exp. Med.*, **185**, 1837–49.

Zheng, L., Fisher, G., Miller, R.E., Peschon, J., Lynch, D.H. and Lenardo, M.J. (1995) Induction of apoptosis in mature T cells by tumour necrosis factor. *Nature*, **377**, 348–51.

5 T Lymphocyte Subpopulations and Immunotoxicology of Drugs of Abuse

Michael P. Holsapple, Norbert E. Kaminski and Stephen B. Pruett

Introduction

There has been a long-standing interest in the effects of drugs of abuse on the immune system. In fact, as reviewed by Holsapple and Munson (1985), some of the earliest studies in the area of immunotoxicology were aimed at characterizing the effects of commonly abused substances, most notably marihuana and ethanol, on immune function. In recent years, interest in characterizing and understanding the consequences of immunotoxicology of drugs of abuse has increased because of two separate phenomena. First, substance abuse has been suggested to play a significant role in the spread of the AIDS epidemic (Hartsock and Genser, 1991). Multiple co-factors that may increase risk are associated with the lifestyle of the drug user, including high risk sexual practises, sharing of needles and the overall health status. An obvious and particularly important co-factor in susceptibility to HIV is the postulated involvement of drug-induced immunological disturbances (Donahoe, 1990), which is the focus of this chapter. Second, the central nervous system (CNS) and the immune system are no longer considered independent and isolated systems; but are now recognized as cooperating systems in constant communication with one another (Dunn, 1989; Chapter 9; Ader *et al.*, 1990). The study of neuroendocrine–immune interactions, also known as psychoneuroimmunology, has undergone an impressive development over the last 5 years. In light of the fact that substances with inherent abuse potential invariably target the CNS and cause a number of measurable changes in neuroendocrine status, several drugs of abuse have proven to be important probes in the characterization of the science of psychoneuroimmunology. Because the mechanism of the immunotoxic actions of ethanol (Han and Pruett, 1996), cocaine (Bayer *et al.*, 1995; Stanulis *et al.*, 1997a) and morphine (Pruett *et al.*, 1992) has been demonstrated to be mediated at least in part via activation

T Lymphocyte Subpopulations in Immunotoxicology. Edited by Ian Kimber and MaryJane K. Selgrade.
 ISBN 0 471 97194 4

of the hypothalamic–pituitary–adrenal (HPA) neuraxis, these substances are correctly classified as 'endocrine disrupters'.

Inclusion of a chapter with a focus on drugs of abuse in a book on the role of T lymphocyte subpopulations in immunotoxicology is appropriate because T cells, especially T helper (Th) cells, have frequently been identified as an important target. In this chapter, the role of Th cells in the immunotoxicity of cannabinoids, ethanol and cocaine are reviewed. As conveyed in the various sections, it is becoming increasingly apparent that disruption in the relative balance between Th1 and Th2 cells may play a role in the mechanism of immunotoxicity of drugs of abuse. It is important to acknowledge that some would consider ethanol and marihuana as contributors to substance abuse; but would not consider them as 'drugs' of abuse. We readily admit that we have taken some liberties with our use of these terms. It is important to emphasize also that if space were available, this chapter could have easily incorporated a section on the immunotoxicity of narcotic opiates, such as morphine and heroin.

Effects of cannabinoids on T lymphocyte subpopulations

Δ^9-Tetrahydrocannabinol (Δ^9-THC) is one of more than 60 structurally related cannabinoid compounds present in the *cannabis sativa* plant. Of the plant-derived cannabinoids, Δ^9-THC is the most potent psychoactive congener and the most extensively characterized with respect to biological activity (Mechoulam, 1970). Over the past 20 years, it has been well established that Δ^9-THC and a number of other plant-derived as well as synthetic cannabinoid congeners produce a broad range of effects on immune function (reviewed: Friedman *et al.*, 1995; Kaminski, 1995). Generally speaking, the plant-derived cannabinoids are relatively weak immunomodulators which have consistently demonstrated immune suppression in the low micromolar range *in vitro* and in the low milligram per kilogram range *in vivo*. More recently, under certain experimental conditions, immune enhancement has been observed at lower cannabinoid concentrations than those that produce immune suppression. Although all three major components of the immune system, cell-mediated, humoral and innate immunity are susceptible to modulation by cannabinoids, the relative sensitivity of various leukocyte cell-types has not yet been well characterized. Inhibition of cell-mediated responses has been demonstrated widely with a variety of cannabinoids including Δ^9-THC following *in vivo* and/or *in vitro* treatment utilizing mixed lymphocyte responses (Schatz *et al.*, 1993), cytotoxic T lymphocyte responses (Fischer-Stenger *et al.*, 1992; Klein *et al.*, 1991), natural killer (NK) cell activity (Kawakami *et al.*, 1988), IL-2 secretion (Condie *et al.*, 1996; Nakano *et al.*, 1993) and proliferative responses to T cell mitogens (Pross *et al.*, 1987; Schatz *et al.*, 1993; Specter *et al.*, 1990). Likewise, Δ^9-THC inhibits the ability of macrophages to elicit an interferon-independent, cell contact-dependent reduction of virus growth in infected cells

(Cabral and Vasquez, 1992), macrophage tumouricidal activity (Burnette-Curley and Cabral, 1995), tumour necrosis factor-alpha (TNFα) secretion (Fischer-Stenger *et al.*, 1993), as well as lipopolysaccharide (LPS)-induced nitric oxide production (Burnette-Curley and Cabral, 1995; Jeon *et al.*, 1996). With respect to humoral immunity, Δ^9-THC produces a marked inhibition of the primary IgM antibody forming cell (AFC) response to sheep erythrocytes (SRBC), a T cell dependent antigen (Kaminski *et al.*, 1992; Schatz *et al.*, 1993). Conversely, Δ^9-THC has no inhibitory effect on the IgM AFC response to DNP-Ficoll, a T cell independent antigen which requires macrophages as accessory cells, or to the polyclonal B cell activator, LPS (Schatz *et al.*, 1993). The profile of activity observed with humoral immune responses to the aforementioned defined antigens has provided some preliminary insights into cell-type sensitivity to Δ^9-THC. Most striking is the fact that B cell effector function (i.e. immunoglobulin secretion) appears to be quite refractory to inhibition by cannabinoids such as Δ^9-THC. Likewise, macrophage accessory help in humoral responses to the soluble antigen, DNP-Ficoll, is also unaffected by Δ^9-THC. Conversely, the observation that antibody responses to T cell dependent antigens are markedly inhibited by a variety of plant-derived and synthetic cannabinoids, as demonstrated with the anti-SRBC responses (Kaminski *et al.*, 1992; Schatz *et al.*, 1993), suggests that Th cells, as well as perhaps Th cell/macrophage interactions, are sensitive to inhibition by cannabinoid compounds. The remainder of this section will focus primarily on the mechanism by which cannabinoids inhibit IL-2 expression, a Th1-derived cytokine, and will address the roles played by specific cannabinoid receptors and by changes in the levels of cAMP. These mechanisms provide insights into potential effects of cannabinoids on Th1 and Th2 cell populations.

Two related forms of GTP-binding protein coupled receptors have been identified and cloned to which cannabinoid compounds bind (see Figure 5.1); cannabinoid receptor type-1 (CB1) (Matsuda *et al.*, 1990) and cannabinoid receptor type-2 (CB2) (Munro *et al.*, 1993). The two forms of cannabinoid receptors are strikingly dissimilar sharing only 44% identity, with this increasing to 68% when exclusively comparing the transmembrane regions (Munro *et al.*, 1993); that portion of the receptor believed to be involved in ligand binding. Equally surprising is that in spite of the structural differences between CB1 and CB2, many of the cannabinoid receptor ligands studied to date exhibit considerable similarity in binding characteristics to both forms of the receptor (Showalter *et al.*, 1996). One of the first cannabinoids found to exhibit differential binding affinity between CB1 and CB2 was the plant-derived cannabinoid, cannabinol, which possesses markedly greater affinity for CB2 than for CB1 (Munro *et al.*, 1993; Showalter *et al.*, 1996; Schatz *et al.*, 1997).

One of the more curious aspects pertaining to cannabinoid receptors is their anatomic distribution. CB1 is primarily expressed within the central nervous system (CNS) with modest expression observed on leukocytes. Northern analysis shows that CB1 is expressed as an approximately 6 kb fragment in human,

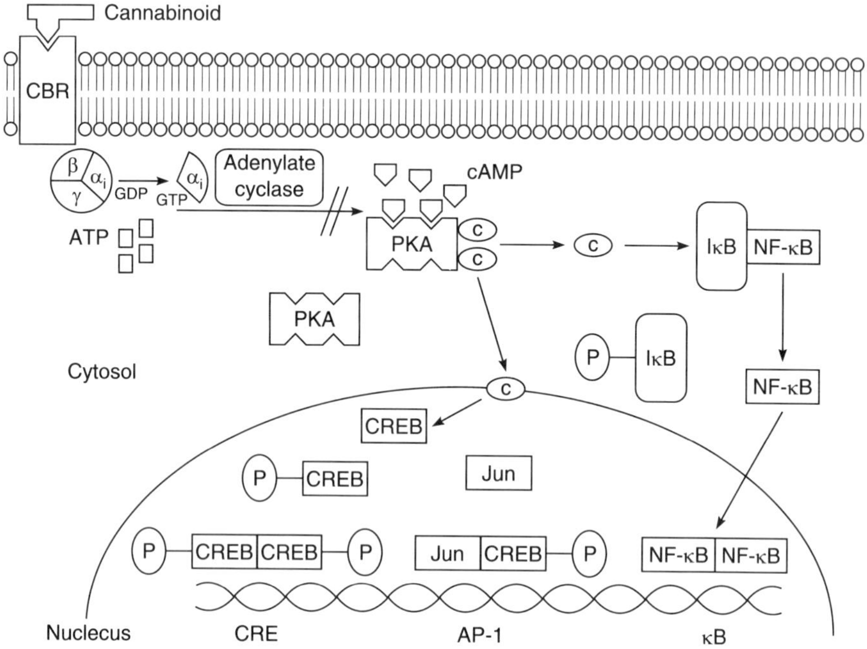

Figure 5.1 Regulation of signal transduction via cannabinoid receptors.

rat and mouse brain (Matsuda *et al.*, 1990). Although little is known about the endogenous ligands for cannabinoid receptors with the exception that several derivatives of arachadonic acid exhibit modest binding activity, CB1 is remarkably conserved phylogenetically, as evidenced by its identification even in fairly primitive life forms such as the sea urchin (Chang *et al.*, 1993). With respect to the immune system, low level expression of CB1 transcripts have been identified in human B cells, T cells and monocytes (Bouaboula *et al.*, 1993), and in mouse spleen (Kaminski *et al.*, 1992) while being below the level of detection in mouse thymus even using RT-PCR (Schatz *et al.*, 1997). Conversely, CB2 which is almost undetectable within the CNS, is the primary form of cannabinoid receptors expressed within the immune system as suggested by radioligand binding, Northern analysis and quantitative RT-PCR (Schatz *et al.*, 1997). Interestingly, CB2 is expressed as a 2.4 kb fragment in human and the rat (Munro *et al.*, 1993) and as an approximately 4 kb transcript in the mouse (Schatz *et al.*, 1997); however, the reason for the difference in transcript size between certain species is presently unknown. CB2 distribution within the immune system appears to be broad based in view of the fact that CB2 transcripts have been identified in rat spleen (Munro *et al.*, 1993; Schatz *et al.*, 1997), mouse spleen and thymus (Schatz *et al.*, 1997), and a number of immune system-derived cell lines including the T cell lines, EL4.IL-2 (mouse; Condie *et al.*, 1996), HPB-ALL (human; Schatz *et al.*, 1997), Jurkat E6-1 (human; Schatz *et al.*, 1997), the monocytic lines, HL60 (human; Munro *et al.*, 1993) and RAW264.7 (mouse; Jeon *et al.*, 1996), the B cell lines, CH12LX (mouse; unpublished observation) and BCL-1 (mouse; unpublished observation) and the mast cell line, RBL-2H3 (rat; Facci *et al.*, 1995). The significance of CB2 tissue distribution is that it provides a target for pharmacologically-based immune modulation by cannabinoids in the absence of undesirable CNS-associated secondary effects.

As presented in the model in Figure 5.1, following ligand binding, both CB1 and CB2 receptors negatively regulate adenylate cyclase as exhibited by a marked decrease in the conversion of ATP to cAMP (Bayewitch *et al.*, 1995; Howlett, 1985; Howlett and Fleming, 1984; Matsuda *et al.*, 1990; Schatz *et al.*, 1992). Of considerable interest has been the observation that cannabinoids concomitantly inhibit immune competence and cAMP formation; an apparent anomaly in light of the long-standing axiom that the cAMP signalling cascade is a negative regulator of leukocyte function. This concept is based on numerous studies that have convincingly demonstrated a negative influence by high concentrations of membrane permeable cAMP analogues (0.2 – 2 mM) on leukocyte function in culture. Under such conditions, immune inhibition has been demonstrated in B cells and T cells (Hsueh and Lai, 1995; Johnson *et al.*, 1988; Tsuruta *et al.*, 1995) and most often has been associated also with an arrest in cell cycle progression (Maraguchi *et al.*, 1984). In contrast, at low concentrations, cAMP analogs ($\leq 100 \mu M$) such as dibutyryl cAMP and 8-bromo-cAMP are immunostimulatory (Kammer, 1988; Kaminski *et al.*, 1994; Koh *et al.*, 1995). Concordantly, the cannabinoid-induced inhibition of many immune responses (e.g. anti-SRBC IgM AFC response, lymphoproliferation and nitric oxide production by macrophages) is abrogated in the presence of low concentrations of cAMP (Jeon *et al.*, 1996; Kaminski *et al.*, 1994). Likewise, glucagon attenuates the inhibition of the anti-sRBC AFC response by Δ^9-THC by increasing intracellular cAMP through glucagon receptors that positively regulate adenylate cyclase (Koh *et al.*, 1996). Similarly, pretreatment of splenocytes with pertussis toxin, an agent which ADP-ribosylates $G\alpha_{inhibitory}$-binding proteins to block the negative regulation of adenylate cyclase in the presence of cannabinoids, attenuates immune inhibition by Δ^9-THC, including suppression of the anti-SRBC IgM AFC response (Kaminski *et al.*, 1994). These results are consistent with a positive role for the cAMP cascade in leukocyte regulation, and suggest that immune suppression by cannabinoids is mediated through an inhibition of the cAMP signal transduction pathway. These mechanisms provide insights into potential effects of cannabinoids on Th1 and Th2 cell responses.

Recent studies aimed at examining the effect of plant derived cannabinoids on Th cell function demonstrated that both Δ^9-THC and cannabinol markedly inhibited IL-2 production by splenic T cells and in the murine thymoma, EL4.IL-2 (Condie *et al.*, 1996). Inhibition of IL-2 production and gene expression in both of the aforementioned cell preparations was demonstrated following treatment with phorbol ester plus calcium ionophore; an activation stimulus independent of macrophage accessory cell function. These findings are consistent with recent evidence suggesting a role for the cAMP signalling cascade in the regulation of IL-2 gene expression. A number of critical observations support this conclusion. First, activation of lymphoid cells by mitogens or phorbol ester plus calcium ionophore produces a rapid (i.e. within 5 min) but transient increase in intracellular cAMP (Hadden *et al.*, 1972; Kaminski *et al.*, 1994; Pepe *et al.*, 1994; Russell, 1978; Smith *et al.*, 1971; Watson *et al.*, 1994) which in the presence of immunosuppressive concentrations of cannabinoids is almost completely inhibited (Kaminski *et al.*, 1994). The sole function of cAMP, as it is presently understood, is to activate protein kinase A (PKA). Consequently, in the presence of cannabinoids, PKA activity in EL4.IL-2 cells (Condie *et al.*, 1996), thymocytes (unpublished observation) and splenocytes (Koh *et al.*, 1997) is dose-dependently inhibited (Figure 5.1). It is notable that cannabinoids have no direct effect on PKA, as evidenced by the restoration of kinase activity by exogenous cAMP in the presence of cannabinoids (Koh *et al.*, 1997). Therefore, inhibition of PKA activity by cannabinoids is mediated indirectly through a decrease in cAMP formation due to a decrease in adenylate cyclase activity. DNA binding to cAMP responsive element (CRE) consensus motifs by CREB/ATF family members represents the terminal event in the cAMP signalling cascade. DNA binding by CREB/ATF proteins is regulated through PKA-mediated phosphorylation of serine residues on these nuclear factors (e.g. Ser 133 on CREB; Gonzalez and Montminy, 1989). A number of recent observations suggest a positive regulatory role for CRE DNA binding in leukocyte regulation, including T cell regulation. For example, anti-CD3 stimulation of splenic T cells (Koh *et al.*, 1997) or thymocytes (unpublished observation) produces a rapid induction of nuclear factor binding to CRE consensus recognition motifs which peaks at approximately 60 min post-stimulation and which then gradually decreases over the next 60–90 min. In the presence of Δ^9-THC, CRE binding activity in the aforementioned preparations is markedly inhibited suggesting that PKA-mediated phosphorylation of CREB/ATF family proteins is decreased by cannabinoid compounds (Condie *et al.*, 1996; Koh *et al.*, 1997). Collectively, these findings suggest that the inhibition of the cAMP signalling cascade is a critical component of the mechanism for immune suppression by cannabinoids.

More recently, studies directed toward elucidating the mechanism for immune modulation by cannabinoids have begun focusing on specific cell-types. As discussed above, one observation from such investigations pertaining specifically to Th cells has been the finding that cannabinoids significantly inhibit IL-2 expression (Condie *et al.*, 1996; Nakano *et al.*, 1993). In spite of the sensitivity of IL-2 production to inhibition by cannabinoids, the minimal essential promoter region for the IL-2 gene possesses no CRE consensus motifs as might be predicted from the discussion above (Novak *et al.*, 1990b). It does, however, possess two AP-1-like sites, the AP-1 proximal (AP-1p) site (-145 bp) and the AP-1 distal (AP-1d) site (-179 bp; Novak *et al.*, 1990b), of which the former is critical to our discussion of IL-2 inhibition by cannabinoids (Figure 5.1). Specifically, CREB proteins have been identified as a constituent of the protein binding complex associated with the AP-1p site of the IL-2 promoter in activated thymocytes. Antibodies specific for Fos, Jun and CREB were used in gel shift assays to identify CREB binding to the AP-1p-site through the formation of 'cross-family' dimers with Fos and Jun (i.e. Fos/CREB and Jun/CREB dimers; Chen and Rothenberg, 1993). Although the formation of CREB/Fos and CREB/Jun 'cross-family' dimers has been previously demonstrated by several laboratories (Hai and Curran, 1991; Ivashkiv *et al.*, 1990), only recently have they been implicated in the regulation of IL-2 gene expression. Consistent with this finding, phorbol ester plus calcium ionophore stimulation of splenocytes (Condie *et al.*, 1996) and thymocytes (Novak *et al.*, 1990a) induces nuclear factor binding to the AP-1p site which is further enhanced by concomitant stimulation with forskolin, a direct activator of adenylate cyclase and thus the cAMP cascade. One explanation for the aforementioned enhancement in AP-1p binding in the presence of forskolin following phorbol ester plus calcium ionophore stimulation is through an increase in CREB phosphorylation, leading to enhanced formation of Fos/CREB and Jun/CREB dimers. A second and concurrent event which likely contributes to enhanced AP-1p-site nuclear factor binding in the presence of forskolin is through an increase in Fos protein expression. It is well established that the regulatory region for the Fos gene contains multiple CRE consensus recognition motifs which helps to regulate positively Fos transcription (Berkowitz *et al.*, 1989). Therefore, by turning on the cAMP cascade, Fos expression is likely to be increased during T cell activation.

Not surprisingly, the cannabinoid-induced inhibition of IL-2 gene transcription by Th cells is more complex than a mechanism simply mediated through a decrease in the phosphorylation and activation of CREB/ATF family members. In the presence of cannabinoids, activated T cells also exhibit a decrease in protein binding to κB motifs (unpublished observation) which are the cognate binding site for NF-κB/cRel family of transcription factors (Figure 5.1). The minimal essential region of the IL-2 promoter possesses a κB consensus motif at -195 bp, which is yet another critical site for the regulation/activation of IL-2 transcription (Novak *et al.*, 1990b). The NF-κB/cRel family of DNA binding proteins are maintained in a quiescent state in the cytosol through their association with the inhibitory protein, IκB. NF-κB/cRel activation is mediated through the phosphorylation of IκB by IκB kinase which marks the inhibitory protein for degradation through the ubiquitin pathway (Chen *et al.*, 1996). IκB kinase appears to be a point of convergence for at least two major signalling pathways, since both PKC and PKA can activate NF-κB. Once activated, NF-κB translocates to the nucleus to bind κB motifs in the regulatory region of a variety of genes (Shirakawa and Mizel, 1989). In light of this, the inhibition of NF-κB/cRel family proteins in the presence of cannabinoids is likely through a decrease in PKA mediated activation of IκB kinase. Collectively, decreased IL-2 expression in the presence of cannabinoids is mediated through decreased formation of Fos/CREB and Jun/CREB dimers which bind to the AP-1p site and decreased activation of NF-κB/Rel family proteins which bind κB sites. The adverse regulation of DNA binding in both cases appears to be through the inhibition of the cAMP cascade. Given the effects of cannabinoids on IL-2, they have the potential to interfere with the development of both Th1 and Th2 responses.

As already mentioned, the accessory function played by Th cells in antibody responses is especially sensitive to inhibition by cannabinoid compounds. This observation is very consistent with recent findings which suggest that the cAMP signalling cascade may serve as a switching mechanism between the Th cell subpopulations, Th1 which facilitate cell-mediated immune responses, and Th2, that subtype which regulates humoral immune responses. Several recent studies indicate that high intracellular cAMP concentrations favor Th2 responses. For example, high concentrations of cAMP (1 mM) activate the binding of nuclear factors to GATA-3 and the conserved lymphokine element (CLE0) to enhance the expression of the B cell differentiation factor, IL-5 (Lee *et al.*, 1995). Moreover, high sustained concentrations of cAMP inhibit IL-2 expression by Th1 cells (Siegel *et al.*, 1995). This, coupled with the fact that T cells show a marked inhibition in a number of responses following cannabinoid treatment, suggests that cAMP is essential for both Th1 and Th2 lymphokine gene expression; however, low and perhaps transient intracellular cAMP concentrations favour the activation of Th1 lymphokines as discussed for IL-2; whereas, high and sustained cAMP concentrations appear to favour the induction of Th2 associated cytokines. Further, the apparent requirement for high cAMP concentrations for IL-5 expression provides an explanation for the marked sensitivity of T cell dependent humoral responses to inhibition by cannabinoids, since these compounds are potent inhibitors of adenylate cyclase.

Effects of ethanol (EtOH) on T lymphocyte subpopulations

It has been known for many years that alcoholism is associated with an increased incidence of infectious disease and cancer (Adams and Jordan, 1984; Breeden, 1984; Nelson *et al.*, 1995). Several studies in humans and experimental animals suggest that the basis for this increase is, at least in part, immunosuppression caused by EtOH (MacGregor, 1986). Recent evidence indicates that acute (binge) as well as chronic EtOH consumption can be immunosuppressive (Carson and Pruett, 1996; Gentilello *et al.*, 1993; Han and Pruett, 1995; Nelson *et al.*, 1989). However, the cellular targets and mechanisms by which EtOH affects the immune system have not been fully delineated.

In assessing the role of T cell subsets as targets of EtOH, it seems reasonable to begin with a review of the effects of EtOH on resistance to diseases for which the role of T cells is known. In a mouse model of chronic EtOH exposure, Saad and colleagues noted that resistance to *Listeria monocytogenes* was decreased significantly in mice given EtOH in a liquid diet preparation (Saad *et al.*, 1993). T cells (Czuprynski *et al.*, 1989), and particularly Th1 cells (Mosmann and Sad, 1996), are known to be important in resistance to *Listeria monocytogenes*. However, neutrophils are also known to be important in resistance to *Listeria monocytogenes* (Czuprynski *et al.*, 1994), and neutrophil chemotaxis can be inhibited by EtOH (Nelson *et al.*, 1989). Similarly, resistance to *Mycobacterium tuberculosis* infection involves innate mechanisms followed by a Th1-mediated immune response (Mosmann and Sad, 1996), and EtOH suppresses resistance to *Mycobacterium tuberculosis* in humans and in animal models (Mendenhall *et al.*, 1990; Nelson *et al.*, 1995). Chronic administration of EtOH in a liquid diet also suppresses host resistance to *Trichinella spiralis*, a parasite for which antibodies and T cell-mediated responses are involved in resistance (Steven *et al.*, 1990). Resistance to *Borrelia burgdorferi* (i.e. the causative agent of Lyme disease) is impaired by EtOH in a similar model, and this effect is associated with a decrease in the delayed type hypersensitivity response to Borrelia antigen (Pavia *et al.*, 1991). Delayed type hypersensitivity is predominantly mediated by Th1 cells (Mosmann and Sad, 1996). These findings are consistent with substantial suppression of Th1 cell function in EtOH-treated animals. However, other immunological parameters are also suppressed in most cases, and it is not clear which immunological decrement contributes most to decreased host resistance.

The available data are consistent also with T cells as a target for EtOH-induced immunosuppression in humans. The infectious diseases most consistently associated with alcoholism in humans are bacterial pneumonia and tuberculosis (Nelson *et al.*, 1995). The major acquired immune defence mechanism against Gram-positive pneumonia is generally considered to be the combined action of antibody, complement, and phagocytic cells (Wood and Davis, 1980). Suppression of Th1 cell activity would not be expected substantially to affect such a response. However, recent evidence from an animal model suggests that alcohol may increase susceptibility to bacterial pneumonia by decreasing the innate antibacterial properties of lung surfactants (Rubins *et al.*, 1996). In contrast, long-term resistance to *Mycobacterium tuberculosis* is dependent on a Th1 response (Mosmann and Sad, 1996), and resistance to this organism is suppressed by EtOH (Nelson *et al.*, 1995). However, it remains possible that suppression of an important innate defence mechanism (e.g., production of TNF*a*) by EtOH may also be involved in decreased resistance to this organism (Nelson *et al.*, 1995).

In a mouse model for alcoholism, Jerrells and colleagues found that T-dependent humoral responses are suppressed (Jerrells *et al.*, 1990a). In contrast, the antibody response to a T-independent antigen was not influenced by

EtOH, suggesting T cells as a major target for EtOH in this system. According to the currently accepted Th1/Th2 paradigm, antibody responses require a Th1/Th2 balance in which Th1 function is not too predominant (Mosmann and Sad, 1996). Thus, a moderate decrease in Th1 function in the absence of a decrease in Th2 function would not be expected to suppress antibody responses substantially. A possible exception would be IgG2a responses, which are Th1-dependent (Schijns *et al.*, 1994). The substantial decreases in the primary (IgM) response to SRBC noted by Jerrells and colleagues suggest that Th1 as well as Th2 function may be decreased by EtOH. Holsapple and colleagues exposed mice to a slightly lower concentration of EtOH in a liquid diet preparation for a longer period of time (up to 42 days) (Holsapple *et al.*, 1993a). Exposure to EtOH under these conditions decreased the primary humoral response to SRBC, but in contrast to the results reported by Jerrells *et al.* (1990a), decreased responses to T cell mitogens were not noted. In addition, EtOH did not directly decrease the humoral response *in vitro*; but serum from EtOH-consuming mice significantly decreased this response. EtOH levels in this serum were negligible, suggesting an indirect effect of EtOH in this system (Holsapple *et al.*, 1993a). Suppression of the primary humoral response to a T-dependent antigen, but not a T-independent antigen, has also been reported following long-term (30 days or more) exposure of rats to EtOH in a liquid diet (Bagasra *et al.*, 1987). Taken together, the available evidence is generally consistent with T cells as an important target of EtOH.

In a rat model in which EtOH is administered by intubation to maintain relatively high blood EtOH levels, both the humoral response to SRBC, and the response to T cell mitogens *in vitro* were suppressed (Jerrells *et al.*, 1986, 1989). Direct administration of EtOH in culture did not suppress responses to T cell mitogens, so indirect mechanisms were again implicated. In addition, adrenalectomy prevented EtOH-induced suppression of the response to T cell mitogens (Jerrells *et al.*, 1989). This suggests the involvement of an EtOH-induced stress response in this suppression of T cell responses to mitogen. Similar effects have been noted in a binge drinking model in mice in which a single dose of EtOH was administered intragastrically (by gavage) (Han and Pruett, 1995). In these studies, EtOH-induced suppression of the primary humoral response to SRBC was partially prevented by the glucocorticoid antagonist, RU 486. This suggests the involvement of glucocorticoids.

Acute and chronic administration of EtOH causes a stress response in rodents characterized by the activation of the hypothalamic–pituitary–adrenal axis (Carson and Pruett, 1996; Jerrells *et al.*, 1990b; see Figure 5.2). A measurable EtOH-induced stress response also occurs in humans (Jenkins and Connolly, 1968; Mendelson *et al.*, 1971; Merry and Marks, 1969; Välimäki *et al.*, 1984). However, precise quantitative comparisons between the rodent and human responses are limited by the dissimilarity of the major glucocorticoid in these species (i.e. corticosterone in rodents and cortisol in humans), and by the small size of most of the human studies. In any case, blood EtOH levels

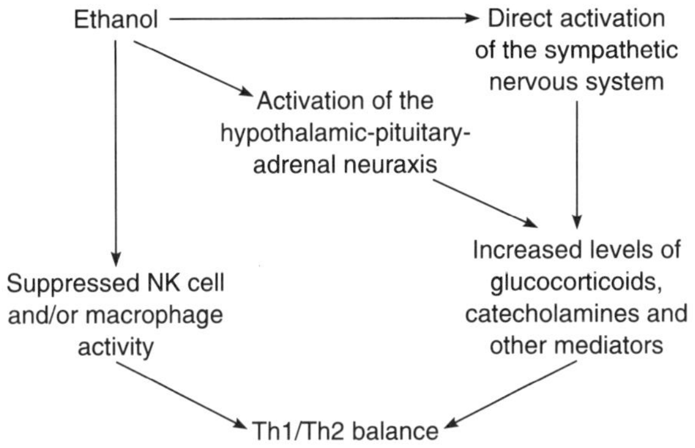

Figure 5.2 Simplified summary of possible pathways by which ethanol could affect the Th1/Th2 balance.

greater than ~0.15% are consistently associated with increases in glucocorticoid levels in humans and rodents (Figure 5.2).

It is widely accepted that glucocorticoids at stress-induced or pharmacological levels have a greater suppressive effect on Th1 cells than on Th2 cells (Chiappelli *et al.*, 1994a; Daynes and Araneo, 1989; Rook *et al.*, 1994; Chapter 10). Thus, it might be expected that EtOH would decrease cell-mediated immune responses to a greater extent than humoral responses or that Th1-dependent immunoglobulins (e.g. IgG2a) would be suppressed to a greater extent than Th2-dependent immunoglobulins (e.g. IgG1 or IgE). However, the conclusion that glucocorticoids suppress Th1 responses to a greater extent than Th2 responses is based on a relatively limited set of data from experiments utilizing purified protein antigens (Stam *et al.*, 1993). Han and colleagues noted that EtOH-induced increases in corticosterone in mice suppressed the induction of increased levels of mRNA for IL-1, IL-2 and IL-4 (Han and Pruett, 1995). Although quantitative RT-PCR was not conducted for all cytokines, there were no obvious differences in the suppression of IL-2 (a Th1 cytokine), and IL-4 (a Th2 cytokine). Furthermore, EtOH suppresses IgG1 production to at least as great an extent as IgG2a production in response to SRBC (Carson and Pruett, 1996). Thus, it is not clear that EtOH-induced stress responses preferentially suppress Th1 cells, but further evidence is needed using a variety of experimental systems (different antigens, mouse and rat strains, EtOH dosing protocols, etc.) before any generalizations are warranted. In addition there have been no studies of possible differences in the effects of EtOH on the major subsets of T cytotoxic cells (Tc1 and Tc2; Mosmann and Sad, 1996).

Acute, or chronic EtOH administration also suppresses basal (Gallucci *et al.*, 1994; Wu *et al.*, 1994) and induced (Gallucci and Meadows, 1996; Wu *et al.*,

1994) NK cell activity in mice. Recent studies indicate that activated NK cells influence isotype switching (presumably through undefined effects on Th cells) (Wilder *et al.*, 1996) and that NK cells are required for the generation of cytotoxic T lymphocytes *in vitro* (Kos and Engleman, 1995). In addition, it has been shown that EtOH strongly affects some macrophage functions (Nelson *et al.*,1989). Because macrophages can influence the Th1/Th2 balance (Mosmann and Sad, 1996), it is conceivable that EtOH could affect the Th1/Th2 balance by altering macrophage function. Thus, there are potentially a number of indirect mechanisms whereby EtOH could alter normal T cell responses (Figure 5.2).

In summary, data from several animal models and from humans indicate that EtOH decreases resistance to microorganisms for which T cells are known to be important in immunity. Several groups have directly demonstrated that T cell responses are suppressed in EtOH-treated animals, and in some cases this suppression is mediated by EtOH-induced activation of the HPA axis. Although stress-induced neuroendocrine mediators may preferentially suppress Th1 cells as compared with Th2 cells, direct evidence for this in EtOH-treated animals or humans is lacking. In addition, no evidence is available regarding the differential effects of EtOH on the $CD8^+$ T cells as compared to $CD4^+$ T cells, or on the major $CD8^+$ subpopulations. Such information will be needed for any future attempts to use pharmacological intervention to prevent stress-induced immunosuppression, and for a full understanding of the normal role of glucocorticoids in homeostasis of the immune system.

Effects of cocaine on T lymphocyte subpopulations

Cocaine is currently one of the most widely abused drugs in the United States (Gawin, 1991), and over the last decade cocaine abuse has reached epidemic proportions. Coincident with the increased abuse of cocaine has been a tremendous increase in the number of infections with the human immunodeficiency virus (HIV) and in the progression of acquired immune deficiency syndrome (AIDS). These two phenomena share a temporal relationship, and are frequently paired within high risk groups for HIV infectivity (Hartsock and Genser, 1991). Cocaine abuse in humans has been reported to be associated with immune abnormalities, including increased incidences of infections by opportunistic pathogens including *Conidiobolus* (Jaffey *et al.*, 1990) and *Candida albicans* (Zagelbaum *et al.*, 1991). There are several ways that cocaine may contribute to the onset and/or progression of AIDS, including the possibility of blood mixing through shared needles and a high-risk lifestyle associated with chronic drug abuse. More direct possibilities include either cocaine-induced changes in HIV infectivity and/or replication, or cocaine-induced changes in immune function. The former possibility was demonstrated by Peterson *et al.* (1991), who showed that direct exposure of human peripheral blood mononuclear cells (PBMC) to cocaine caused an increase in the replication of HIV, as evidenced by the expression of the P24 antigen. Interestingly, this effect by cocaine was noted at low (10^{-9} M) but not high (10^{-6} M) concentrations of cocaine, and was blocked by co-treatment with an antibody against transforming growth factor-β (TGF-β). The authors speculated that the latter observation was due to the release of this pleiotropic growth factor from lymphocytes caused by exposure to cocaine, although no attempt was made to confirm this possibility. It is interesting to speculate that cocaine-induced changes in immunocompetence in either direction could act as a co-factor in AIDS. Cocaine-induced activation of mononuclear cells could increase the infectivity of these cells. As indicated below, some studies have suggested that exposure to cocaine may enhance the activation of immunocompetent cells or increase the magnitude of an immune response. A number of studies have demonstrated that cocaine suppresses immunocompetence, which could be an important co-factor by acting in concert with the virus to impair or destroy the ability to combat the various opportunistic pathogens which are associated with AIDS.

Studies of cocaine's effects on immunocompetence have included both direct exposure to lymphocytes *in vitro*, and *in vivo* (i.e. whole animal) exposure. The aim of *in vitro* studies is generally to address the direct effects of cocaine on immune function. Before summarizing the *in vitro* studies with cocaine, it is important to discuss briefly what is known about plasma and tissue levels of cocaine. Cone *et al.* (1988) reported that peak plasma concentrations for cocaine ranged from 2.84×10^{-7} to 1.73×10^{-6} M. Cocaine levels in the blood over 2 mg/l (i.e. 2 μg/ml or 6×10^{-6} M) are usually associated with fatalities (Smart and Anglin, 1987).

Two studies using human peripheral blood lymphocytes have failed to detect any direct effects by cocaine. First, Martinez and Watson (1990) showed that concentrations of cocaine from 0.01 to 100 μM had no effect on pokeweed mitogen (PWM)-stimulated IgG synthesis *in vitro*. The second paper by Luo *et al.*, (1992) showed that exposure to cocaine from 100 to 3200 ng/well had no effect on mitogen-induced T cell proliferation (i.e. either PHA or Con A) by either mouse or human lymphocytes. Since each well contained 0.1 ml, the effective concentration range of cocaine was from 1 to 32 μg/ml. Although Klein *et al.* (1988) showed that direct exposure to cocainc suppressed T cell proliferation (i.e. Con A); but had no effect on B cell proliferation (i.e. LPS) by murine splenocytes, the significance of the former results is questionable because concentrations of 100–200 μg/ml (3–6 $\times$ 10^{-4} M) were required. They showed a similar profile for human lymphocytes, in that the proliferation in response to PHA was unaffected by concentrations of cocaine up to 100 μg/ml. Similar results were reported in a second study using human peripheral blood lymphocytes, as Delafuente and DeVane (1991) showed significant suppression of mitogen-induced proliferation; but only when 100 μg/ml cocaine was added daily to the cultures. The rationale for the daily administration of cocaine was based on *in vitro* kinetics studies which indicated that 90% of the cocaine was degraded within 72 h under tissue culture conditions, and that the major degradation product was benzoylecgonine. Delafuente and DeVane (1991) also tested the daily administration of benzoylecgonine, ecgonine and ecgonine methyl ester (i.e. all metabolites resulting from the hydrolysis of cocaine) and observed no suppression. More recently, Jeong *et al.* (1995a) reported similar results with the T-dependent antibody response to SRBC in that benzoyecgonine was devoid of activity at concentrations up to 10^{-4} M, while cocaine caused a modest but nonsignificant suppression at the same concentration. Previous results by Holsapple *et al.* (1993b) confirmed this effect at high concentrations of cocaine and showed an identical profile of activity with procaine, which suggested that this suppression may be a consequence of the local anaesthetic properties by these two drugs. Support for this possibility came from a report by Klein *et al.* (1993) that showed that high concentrations of cocaine suppressed the proliferation of human T cells in response to PHA, under conditions of suboptimum proliferation, and that the effects on proliferation were associated with a decrease in the mobilization of cytosolic free calcium. Interestingly, the studies by Jeong *et al.* (1995a) showed that direct exposure to norcocaine at concentrations as low as 10^{-5} M caused a significant suppression (>60%) of the antibody response. Norcocaine is the initial product of the *N*-demethylation of cocaine by the P-450 pathway, and these results were the first to suggest a role for P-450 metabolism, which is normally a minor pathway for the degradation of cocaine. Taken together, the available evidence from *in vitro* studies does not provide compelling evidence that changes in immunocompetence associated with exposure to cocaine are mediated by direct effects on lymphocytes. This conclusion must be qualified based on some recent studies demonstrating a direct effect by cocaine at reasonable concentrations on cytokine production, as discussed below.

The literature on *in vivo* studies presents a confusing picture of sometimes contradictory effects on a variety of immune function parameters. The profile of cocaine immunotoxicity is characterized by increases, decreases and the absence of effect. The difficulties in interpretation and comparison are due in part to the different exposure regimens (including dose, route, time-course, etc.), sex, strain, and assay methods. Results from an immunotoxicology validation study, conducted as part of the National Toxicology Program (NTP), indicated that the best correlations and the greatest degree of reproducibility were achieved when studies were conducted under similar conditions (Luster *et al.*, 1988).

Watson *et al.* (1983) reported that five daily exposures to cocaine HCl from 15 to 60 mg/kg per day produced a non-dose-related suppression (~30%) of the delayed type hypersensitivity response to the contact sensitizer, dinitrofluorobenzene in male ICR mice. The antibody response to SRBC proved to be more sensitive, as exposure to cocaine HCl from 3 mg/kg per day to 30 mg/kg per day produced a dose-related suppression of 8–46%. Holsapple and Munson (1985) reported that the effects of cocaine on immune function were quite modest, in spite of the fact that behaviorally active doses of cocaine were tested (i.e. 25 and 50 mg/kg per day for 14 days in male CD-1 mice – all mice became hyperactive, and a number of mice convulsed and died at the high dose). Delayed type hypersensitivity to SRBC was unaffected, while the antibody response to the same antigen was suppressed in a non-dose-related fashion with significance only noted at the lowest dose (35%). Havas *et al.* (1987) studied the effects of cocaine in BALB/c mice using a complex exposure paradigm (i.e. mice received either three or five i.p. injections of cocaine from 20 to 50 mg/kg for 1 day, or two i.p. injections per day of cocaine from 30 to 60 mg/kg for 5 consecutive days) and reported increases in the antibody responses to both T-independent and T-dependent antigens, and no effect on the infection to *S. pneumoniae* or on the resistance to an injection with a solid tumour. In contrast, Ou *et al.* (1989) demonstrated a non-dose-related increase in the size of an injected tumour in female BALB/c mice after 10 days of treatment (i.e. daily i.m. injections of cocaine from 0.05 to 5.0 mg/kg per day). This study also showed that exposure to cocaine caused a 34% decrease in the antibody response to SRBC at the highest doses tested. Bagasra and Forman (1989) characterized the effects of cocaine (i.e. daily doses of 1.25–5 mg/kg cocaine for 10 days) in male Fisher rats and demonstrated an elevation in the antibody responses to both a T-independent (pneumococcal polysaccharide type III) and a T-dependent (SRBC) antigen at the low doses, and a suppression at the high dose. The elevated antibody responses were attributed to the combined effects of a loss of T suppressor cells, a direct stimulation of B cells and a stimulation of T helper cells, although it is important to emphasize that definitive studies to test and confirm this profile of activity were not reported. Starec *et al.* (1991) showed that male DBA/2 mice exposed to i.p. injections of cocaine for 5 days at 50 mg/kg (two per day) followed by single daily injections of 30 mg/kg for 14 days had decreased resistance to infection with Friend virus, decreased total immunglobulin titers following immunization with SRBC, but no effect on the delayed type hypersensitivity response to the contact sensitizer, oxazalone. Matulka *et al.* (1996) compared the effects of repeated exposure to cocaine (i.e. daily exposure to cocaine at 5–60 mg/kg per day for 7 or 14 days) in male and female B6C3F1 and DBA/2 mice using the antibody response to SRBC. They observed a sex-dependent (i.e. male > female) and strain-dependent (i.e. B6 > D2) difference in immunosuppression by cocaine, which was consistent with a profile of activity mediated via metabolism of cocaine by the cytochrome P-450 pathway. Additional support for the interpretation that the suppression by high doses of cocaine was dependent on the generation of a P-450 metabolite came from studies in which the immunotoxicity by cocaine was increased by the co-administration of either an inducer of cytochrome P-450 or an inhibitor of the esterase pathway (Jeong *et al.*, 1995b, 1996; and Kump *et al.*, 1996). The studies suggesting a role by P-450 metabolism in the immunotoxicity by cocaine (Jeong *et al.*, 1995b, 1996; Kump *et al.*, 1996; Matulka *et al.*, 1996) differed from most of the previous studies because the magnitude of suppression of the antibody response to SRBC was frequently >50%. These studies also showed that the conditions necessary for cocaine-induced immunosuppression were generally associated with liver damage and the authors speculated that the effects on the immune system were an indirect consequence of the damage to the liver. Support for this speculation came from recent studies which showed that cocaine-induced hepatotoxicity was associated with an elevation in the serum levels of TGF-β, and that the immunosuppression by cocaine was reversed by the administration of an antibody against TGF-β (Matulka *et al.*; manuscript in preparation).

The studies by Matulka *et al.* (1996) are also significant because they observed a significant enhancement in the antibody response at low doses of cocaine in B6C3F1 mice; but not in DBA/2 mice. This profile of activity (i.e. enhancement at low doses, and suppression at high doses) is identical to previous results by Bagasra and Forman (1989). More recent results by Stanulis *et al.* (1997a) showed that acute exposure to a behaviourally active dose of cocaine also caused a significant enhancement in the antibody response to SRBC. Interestingly, a subsequent study by the same group indicated that exposure to cocaine under conditions which elevated the T-dependent antibody response to SRBC had no effect on either the delayed type hypersensitivity response to the same antigen, or a T-independent antibody response (Stanulis *et al.*, 1997b). The latter observation is in contrast with previous results by Havas *et al.* (1987) and by Bagasra and Forman (1989), who demonstrated significant increases in antibody responses to both T-dependent and T-independent antibody responses.

Although the results to-date present a complex picture of the profile of cocaine immunotoxicity, they also allow a number of important observations to be made. First, the results indicate that delayed type hypersensitivity, a Th1-dependent immune parameter, appears to be relatively refractory to the effects of cocaine, whereas humoral immunity, a Th2-dependent immune parameter (i.e. at least under conditions where T-dependence has been demonstrated), is sensitive. Second, the dosimetry for the effects of cocaine on the immune system is complex, with enhancement noted at low doses and suppression noted at high doses. Because the low doses are also associated with the onset of behavioural activity, it is reasonable to suggest that there may be a relationship to cocaine's effects on neuroendocrine processes. In this regard, it is imporant to note that the most comprehensive review to-date on the effects of cocaine on immunocompetence was written from the perspective that the immunological effects of cocaine are indirectly mediated through the neuroendocrine effects of this drug (Watzl and Watson, 1990). More recent results are consistent with the involvement of a central nervous system pathway in the modulation of the immune system by cocaine (Bayer *et al.*, 1995, 1996). As noted above, the evidence to date suggests that the suppression associated with high doses of cocaine are an indirect consequence of the liver damage caused by the generation of a reactive P-450 metabolite (Jeong *et al.*, 1995a,b, 1996;

Kump *et al.*, 1996; Matulka *et al.*, 1996), possibly mediated through an increase in the serum levels of TGF-β (Matulka *et al.*, manuscript in preparation). It is interesting to speculate that the absence of effects on the immune system by intermediate doses of cocaine may actually reflect the complex interactions by two opposing actions of neuroendocrine changes and liver damage, and are really not indicative of the absence of activity by cocaine.

The remainder of this section will focus on the effects by low behaviourally active doses of cocaine because these effects have been the most extensively characterized from a mechanistic perspective, and because these doses are perhaps the easiest to extrapolate to humans. Stanulis *et al.* (1997b) concluded that the increase in the antibody response to SRBC was due to an effect on Th2 cells because neither a T-independent antibody response nor the delayed type hypersensitivity response to SRBC were affected by acute exposure to cocaine. More definitive evidence came from the analysis of cytokine levels in anti-CD3-stimulated splenocyte suspensions from cocaine-treated mice. They observed a significant increase in Th2 cytokines, IL-4 and IL-10, and no effect on Th1 cytokines, IL-2 and interferon-gamma (IFN-*g*; Stanulis *et al.*, 1997b). The mechanism for this profile of activity proved to be mediated by an increase in the serum levels of corticosterone based on the following results: acute exposure to cocaine caused an increase in serum corticosterone; acute exposure to intermediate doses of corticosterone (i.e. 50 and 100 mg/kg) mimicked cocaine and caused an increase in the antibody response; the cocaine-induced increase in the antibody response was completely blocked by adrenalectomy and by the administration of a corticotropin releasing factor (CRF) antagonist, was partially blocked by the administration of the glucocorticoid antagonist, RU 486, and was unaffected by adrenal demedullation (i.e. thereby ruling out a contribution by a sympathetic component mediated by the release of adrenal catecholamines). Cocaine had previously been shown to activate the HPA axis in the hypothalamus causing the release of CRF, which acted on the pituitary to release adrenocorticotropin (ACTH), which acted on the adrenal cortex to release corticosterone (Rivier and Vane, 1987; Sarnyai *et al.*, 1992).

The apparent involvement of corticosterone in an enhanced antibody response may seem odd, because the glucocorticoids have generally been associated with suppression of immune parameters. The initial suspicion was that corticosterone suppressed Th1 cells and caused a shift in the Th-cell balance, which favoured Th2 cell activity and indirectly resulted in an enhanced antibody response. This speculation was consistent with a recent study in humans by Elenkov *et al.* (1996), which showed that dexamethasome inhibited the secretion of IL-12, a cytokine produced by macrophages and B cells that is known to affect Th1-dependent responses, but not IL-10, a Th2-derived cytokine; and caused a shift toward a Th2 type of cytokine pattern and immune responsiveness. Stanulis *et al.* (1997b) did demonstrate that Th1 cytokines (i.e. splenocyte cultures from naive B6C3F1 mice stimulated with anti-CD3) were suppressed by direct exposure to corticosterone in a dose-related fashion with significance noted at 0.1 μM for IL-2 and at 0.03 μM for IFN-*g*. However, they also surprisingly showed that direct addition of corticosterone resulted in a biphasic effect on Th2 cytokines, IL-4 and IL-10, with a significant enhancement noted at 0.01 μM (i.e. the lowest concentration tested) and a significant suppression only noted at 0.3 μM (i.e. the highest concentration tested). Additional support for a differential profile of activity by corticosterone on Th1 and Th2 cells comes from recent work which demonstrated a dose-related suppression of the proliferation of the Th1 cell line, 3DO-54.8, with significance noted at 0.03 μM; and a biphasic effect on the proliferation of the Th2 cell line, CDC.35, with a significant enhancement noted at 0.03 and 0.1 μM and only a modest, non-significant suppression noted at 3.0 μM (Kump *et al.*; manuscript in preparation). These results are consistent with previous studies which indicated that glucocorticoids select for Th2-type immune responses by inducing the production of IL-4 (Daynes and Araneo, 1989), and by inhibiting transcription of IFN-*g* and IL-2 genes (Daynes and Araneo, 1989; Serfling *et al.*, 1989; Vacca *et al.*, 1992). The specific mechanism for this differential regulation of the Th1/Th2 system by glucocorticoids is not known.

A number of other studies have characterized the effects of cocaine on cytokine production in both mice and humans after both *in vivo* and *in vitro* exposure, as summarized in Table 5.1. Some studies have characterized the effects of exposure to cocaine on the levels of pro-inflammatory cytokines, including IL-6 (Wang *et al.* 1994), IL-8 (Mao *et al.*, 1996) and TNFα (Wang *et al.*, 1994). However, the focus of this section will be on Th-derived cytokines and/or cytokines known to affect Th1/Th2 balance. In contrast with the studies by Stanulis *et al.* (1997b), DiFrancesco *et al.* (1992) showed that acute administration of cocaine suppressed the production of both Th1 cytokines, IFN-*g* and IL-2, and a Th2 cytokine, IL-4, by splenocytes from BALB/c mice. Adding to the confusion is a report by Wang *et al.* (1994), who compared the effects of *in vivo* and *in vitro* exposure to cocaine on a number of Th1 and Th2 cytokines in C57BL/6 mice. Repeated exposure to cocaine increased the secretion of IL-2 and IL-5, decreased the production of IL-4 and IL-10, and had no effect on the production of IFN-*g*. Surprisingly, the authors offered no explanation for this complex profile of activity. They did not even speculate about the significance of opposite effects on various Th1 (i.e., an increase in IL-2, and no effect on IFN-*g*) and Th2 cytokines (i.e. increase in IL-5, and decreases in IL-4 and IL-10). It is possible that the different profiles of activity observed in these three papers are due in part to the study designs which employed different exposure conditions and mouse strains. The latter point is particularly relevant because different mouse strains are known to possess different propensities to express either a Th1 or a Th2 phenotype (McLeod *et al.*, 1995). In this regard, it is important to recall that Matulka *et al.* (1996) observed the enhancement in the antibody response to SRBC following exposure to low, behavorially active doses of cocaine in B6C3F1 mice (i.e. a Th2-predominant strain), but not in DBA/2 mice (i.e. a Th1-predominant strain).

Table 5.1 Summary of the effects of cocaine on cytokines involved in the Th1/Th2 balance

Study design			Cytokines – Th1 system			Cytokines – Th2 system			
Species	Exposure	Parameter	IL-2	IFN-g	IL-12	IL-4	IL-5	IL-10	Reference
Mice (BALB/c)	*in vivo* (acute)	SPLC – Con A	↓	↓		↓			DiFrancesco *et al*. (1992)
Mice (C57BL/6)	*in vivo* (repeated)	SPLC – Con A	↑	No effect		↓	↑	↓	Wang *et al*. (1994)
Mice (B6C3F1)	*in vivo* (acute)	SPLC – antiCD3	No effect	No effect		↑		↑	Stanulis *et al*. (1997b)
Human (addicts)	*in vivo*	Serum		↑	No effect			↓	Fiala *et al*. (unpublished)
Human (addicts)	*in vivo*	PBMC – PHA		↑	↑			↓	Fiala *et al*. (unpublished)
Mice (C57BL/6)	*in vitro* 0.33–333.0 μM	SPLC – Con A	↓	↓		↓↓	↓↓	↓↓	Wang *et al*. (1994)
Human	*in vivo* 0.09–96.0 μM	T cells – antiCD3	↑						Matsui *et al*. (1992)
Human	*in vitro* 0.1–10.0 μM	PBMC – Con A	↑ / ↓						Chiappelli *et al*. (1994b)
Human	*in vitro* 0.003–3.3 μM	PBMC – IL-2		↓		No effect	No effect	No effect	Mao *et al*. (1996)

Milan Fiala and co-workers from the UCLA School of Medicine studied the effects of acute i.v. cocaine in cocaine-addicted human subjects (personal communication). Post-cocaine serum levels of IL-10 were reduced in four of four subjects, while those of IFN-*g* were moderately increased, and the levels of IL-12 were unaffected. When peripheral blood mononuclear cells (PBMC) were cultured without stimulation, a reduction in the levels of IL-10 was noted in post-cocaine supernatants relative to pre-cocaine supernatants, while IL-12 was unaffected. When PBMC cultures were stimulated with PHA, there was an increase in the post-cocaine levels of IFN-*g* and IL-12, while the levels of IL-10 were suppressed following treatment with cocaine. It is interesting that these investigators also showed that six of seven subjects showed an increase in serum cortisol after treatment with cocaine. These results suggest that the stimulation of the HPA axis by cocaine in humans is associated with a decrease the levels of a Th2 cytokine (IL-10) and a possible increase in the levels of a Th1 cytokine (IFN-*g*). These results are opposite to those reported by Stanulis *et al.* (1997a,b) following acute exposure to cocaine in mice. While the different results may reflect inherent differences between humans and mice, it is also important to emphasize that the studies by Fiala and co-workers were conducted in cocaine addicts, and that it is impossible to determine the contribution by the previous chronic exposure to cocaine.

The *in vitro* studies with cocaine on cytokine production also offer a puzzling profile of activity (Table 5.1). In studies with splenocytes from C57BL/6 mice, Wang *et al.* (1994) reported that direct exposure to cocaine inhibited the production of a number of cytokines (IL-2, IFN-*g*, IL-4, IL-5 and IL-10), with a greater degree of suppression of the Th2 cytokines than the Th1 cytokines. In contrast, Matsui *et al.* (1992) showed that cocaine in the range of 0.09–96.0 μM increased the production of IL-2 by human T-cells stimulated with anti-CD3 antibody. Interestingly, the proliferative response by these T-cell cultures, and the level of cytosolic free calcium mobilization were also increased over a comparable concentration range. A third study by Chiappelli *et al.* (1994b) characterized the effects of cocaine on Con A-induced IL-2 production in human PBMC. These authors reported a biphasic effect with increased levels of IL-2 noted at 0.1 and 1.0 μM cocaine, and a significant reduction noted at 10.0 μM cocaine. Again, the range of effects and different profiles of results may be due to either inherent differences between humans and mice and/or differences in the study design. However, because all of these studies used relatively high concentrations of cocaine, the significance of these results is suspect, as emphasized above.

The same concern cannot by made for a recent report by Mao *et al.* (1996), which characterized the direct effect of cocaine at concentrations from 1 to 1000 ng/ml on IL-2-stimulated production of IL-4, IL-5, IL-10 and IFN-*g* by human peripheral blood lymphocytes. Cocaine did not affect IL-4, IL-5 or IL-10; but directly suppressed the IL-2-stimulated production of IFN-*g* at concentrations as low as 1 ng/ml, as measured by cytokine-specific ELISA.

Unfortunately, the opposite effects by cocaine on IFN-*g* production in the two human studies (i.e. increased following acute exposure as demonstrated by Gan *et al.*, 1997; and decreased following direct exposure as demonstrated by Mao *et al.*, 1996) cannot be resolved at the present time. The studies by Mao *et al.* (1996) did show that direct exposure to cocaine decreased the expression of IFN-*g* mRNA, as measured by Northern blot and slot blot analysis; but had no effect on the stability of the specific mRNA. Additional studies with nuclear run-on assays revealed that cocaine down-regulated the rate of IFN-*g* transcription, although the specific mechanism was not elucidated.

Taken together, although the available results indicate that cocaine produces a complex profile of activity as it relates to cytokine production, there is enough compelling evidence to suggest that the cytokine crosstalk between Th1 and Th2 cells is a fruitful area for further study into the mechanism of immunotoxicity by this drug. However, the complex profile of activity for the effects of cocaine on cytokine expression may also indicate a potential effect on the third type of Th cells, Th0 cells. Although there is considerable debate concerning the role of Th0 cells in Th cell crosstalk and/or development, it is reported that these cells produce both Th1-type and Th2-type cytokine profiles (Del Prete, 1992). Future studies on the modulation of the expression of cytokine genes will need to address the possible direct effects of cocaine, as reported by Mao *et al.* (1996), as well as possible indirect effects mediated through the release of glucocorticoids. The latter effects will have to consider the possibility that glucocorticoids suppress cytokine gene expression as reported by Daynes and Araneo (1989), Serfling *et al.* (1989), and Vacca *et al.* (1992) and the possibility for a glucocorticoid-induced activation of cytokine gene expression, as reported by Daynes and Araneo (1989). The complexity of the interaction between cocaine and T lymphocyte subpopulations is perhaps best exemplified by one of the earliest studies to characterize the immunocompetence in human addicts. Donahoe *et al.* (1986) showed that the severe effects on T cells observed in heroin addicts were reversed in heroin addicts who also used cocaine. Clearly, this observation can not be easily explained on the basis of a profile of activity by cocaine which was simply limited to a generalized immunosuppression, and may reflect an important action mediated at the level of Th1/Th2 balance.

Conclusions

The results reviewed in this chapter indicate that a number of different classes of drugs of abuse can affect the immune system and that T lymphocyte subpopulations, especially Th cells, are a primary target for these effects. It has been suggested that selective modulation of one Th subtype over the other may determine the type of response that is elicited (Bottomly *et al.*, 1989), and the assertion has been made that control at the level of the Th cell network may determine whether Th1 or Th2 responses are produced (Powrie *et al.*, 1994).

Moreover, it has been proposed that a Th1 to Th2 switch is a critical step in the onset and/or progression of HIV (Clerici and Shearer, 1993). As such, a disruption in the balance between Th1 and Th2 phenotypes following the exposure to cannabinoids, ethanol or cocaine could contribute to this switch, and thereby contribute to the onset and/or progression of AIDS.

The mechanisms responsible for the effects of drugs of abuse on the Th cell network include both direct effects, as clearly demonstrated for the cannabinoids and suggested for cocaine, and indirect effects, as demonstrated for ethanol and cocaine. The latter effects are mediated, at least in part via an activation of the HPA axis. Glucocorticoids have clearly been demonstrated to be an important endogenous modulator of the Th cell network (Chiapelli *et al.*, 1994; Chapter 10). Glucocorticoids have also been shown to contribute to the immunosuppression associated with chemical-induced stress as a general model of indirect immunotoxicity (Pruett *et al.*, 1993). It is anticipated that future studies characterizing the immunotoxicity of drugs of abuse will provide new insights into our understanding of the role of glucocorticoids in homeostasis of the immune system, and of the modulation of the Th cell network associated with stress.

Acknowledgements

Work on the cannabinoid project was supported in part by funds from NIDA Grants DA09789 and DA07908. The authors recognize the following individuals for their contributions to the cannabinoid research: Dr Mary Abood, Robin Condie, Faye Kessler, Sung Hyun Han, Amy Herring, Dr Seong Gu Huong, Dr Young Jin Jeon, Dr Woo Suk Koh, Dr Michael Lee and Dr Anthony Schatz.

Work on the ethanol project was supported in part by funds from NIAAA Grant AA09505. Dr Stephen Pruett is also supported by a Research Scientist Development Award (K02 AA00201) from NIAAA. The authors recognize the following individuals for their contributions to the ethanol research: Ed Carson, Stephanie Collier, Yun-Cheng Han, Deborah Kiel, Paul Weiss and Wen-Jun Wu.

Work on the cocaine project was supported in part by funds from NIDA Grant DA08161. The authors recognize the following individuals for their contributions to the cocaine research: Dr Tae Cheon Jeong, Steve Jordan, Denton F. Kump, Dr Ray Matulka, Dr John Rosecrans and Dr Eric Stanulis.

References

Adams, H.G. and Jordan, C. (1984) Infections in the alcoholic. *Med. Clin. N. Am.*, **68**, 179–200.

Ader, R., Felton, D. and Cohen, N. (1990) Interaction between the brain and the immune system. *Annu. Rev. Pharmacol. Toxicol.*, **30**, 561–602.

Bagasra, O. and Forman, L. (1989) Functional analysis of lymphocytes subpopulations in experimental cocaine abuse. I. Dose-dependent activation of lymphocyte subsets. *Clin. Exp. Immunol.*, **77**, 289–93.
Bagasra, O., Howeedy, A., Dorio, R. and Kajdacsy-Balla, A. (1987) Functional analysis of T-cell subsets in chronic experimental alcoholism. *Immunology*, **61**, 63–9.
Bayer, B.M., Mulroney, E., Hernandez, M.C. and Ding, X.Z. (1995) Acute infusions of cocaine result in time- and dose-dependent effects of lymphocyte responses and corticosterone secretion. *Immunopharmacology*, **29**, 19–28.
Bayer, B.M., Hernandez, M.C., Dunn, K.L., Melton, R.D. and Pellegrino, T.C. (1996) Central pathways are involved in morphine and cocaine modulation of the immune system. *J. Neuroimmunol.*, **69**, 33–5.
Bayewitch, M., Avidor-Reiss, T., Levy, R., Barg, J. and Mechoulam, R. (1995) The peripheral cannabinoid receptor: adenylate cyclase inhibition and G protein coupling. *FEBS Lett.*, **375**, 143–7.
Berkowitz, L.A., Riabowol, K.T. and Gilman, M.Z. (1989) Multiple sequence elements of a single functional class are required for cyclic AMP responsiveness of the mouse c-fos promoter. *Mol. Cell. Biol.*, **9**, 4272–81.
Bouaboula, M., Rinaldi, M., Carayon, P., Carillon, C., Delpech, B., Shire, D., Le Fur, G. and Casellas, P. (1993) Cannabinoid-receptor expression in human leukocytes. *Eur. J. Biochem.*, **214**, 173–80.
Bottomly, K., Lugman, M., Greenbaum, L. and Carding, S. (1989) A monoclonal antibody to murine CD45R distinguishes CD4 T cell populations that produce different cytokines. *Eur. J. Immunol.*, **19**, 617–23.
Breeden, J.H. (1984) Alcohol, alcoholism, and cancer. *Med. Clin. N. Am.*, **68**, 163–78.
Burnette-Curley, D. and Cabral, G.A. (1995) Differential inhibition of RAW264.7 macrophage tumoricidal activity by delta-9-tetrahydrocannabinol. *Proc. Soc. Exp. Biol. Med.*, **210**, 64–76.
Cabral, G.A. and Vasquez, R. (1992) Delta 9-tetrahydrocannabinol suppresses macrophage extrinsic antiherpes virus activity. *Proc. Soc. Exp. Biol. Med.*, **199**, 255–63.
Carson, E. and Pruett, S.B. (1996) Development and characterization of a binge drinking model in mice for evaluation of the immunological effects of ethanol. *Alcohol. Clin. Exp. Res.*, **20**, 132–8.
Chang, M.C., Berkery, D., Schuel, R., Laychock, S.G., Zimmerman, A.M., Zimmerman, S. and Schuel, H. (1993) Evidence for a cannabinoid receptor in sea urchin sperm and its role in blockade of the acrosome reaction. *Mol. Reprod. Dev.*, **36**, 507–16.
Chen, D. and Rothenberg, E. (1993) Molecular basis for developmental changes in interleukin-2 gene inducibility. *Mol. Cell. Biol.*, **13**, 228–37.
Chen, Z.J., Parent, L. and Maniantis, T. (1996) Site specific phosphorylation of IkBa by a novel ubiquitination-dependent kinase activity. *Cell*, **84**, 853–62.
Chiappelli, F., Manfrini, E., Franceschi, C., Cossarizza, A. and Black, K.L. (1994a) Steroid regulation of cytokines. Relevance for Th1-to-Th2 shift? *Ann. N.Y. Acad. Sci.*, **746**, 204–15.
Chiappelli, F., Frost, P., Manfrini, E., Lee, P., Pham, L., Garcia, C., Daley, S., Kung, M. and Villanueva, P. (1994b) Cocaine blunts human CD4+ cell activation. *Immunopharmacology*, **28**, 233–40.
Clerici, M. and Shearer, G.M. (1993) A Th1 to Th2 switch is a critical step in the etiology of HIV infection. *Immunol. Today*, **14**, 107–10.
Condie, R., Herring, A., Koh, W.S., Lee, M. and Kaminski, N.E. (1996) Cannabinoid inhibition of adenylate cyclase-mediated signal transduction and IL-2 expression in the murine T-cell line, EL4.IL-2. *J. Biol. Chem.*, **271**, 13175–83.

Cone, E.J., Kumor, K., Thompson, L.K. and Sherer, M. (1988) Correlation of saliva cocaine levels with plasma levels and with pharmacologic effects after intravenous cocaine administration in human subjects *J. Anal. Toxicol.*, **12**, 200–6.

Czuprynski, C.J., Brown, J.F., Young, K.M. and Cooley, A.J. (1989) Administration of purified anti-L3T4 monoclonal antibody impairs the resistance of mice to *Listeria monocytogenes* infection. *Infect. Immun.*, **57**, 100–9.

Czuprynski, C.J., Brown, J.F., Maroushek, N., Wagner, R.D. and Steinburg, H. (1994) Administration of anti-granulocyte mAb RB6-8C5 impairs the resistance of mice to *Listeria monocytogenes* infection. *J. Immunol.*, **152**, 1836–46.

Daynes, R.A. and Araneo, B.A. (1989) Contrasting effects of glucocorticoids on the capacity of T cells to produce the growth factors interleukin 2 and interleukin 4. *Eur. J. Immunol.*, **19**, 2319–25.

Delafuente, J.C. and DeVane, C.L. (1991) Immunologic effects of cocaine and related alkaloids. *Immunopharmacol. Immunotoxicol.*, **13**, 11–23.

Del Prete, G. (1992) Human Th1 and Th2 lymphocytes: their role in the pathophysiology of atopy. *Allergy*, **47**, 450–5.

DiFrancesco, P., Marini, S., Pica, F., Favalli, C., Tubaro, E. and Garaci, E. (1992) *In vivo* cocaine administration influences lymphokine production and humoral immune response. *Immunol. Res.*, **11**, 74–9.

Donahoe, R.M. (1990) Drug abuse and AIDS: Causes for the connection, in *Drugs of Abuse: Chemistry, Pharmacology, Immunology and AIDS* (eds P.T.K. Pham and K. Rice), NIDA Research Monograph 96, GPO, Washington, DC.

Donahoe, R.M., Nicholson, J.K.A., Madden, J.J., Donahoe, F., Shafer, D.A., Gordon, D., Bokos, P. and Falek, A. (1986) Coordinate and independent effects of heroin, cocaine, and alcohol abuse on T-cell E-rosette formation and antigenic marker expression. *Clin. Immunol. Immunopathol.*, **41**, 254–64.

Dunn, A.J. (1989) Psychoneuroimmunology for the psychoneuroendocrinologist: a review of animal studies of nervous system–immune system interactions. *Psychoneuroendocrinology*, **14**, 251–74.

Elenkov, I.J., Papanicolaou, D.A., Wilder, R.L. and Chrousos, G.P. (1996) Modulatory effects of glucocorticoids and catecholamines on human interleukin-12 and interleukin-10 production: clinical implications. *Proc. Assoc. Am. Physiol.*, **108**, 374–85.

Facci, L., Dal Toso, R., Romanello, S., Buriani, A., Skaper, D. and Leon, A. (1995) Mast cells express a peripheral cannabinoid receptor with differential sensitivity to anandamide and palmitoylethanolamide. *Proc. Natl Acad. Sci. USA*, **92**, 3376–80.

Fischer-Stenger, K., Updegrove, A.W. and Cabral, G.A. (1992) Delta-9-tetrahydrocannabinol decreases cytotoxic T lymphocyte activity to herpes simplex virus type 1-infected cells. *Proc. Soc. Exp. Biol. Med.*, **200**, 422–30.

Fischer-Stenger, K., Dove-Pettit, D.A. and Cabral, G.A. (1993) Delta 9-tetrahydrocannabinol inhibition of tumour necrosis factor-alpha: suppression of post-translational events. *J. Pharmacol. Exp. Ther.*, **267**, 1558–65.

Friedman, H., Shrivers, S.C. and Klein, T.W. (1995) Drugs of abuse and the immune system, in *Immunotoxicology and Immunopharmacology* (eds J.H. Dean, M.I. Luster, A.E. Munson and I. Kimber), Raven Press, New York, pp. 303–22.

Gallucci, R.M. and Meadows, G.G. (1996) Ethanol consumption suppresses the IL-2 induced proliferation of NK cells. *Toxicol. Appl. Pharmacol.*, **138**, 90–7.

Gallucci, R.M., Pfister, L.J. and Meadows, G.G. (1994) Effects of ethanol consumption on enriched natural killer cells from C57BL/6 mice. *Alcohol. Clin. Exp. Res.*, **18**, 625–31.

Gawin, F.H. (1991) Cocaine addiction: psychology and neurophysiology. *Science*, **251**, 1580.

Gentilello, L.M., Cobean, R.A., Walker, A.P., Moore, E.E., Wertz, M.J. and Dellinger, E.P. (1993) Acute ethanol intoxication increases the risk of infection following penetration abdominal trauma. *J. Trauma*, **34**, 669–75.

Gonzalez, G.A. and Montminy, M.R. (1989) Cyclic AMP stimulates somatostatin gene transcription by phosphorylation of CREB at serine 133. *Cell*, **59**, 675–80.

Hadden, J.W., Hadden, E.M., Haddox, M.K. and Goldberg, N.D. (1972) Guanosine 3′:5′-cyclic monophosphates: a possible intracellular mediator of mitogenic influences in lymphocytes. *Proc. Natl. Acad. Sci. USA*, **69**, 3024–7.

Hai, T.-Y. and Curran, T. (1991) Cross-family dimerization of transcription factors Fos: Jun and ATF/CREB alters DNA binding specificity. *Proc. Natl Acad. Sci. USA*, **88**, 3720–4.

Han, Y.-C. and Pruett, S.B. (1995) Mechanisms of ethanol-induced suppression of a primary antibody response in a mouse model for binge drinking. *J. Pharmacol. Exp. Ther.*, **275**, 950–7.

Hartsock, P. and Genser, S.G. (1991) Longitudinal studies of HIV infection in intravenous drug users. *NIDA Research Monograph* #109, GPO, Washington, DC.

Havas, H.F., Dellaria, M., Schiffman, G., Geller, E.B. and Adler, M.W. (1987) Effect of cocaine on the immune response and host resistance in BALB/c mice. *Int. Arch. Allergy Appl. Immun.*, **83**, 377–83.

Holsapple, M.P. and Munson, A.E. (1985) Immunotoxicology of abused drugs, in *Immunopharmacology and Immunotoxicology* (eds J.H. Dean, M.I. Luster, A.E. Munson and H. Amos), Raven Press, New York, pp. 381–92.

Holsapple, M.P., Eads, M., Stevens, W.D., Wood, S.C., Kaminski, N.E., Morris, D.L., Poklis, A., Kaminski, E.J. and Jordan, S.D. (1993a) Immunosuppression in adult female B6C3F1 mice by chronic exposure to ethanol in a liquid diet. *Immunopharmacology*, **26**, 31–51.

Holsapple, M.P., Matulka, R.A., Stanulis, E.D. and Jordan, S.D. (1993b) Cocaine and immunocompetence: possible role of reactive metabolites, in *Drugs of Abuse, Immunity, and AIDS* (eds H. Friedman and T. Klein), Plenum Publishing Co., New York, pp. 121–6.

Howlett, A.C. (1985) Cannabinoid inhibition of adenylate cyclase. Biochemistry of the response in neuroblastoma cell membranes. *Mol. Pharmacol.*, **27**, 429–36.

Howlett, A.C. and Fleming, R.M. (1984) Cannabinoid inhibition of adenylate cyclase. Pharmacology of the response in neuroblastoma cell membranes. *Mol. Pharmacol.*, **26**, 532–8.

Hsueh, Y.P. and Lai, M.Z. (1995) c-Jun N-terminal kinase but not mitogen-activated protein kinase is sensitive to cAMP inhibition in T lymphocytes. *J. Biol. Chem.*, **270**, 18094–8.

Ivashkiv, L.B., Liou, H.C., Kara, C.J., Lamph, W.W., Verma, I.M. and Glimcher, L.H. (1990) mXBP/CRE-BP2 and c-jun form a complex which binds cAMP, but not to the 12-O-tetradecanoyl phorbol-13-acetate response element. *Mol. Cell. Biol.*, **10**, 1609–21.

Jaffey, P.B., Haque, A.K., El-Zaatari, M., Pasarelli, L. and McGinnis, M.R. (1990) Disseminated *Conidiobolus* infection with endocarditis in a cocaine abuser. *Arch. Pathol. Lab. Med.*, **114**, 1276–8.

Jenkins, J.S. and Connolly, J. (1968) Adrenocortical response to ethanol in man. *Br. Med. J.*, **2**, 804–5.

Jeon, Y.J., Yang, K.H., Pulaski, J.T. and Kaminski, N.E. (1996) Attenuation of iNOS gene expression by Δ^9-tetrahydrocannabinol is mediated through the inhibition of NF-kB/Rel activation. *Mol. Pharmacol.*, **50**, 334–41.

Jeong, T.C., Matulka, R.A., Jordan, S.D., Yang, K.H. and Holsapple, M.P. (1995a) Role of metabolism in cocaine-induced immunosuppression in splenocyte cultures from B6C3F1 female mice. *Immunopharmacology*, **29**, 37–46.

Jeong, T.C., Jordan, S.D., Matulka, R.A., Stanulis, E.D., Kaminski, E.J. and Holsapple, M.P. (1995b) Role of metabolism by esterase and cytochrome P-450 in cocaine-induced suppression of the antibody response. *J. Pharmacol. Exp. Ther.*, **272**, 407–16.

Jeong, T.C., Jordan, S.D., Matulka, R.A., Stanulis, E.D., Park, S.S. and Holsapple, M.P, (1996) Immunosuppression induced by acute exposure to cocaine is dependent on metabolism by cytochrome P-450. *J. Pharmacol. Exp. Ther.*, **276**, 1257–65.

Jerrells, T.R., Marietta, C.A., Eckardt, M.J., Majchrowicz, E. and Weight, F.F. (1986) Effects of ethanol administration on parameters of immunocompetency in rats. *J. Leuk. Biol.*, **39**, 499–510.

Jerrells, T.R., Peritt, D., Marietta, C. and Eckardt, M.J. (1989) Mechanisms of suppression of cellular immunity induced by ethanol. *Alcohol. Clin. Exp. Res.*, **13**, 490–3.

Jerrells, T.R., Smith, W. and Eckardt, M.J. (1990a) Murine model of ethanol-induced immunosuppression. *Alcohol. Clin. Exp. Res.*, **14**, 546–50.

Jerrells, T.R., Marietta, C.A., Weight, F.F. and Eckardt, M.J. (1990b) Effect of adrenalectomy on ethanol-associated immunosuppression. *Int. J. Immunopharmacol.*, **12**, 435–42.

Johnson, K.W., Davis, B.H. and Smith, K.A. (1988) cAMP antagonizes interleukin 2-promoted T-cell cycle progression at a discrete point in early G_1. *Proc. Natl Acad. Sci. USA*, **83**, 6072–6.

Kaminski, N.E. (1995) Mechanisms of immune modulation by cannabinoid compounds, in *Immunotoxicology and Immunopharmacology* (eds J.H. Dean, M.I. Luster, A.E. Munson and I. Kimber), Raven Press, New York, pp. 349–62.

Kaminski, N.E., Abood, M.E., Kessler, F.K., Martin, B.R. and Schatz, A.R. (1992) Identification of a functionally relevant cannabinoid receptor on mouse spleen cells that is involved in cannabinoid-mediated immune modulation. *Mol. Pharmacol.*, **42**, 736–42.

Kaminski, N.E., Koh, W.S., Lee, M., Yang, K.H. and Kessler, F.K. (1994) Suppression of the humoral immune response by cannabinoids is partially mediated through inhibition of adenylate cyclase by a pertussis toxin-sensitive G-protein coupled mechanism. *Biochem. Pharmacol.*, **48**, 1899–908.

Kammer, G.M. (1988) The adenylate cyclase-cAMP-protein kinase A pathway and regulation of the immune response. *Immunol. Today*, **9**, 222–9.

Kawakami, Y., Klein, T.W., Newton, C., Djeu, J.Y., Specter, S. and Friedman, H. (1988) Suppression by delta-9-tetrahydrocannabinol of interleukin 2-induced lymphocyte proliferation and lymphokine-activated killer cell activity. *Int. J. Immunopharmacol.*, **10**, 485–8.

Klein, T.W., Newton, C.A. and Friedman, H. (1988) Suppression of human and mouse lymphocyte proliferation by cocaine. *Adv. Biochem. Psychopharmacol.*, **44**, 139–43.

Klein, T., Kawakami, Y., Newton, C. and Friedman, H. (1991) Marijuana components suppress induction and cytolytic function of murine cytotoxic T cells *in vitro* and *in vivo*. *J. Toxicol. Envir. Health*, **32**, 465–77.

Klein, T.W., Matsui, K., Newton, C.A., Young, J., Widen, R.E. and Friedman, H. (1993) Cocaine suppresses proliferation of phytohemagglutinin-activated human peripheral blood T-cells. *Int. J. Immunopharmacol.*, **15**, 77–86.

Koh, W.S., Yang, K.H. and Kaminski, N.E. (1995) Cyclic AMP is an essential factor in immune responses. *Biochem. Biophys. Res. Commun.*, **206**, 703–9.

Koh, W.S., Lee, M., Yang, K.H. and Kaminski, N.E. (1996) Expression of functional glucagon receptors on mouse spleenocytes. *Life Sci.*, **58**, 741–51.

Koh, W.S., Crawford, R.B. and Kaminski, N.E. (1997) Δ^9-Tetrahydrocannabinol inhibits the activation and binding of CRE-specific transcription factors. A putative mechanism of cannabinoid-induced immune modulation. *Biochem. Pharmacol.*, in press.

Kos, F.J. and Engleman, E.G. (1995) Requirement for natural killer cells in the induction of cytotoxic T cells. *J. Immunol.*, **155**, 578-84.

Kump, D.F., Matulka, R.A., Burton, G.F., Jordan, S.D. and Holsapple, M.P. (1996) Alterations in splenocyte and thymocyte subpopulations in B6C3F1 mice exposed to cocaine plus diazinon. *J. Pharmacol. Exp. Ther.*, **277**, 1477–85.

Lee, H.J., Masuda, E.S., Arai, N., Arai, K. and Yokota, T. (1995) Definition of *cis*-regulatory elements of the mouse interleukin-5 gene promoter. *J. Biol. Chem.*, **270**, 17541–50.

Luo, Y.D., Patel, M.K., Wiederhold, M.D. and Ou, D.W. (1992) Effects of cannabinoids and cocaine on the mitogen-induced transformations of lymphocytes of human and mouse origins. *Int. J. Immunopharmacol.*, **14**, 49–56.

Luster, M.I., Munson, A.E., Thomas, P.T., Holsapple, M.P., Fenters, J., White, K.L. Jr, Lauer, L.D., Germolec, D.R., Rosenthal, G.J. and Dean, J.H. (1988) Development of a testing battery to assess chemical-induced imunotoxicity: National Toxicology Program's guidelines for immunotoxicity evaluation in mice. *Fundam. Appl. Toxicol.*, **10**, 2–19.

MacGregor, R.R. (1986) Alcohol and immune defense. *JAMA*, **256**, 1474–9.

Mao, J.T., Huang, M., Wang, J., Sharma, S., Tashkin, D.P. and Dubinett, S.M. (1996) Cocaine down-regulates IL-2-induced peripheral blood lymphocyte IL-8 and IFN-*g* production. *Cell. Immunol.*, **172**, 217–23.

Maraguchi, A., Miyazaki, K. and Fauci, A.S. (1984) Inhibition of human B cell activation by diterpine forskolin: interference with B cell growth factor-induced G1 to S transition of the B cell cycle. *J. Immunol.*, **133**, 1283–7.

Martinez, F. and Watson, R.R. (1990) Effects of cocaine and morphine on IgG production by human peripheral blood lymphocytes *in vitro*. *Life Sci.*, **47**, 59–64.

Matsuda, L.A., Lolait, S.J., Brownstein, M.J., Young, A.C. and Bonner, T.I. (1990) Structure of a cannabinoid receptor and functional expression of the cloned cDNA. *Nature*, **346**, 561–4.

Matsui, K., Friedman, H. and Klein, T.W. (1992) Cocaine augments proliferation of human peripheral blood T-lymphocytes activated with anti-CD3 antibody. *Int. J. Immunopharmacol.*, **14**, 1213–20.

Matulka, R.A., Jordan, S.D., Stanulis, E.D. and Holsapple, M.P. (1996) Evaluation of sex- and strain-dependency of cocaine-induced immunosuppression in B6C3F1 and DBA/2 mice. *J. Pharmacol. Exp. Ther.*, **279**, 12–17.

McLeod, R., Buschman. E., Arbuckle, L.D. and Skamette, E. (1995) Immunogenetics in the analysis of resistance to intracellular pathogens. *Curr. Opin. Immunol.*, **7**, 539–52.

Mechoulam, R. (1970) Marihuana chemistry. *Science*, **168**, 1159–60.

Mendelson, J.H., Ogata, M. and Mello, N.K. (1971) Adrenal function and alcoholism. I. Serum cortisol. *Psychosom. Med.*, **33**, 145–57.

Mendenhall, C.L., Grossman, C.J., Roselle, G.A., Ghosn, S., Gartside, P.S., Rouster, S.D., Chalasani, P.V.R.K., Schmitt, G., Martin, K. and Lamping, K. (1990) Host response to mycobacterial infection in the alcoholic rat. *Gastroenterology,* **99**, 1723–6.

Merry, J. and Marks, V. (1969) Plasma hydrocortisone response to ethanol in chronic alcoholics. *Lancet*, **1**, 921–3.

Mosmann, T.R. and Sad, S. (1996) The expanding universe of T-cell subsets: Th1, Th2 and more. *Immunol. Today*, **17**, 138–46.

Munro, S., Thomas, K.L. and Abu-Shaar, M. (1993) Molecular characterization of a peripheral receptor for cannabinoids. *Nature*, **365**, 61–5.

Nakano, Y., Pross, S. and Friedman, H. (1993) Contrasting effect of delta-9-tetrahydrocannabinol on IL-2 activity in spleen and lymph node cells of mice of different ages. *Life Sci.*, **52**, 41–51.

Nelson, S., Bagby, G.J., Bainton, B.G. and Summer, W.G. (1989) The effects of acute and chronic alcoholism on tumor necrosis factor and the inflammatory response. *J. Infect. Dis.*, **160**, 422–9.

Nelson, S., Mason, C., Bagby, G. and Summer, W. (1995) Alcohol, tumor necrosis factor, and tuberculosis. *Alcohol. Clin. Exp. Res.*, **19**, 17–23.

Novak, T.J., Chen, D. and Rothenberg, E.V. (1990a) Interleukin-1 synergy with phosphoinositide pathway agonists for induction of interleukin-2 gene expression: molecular basis of costimulation. *Mol. Cell. Biol.*, **10**, 6325–34.

Novak, T.J., White, P.M. and Rothenberg, E.V. (1990b) Regulatory anatomy of the murine IL-2 gene. *Nucl. Acids Res.*, **18**, 4523–33.

Ou, D.W., Shen, M.L. and Luo, Y.D. (1989) Effects of cocaine on the immune system of BALB/C mice. *Clin. Immunol. Immunopathol.*, **52**, 305–12.

Pavia, C.S., Bittker, S. and Cooper, D. (1991) Immune response to Lyme spirochete *Borrelia burgdorferi* affected by ethanol consumption. *Immunopharmacology*, **22**, 165–74.

Pepe, S., Ruggiero, A., Tortora, G., Ciaardiello, F., Garbi, C., Yokozaki, H., Cho-Chung, Y.S., Clair, T., Skalhegg, B.S. and Bianco, A.R. (1994) Flow cytometric detection of the RI alpha subunit of type-I cAMP-dependent protein kinase in human cells. *Cytometry*, **15**, 73–9.

Peterson, P.K., Gekker, G., Chao, C.C., Schut, R., Molitor, T.W. and Balfour, H.H. Jr (1991) Cocaine potentiates HIV-1 replication in human peripheral blood mononuclear cell cocultures: Involvement of transforming growth factor-β. *J. Immunol.*, **146**, 81–4.

Powrie, F., Correa-Oliveira, R., Mauze, S. and Coffman, R.L. (1994) Regulatory interactions between CD45RB^{high} and CD45RB^{low} CD4^{+} T cells are important for the balance between protective and pathogenic cell-mediated immunity. *J. Exp. Med.*, **179**, 598–600.

Pross, S., Newton, C., Klein, T. and Friedman, H. (1987) Age-associated differences in cannabinoid-induced suppression of murine spleen, lymph node and thymus cell blastogenic responses. *Immunopharmacology*, **14**, 159–68.

Pruett, S.B., Han, Y.C. and Fuchs, B.A. (1992) Morphine suppresses primary humoral immune responses by a predominately indirect mechanism. *J. Pharmacol. Exp. Ther.*, **262**, 923–8.

Pruett, S.B., Ensley, D.K. and Crittenden, P.L. (1993) The role of chemical-induced stress responses in immunosuppression: a review of quantitative associations and cause-effect relationships between chemical-induced stress responses and immunosuppression. *J. Toxicol. Environ. Health*, **39**, 163–92.

Rivier, C. and Vale, W. (1987) Cocaine stimulates adrenocorticotropin (ACTH) secretion through a corticotropin-releasing factor (CRF) mediated mechanism. *Brain Res.*, **422**, 403–6.

Rook, G.A.W., Hernandez-Pando, R. and Lightman, S.L. (1994) Hormones, peripherally activated prohormones and regulation of the Th1/Th2 balance. *Immunol. Today*, **15**, 301–3.

Rubins, J.B., Charboneau,D., Prigge, W. and Mellencamp, M.A. (1996) Ethanol ingestion reduces antipneumococcal activity of rat pulmonary surfactant. *J. Infect. Dis.*, **174**, 507–12.

Russell, D.H. (1978) Type I cyclic AMP-dependent protein kinase as a positive effector of growth. *Adv. Cyclic Nucl. Res.*, **9**, 493–506.

Saad, A.J., Domiati-Saad, R. and Jerrells, T.R. (1993) Ethanol ingestion increases susceptibility of mice to *Listeria monocytogenes*. *Alcohol. Clin. Exp. Res.*, **17**, 75–85.

Sarnyai, Z., Biro, E., Penke, B. and Telegdy, G. (1992) The cocaine-induced elevation of plasma corticosterone is mediated by endogenous corticotropin-releasing factor (CRF) in rats. *Brain Res.*, **589**, 154–6.

Schatz, A.R., Kessler, F.K. and Kaminski, N.E. (1992) Inhibition of adenylate cyclase by Δ^9-tetrahydrocannabinol in mouse spleen cells: a potential mechanism for cannabinoid-mediated immunosuppression. *Life Sci.*, **51**, 25–30.

Schatz, A.R., Koh, W.S. and Kaminski, N.E. (1993) Δ^9-tetrahydrocanabinol selectively inhibits T-cell dependent humoral immune responses through direct inhibition of accessory T-cell function. *Immunopharmacology*, **26**, 129–37.

Schatz, A.R., Lee, M., Condie, R., Pulaski, J. and Kaminski, N.E. (1997) Cannabinoid receptors CB1 and CB2: A characterization of expression and adenylate cyclase modulation within the immune system. *Toxicol. Appl. Pharmacol.*, **142**, 278–87.

Schijns, V.E.C.J., Haagmans, B.L., Rijke, E.O., Huang, S., Aguet, R. and Horzinek, M.C. (1994) IFN-g receptor-deficient mice generate antiviral Th1-characteristic cytokine profiles but altered antibody responses. *J. Immunol.*, **153**, 2029–36.

Serfling, E.R., Barthelmas, R. and Pfeuffer, A. (1989) Ubiquitous and lymphocyte specific factors are involved in the induction of the mouse interleukin-2 gene on T lymphocytes. *EMBO J.*, **8**, 465–72.

Shirakawa, F. and Mizel, S.B. (1989) *In vitro* activation and nuclear translocation of NF-kB catalyzed by cyclic AMP-dependent protein kinase and protein kinase C. *Mol. Cell. Biol.*, **9**, 2424–30.

Showalter, V.M., Compton, D.R., Martin, B.R. and Abood, M.E. (1996) Evaluation of binding in a transfected cell line expressing a peripheral cannabinoid receptor (CB2): identification of cannabinoid receptor subtype selective ligands. *J. Pharmacol. Exp. Ther.*, **278**, 989–99.

Siegel, M.D., Zhang, D.H., Ray, P. and Ray, A. (1995) Activation of the interleukin-5 promoter by cAMP in murine EL-4 cells requires the GATA-3 and CLE0 elements. *J. Biol. Chem.*, **270**, 24548–55.

Smart, R.G. and Anglin, L. (1987) Do we know the lethal dose of cocaine? *J. Forens. Sci.*, **32**, 303–12.

Smith, J.W., Steiner, A.L., Newberry, W.M,, and Parker, C.W. (1971) Cyclic adenosine 3′,5′-monophosphate in human lymphocytes. Alteration after phytohemagglutinin. *J. Clin. Invest.*, **50**, 432–41.

Specter, S., Lancz, G. and Hazelden, J. (1990) Marijuana and immunity: tetrahydrocannabinol mediated inhibition of lymphocyte blastogenesis. *Int. J. Immunopharmacol.*, **12**, 261–7.

Stam, W.B., Van Oosterhout, A.J.M. and Nijkamp, F.P. (1993) Pharmacologic modulation of Th1- and Th2-associated lymphokine production. *Life Sci.*, **53**, 1921–34.

Stanulis, R.A., Matulka, R.A., Jordan, S.D., Rosecrans, J.A. and Holsapple, M.P. (1997a) Role of corticosterone in the enhancement of the antibody response after acute cocaine administration. *J. Pharmacol. Exp. Ther.*, in press.

Stanulis, R.A., Jordan, S.D., Rosecrans, J.A. and Holsapple, M.P. (1997b) Disruption of Th1/Th2 cytokine balance by cocaine is mediated by corticosterone. *Immunopharmacology*, **37**, 25–33.

Starec, M.V., Rouveix, B., Sinet, M., Chau, F., Desforges, B., Pocidalo, J.J. and Lechat P. (1991) Immune status and survival of opiate- and cocaine-treated mice infected with Friend virus. *J. Pharmacol. Exp. Ther.*, **259**, 745–50.

Steven, W.M., Kumar, S.N., Stewart, G.L., Seeing, J. and Leonard, L. (1990) The effects of ethanol consumption on the expression of immunity to *Trichinella spiralis* in rats. *Alcohol. Clin. Exp. Res.*, **14**, 87–91.

Tsuruta, L., Lee, H., Masuda, E., Koyano-Nakagawa, N., Arai, N., Arai, K. and Yokota, T. (1995) Cyclic AMP inhibits expression of the IL-2 gene through the nuclear factor of activated T cells (NF-AT) site, and transfection of NF-AT cDNAs abrogates the sensitivity of EL-4 cells to cyclic AMP. *J. Immunol.*, **154**, 5255–64.

Vacca, A., Felli, M.P. and Fatina, A.R. (1992) Glucocorticoid receptor mediated suppression of the interleukin-2 gene expression through impairment of the cooperativity between nuclear factors of activated T cells and AP 1 enhancer elements. *J. Exp. Med.*, **175**, 637–46.

Välimäki, M.J., Härkönen, M., Eriksson, C.J.P. and Ylikahri, R.H. (1984) Sex hormones and adrencortical steroids in men acutely intoxicated with ethanol. *Alcohol*, **1**, 89–93.

Wang, Y., Huang, D.S. and Watson, R.R. (1994) *In vivo* and *in vitro* cocaine modulation on production of cytokines in C57BL/6 mice. *Life Sci.*, **54**, 401–10.

Watson, E.S., Murphy, J.C., ElSohly, H.N., ElSohly, M.D. and Turner, C.E. (1983) Effects of the administration of coca alkaloids on the primary immune responses of mice: interaction with delta-9-tetrahydrocannabinol and ethanol. *Toxicol. Appl. Pharmacol.*, **71**, 1–13.

Watson, P., Krupinski, J., Kempinski, A. and Frankenfield, C. (1994) Molecular cloning and characterization of the type VII isoform of mammalian adenylyl cyclase expressed widely in mouse tissues and in S49 mouse lymphoma cells. *J. Biol. Chem.*, **269**, 893–8.

Watzl, B. and Watson, R.R. (1990) Immunomodulation by cocaine – a neuroendocrine mediated response. *Life Sci.*, **46**, 1319–29.

Wilder, J.A., Koh, C.Y. and Yuan, D. (1996) The role of NK cells during *in vivo* antigen-specific antibody responses. *J. Immunol.*, **156**, 146–52.

Wood, W.B. and Davis, B.D. (1980). In *Microbiology* (eds B.D. Davis, R. Dulbecco, H.N. Eisen and H.S. Ginsberg), Harper and Row, Philadelphia, PA., pp. 552–84.

Wu, W.J., Wolcott, R.M. and Pruett, S.B. (1994) Ethanol decreases the number and activity of splenic natural killer cells in a mouse model for binge drinking. *J. Pharmacol. Exp. Ther.*, **271**, 722–9.

Zagelbaum, B.M., Tannenbaum, M.H. and Hersh, P.S. (1991) *Candida albicans* corneal ulcer associated with crack cocaine. *Am. J. Ophthalmol.*, **11**, 248–9.

6 Oxidative Stress and Heavy Metal Modification of T Lymphocyte Subsets

J.A. Krieger, Y. Heo and D.A. Lawrence

Introduction

A mammalian cell is complex, in that it possesses multilayered, multifactorial methods of self regulation. A major influence on the control of these processes is the net oxidation state of proteinous cellular constituents. The redox positionings of thiol groups on proteins are posited to be critical intrinsic factors in deciding the fate of cells. Environmental oxidants, heavy metals, and pathological conditions, inducing oxidative stress, can alter thiols or directly or indirectly other nucleophilic moieties on critical proteins. Hence, oxidative stress has the potential to affect cell growth, differentiation, and function, all of which will have an impact on the homeostasis of the whole organism. The immune system is analogous to a symphony; every musician must be ready to play together and on cue. Immune homeostasis relies on the proper distribution and talent of these players. As detailed in Chapters 2 and 3 of this book, essential sections of our immune orchestra are the $CD4^+$ and $CD8^+$ T lymphocytes and their subsets (e.g. Th1 and Th2 and Tc1 and Tc2 cells). Events which disturb the fidelity of the balance between the regulatory and effector T cells are expected to correlate with an increased incidence of infection, autoimmune disease, or cancer. In this chapter, we will discuss how oxidative conditions and heavy metal exposures may alter immune function via modification of T cell subset distribution. Thiol-reactive metals such as mercury may directly induce oxidative stress by elimination of cellular thiols and indirectly by inhibition of enzymes necessary for the production of cellular thiols such as glutathione.

T Lymphocyte subpopulations in Immunotoxicology. Edited by Ian Kimber and MaryJane K. Selgrade. Published in 1998 by John Wiley & Sons Ltd. ISBN 0 471 97194 4

Effects of oxidative stress on lymphocyte activation and proliferation

In addition to overt denaturation of cellular proteins by oxidation resulting in transformation or cell death, sublethal oxidant conditions may alter many regulatory cellular proteins critical to the development, functioning and/or survival of lymphocytes. Lymphocytes normally undergo oxidative stress during progression through the cell cycle or when stimulated or activated. Numerous membrane, cytosolic, and nuclear proteins important in signal transduction and consequently cell activation have proved to be molecular targets of oxidation (see Table 6.1). For example, the activities of certain protein tyrosine kinases (Kanner *et al.*, 1992; Scheiven *et al.*, 1993), phosphatases (Hecht and Zink, 1992), and protein kinase C isozymes (Kass *et al.*, 1989) are redox regulated. Regulatory nuclear redox-sensitive proteins involved in numerous activations include Fos, Jun, AP-1, and NF- κB (Abate *et al.*, 1990; Beiqing *et al.*, 1996; Meyer *et al.*, 1994); p21ras has been implicated in the redox activation of NF-κB (Lander *et al.*, 1995). Mild oxidative stress via oxidant treatment or GSH depletion has been shown to inhibit lymphocyte proliferation and signalling (Messina and Lawrence, 1989; Flescher *et al.*,

Table 6.1 Cellular thiol-reactive proteins and lymphocyte activation and proliferation*

Cellular compartment	Protein	Activity	Reference
Plasma membrane	PLCγ1	phospholipase	Kanner *et al.* (1992); Kavanagh *et al.* (1993)
	PLD	phospholipase	Dai *et al.* (1992); Natarajan *et al.* (1993)
	CD45 PTPase	phosphatase	Hecht and Zick (1992)
Cytosol	p56lck	Tyrosine kinase	Kanner *et al.* (1992)
	Protein kinase C	Ser/Thr kinase	Kass *et al.* (1989); Taher and Abd-Elfattah (1996)
	ZAP-70	Tyrosine kinase	Schieven *et al.* (1994)
	Ref-1	Transcription factor	Hirota *et al.* (1997); Qin *et al.* (1996); Xanthoudakis *et al.* (1994)
	Jun/fos	Transcription factor	Abate *et al.* (1990); Handel *et al.* (1995); Meyer *et al.* (1993)
	NF-κB	Transcription factor	Lin *et al.* (1995); Mahon and O'Neill (1995); Meyer *et al.*, (1993)
	NF-AT	Transcription factor	Beiqing *et al.* (1996)

* Partial list of some cellular proteins with reactivities that can control lymphocyte activation or proliferation and are influenced by their expression of sulphydryls.

1994), even though a slight, brief oxidative stress may be requisite for lymphocyte activation and proliferation (Fidelus, 1988; Los *et al.*, 1995; Sekkat *et al.*, 1988; Novogrodsky *et al.*, 1982). Hence, different degrees of oxidative stress modulating differential lymphocyte activity could be a basis for subset skewing by environmental and pathological conditions. Thiols, chemical antioxidants, or low pO_2 conditions can augment proliferation, retard cell death (both apoptotic and necrotic), and modify lymphocyte function in a cell type-specific manner (Broom and Jeng, 1973; Dröge *et al.*, 1994; Kohno *et al.*, 1996; Krieger *et al.*, 1997; Lawrence, 1981a; Mishell and Dutton, 1967). Our previous reports have shown that pO_2 can influence cytokine production and lymphocyte proliferation (Krieger *et al.*, 1997) as well as the sensitivity of cell lines to hydrogen peroxide (Lawrence *et al.*, 1996a), but until recently, few studies have attempted to delineate which lymphoid subsets and subcellular mechanisms are affected by redox modulation.

It has been established that lymphocyte thiol status is altered in immune diseases such as human immunodeficiency virus (HIV) infection (reviewed by Stall *et al.*, 1990). T cells from HIV patients have been shown to be depleted of reduced cellular thiols including glutathione (GSH), the most abundant cytosolic sulphydryl compound, and extracellular protein thiols (Stall *et al.*, 1992). Membrane thiols comprise only about 5% of the total thiol base, but appear to be critical regulators of cell activation (Noelle and Lawrence, 1981a; Duncan and Lawrence, 1988). This has been evidenced by reduced cellular functioning and inhibition of cell cycle progression in lymphocytes treated *in vitro* with the impermeant thiol reactive agents as well as compounds that diminish total reduced thiols (Noelle and Lawrence, 1981b; Ristow *et al.*, 1985; Duncan and Lawrence 1988, 1989; Levy *et al.*, 1992). This loss of lymphoproliferative capacity *in vitro* may be similar to that associated with progressing HIV disease. Moreover, it has been shown that there is a skewing toward a type-2 lymphoid predominance (Th2 and Tc2 cells) in HIV-infected individuals (reviewed by Clerici and Shearer, 1994), suggesting that type-1 responses may be more sensitive to oxidative stress.

Similar to HIV infection, a skewing toward type-2 reactivities in the elderly as been demonstrated (Weksler, 1993; Meydani *et al.*, 1995). The decrease in ability to cope with infectious agents (normally handled by type-1 driven cell-mediated immunity) that occurs with ageing also supports an age-dependent skewing (Albright and Albright, 1994; Horvath-Arcidiacono *et al.*, 1996; Mbawuike *et al.*, 1996). Any skewing with age could not be accounted for by a differential ability of IL-2 versus IL-4 transcription factors to translocate, but differences in costimulatory signals (e.g. thru CD28) were observed (Albright *et al.*, 1995). Paganelli *et al.* (1996) cloned T cells from two centenarians and compare the results with clones from three young subjects. Interestingly, a higher percentage of T cell clones were $CD4^+$ (46%) from the aged subjects in comparison to the young subjects (35%), and 32% fewer clones from the older population produced interferon (IFN)γ (a type-1 cytokine). Although the numbers of classical type-2 clones were identical, the centenarians had double the number of clones producing IFNγ and IL-4, thought to be Th0 cells. With ageing, there is a decline in endogenous antioxidants including glutathione (Masoro, 1990; Richie, 1992; Kretzschmar and Muller, 1993). Lower cellular thiols could account for the oxidative stress and the consequential loss of T cell responses observed in older individuals, as discussed earlier for AIDS patients. In addition, $CD8^+$ T cells have been suggested to be more sensitive to oxidative stress than $CD4^+$ T cells (Lawrence, 1981a; Lawrence *et al.*, 1996b) which would also account for the cloning data (Paganelli *et al.*, 1996).

Effects of oxidative stress on apoptosis

In addition to effects on activation and proliferation, oxidative stress may skew T cell responses as a result of effects on differential apoptosis. *In vitro* treatment of human lymphocytes with chemical or physical (UV or γ-radiation) oxidants has been shown to induce apoptosis, or programmed cell death, and associated mechanisms are related to redox signalling pathways (reviewed by Buttke and Sandstrom, 1995). Dexamethasone or γ-irradiation appear to induce apoptosis via oxidation of critical mitochondrial thiols, whereas anti-Fas-induced apoptosis involves cellular thiols other than those associated with the mitochondrial membrane (Marchetti *et al.*, 1997). In a fashion similar to the oxidant-generated signalling induced by TNF-α which results in NF-κB activation, apoptosis induced via Fas (CD95) is dependent on generation of oxygen radicals (Kohno *et al.*, 1996). As a further example, it has been established that elevated intracellular calcium levels may be requisite for and precede apoptosis in lymphoid cells (reviewed by Duvall and Wyllie, 1986). Micromolar concentrations of a calcium ionophore or oxidative compounds such as hydrogen peroxide or *t*-butylhydroquinone (TBHQ), are known to induce calcium fluxes and apoptosis in lymphocytes (reviewed by Orrenius, 1993). Depletion of glutathione is an early event in thymocyte apoptosis (Macho *et al.*, 1997). It has been proposed that oxidants produced endogenously, or those applied exogenously, interact with certain enzymes, such as phosphatases, kinases, ATPases, etc. Each of these enzyme systems plays a critical role in lymphocyte calcium homeostasis modulating intracellular calcium concentrations. An oxidative stress placed on one or all of these signalling pathways may lead to altered cellular processes or apoptosis (Duncan and Lawrence, 1989). Ligation of monoclonal antibodies to the T cell receptor or activation by lectins causes intracellular calcium levels to become elevated and may induce a form of apoptosis in primed lymphocyte subsets. This process has been termed activation-induced apoptosis (AIA; reviewed by Green and Scott, 1994). As mentioned earlier, the degree of oxidative stress is probably the determinant in whether the lymphocytes are activated by an oxidant or killed by it. Just as calcium flux has been implicated in oxidant-induced apoptosis, it has also been reported that the oxidant hydrogen peroxide can signal in a manner similar to anti-TCR by induction of calcium flux, PTK activity and AP-1 binding (Liu *et al.*, 1996).

There have also been reports of 'calcium-independent' receptor/ligand-induced apoptosis scenarios that appear to rely on oxidative intermediates, since cell death was blocked by the addition of anti-oxidants. These include, but are not limited to TNF/TNFR and FasL/CD95. Interestingly, TNF mediates death of murine $CD8^+$ cells, whereas FasL induces apoptosis of primarily $CD4^+$ lymphocytes (Kohno *et al.*, 1996; Zheng *et al.*, 1995). Additionally, it had been demonstrated that cloned Th1 cells express high levels of FasL and more readily undergo AIA than their FasL-low Th2 counterpart (Ramsdell *et al.*, 1994). AIA induced by lectins, phorbol esters, etc. may involve the induction of CD95 and FasL expression leading to a disproportionate deletion of $CD4^+$ lymphocytes in a mixed cell population, such as human peripheral blood mononuclear cells (PBMC).

Prevention of T cell apoptosis has been shown to occur when a normally tolerogenic signal is provided during an inflammatory response (Röcken *et al.*, 1992; Vella *et al.*, 1996). Tolerogens can block immune responses by inducing anergy or clonal deletion, probably via apoptosis. Blockage of apoptosis is a critical observation in that induction of peripheral T cell unresponsiveness or tolerance to a self constituent is the mechanism to prevent an autoimmune disease, which can be associated with a regulatory type-1:type-2 imbalance. Inflammatory responses induce numerous cytokines as well as oxidants which could be responsible for prevention of the tolerogenic signal. Furthermore, it has been suggested that Th2 cells are preferentially rescued by infection (Röcken *et al.*, 1994) although rescue by LPS (and TNFα) could not confirm differential Th1 versus Th2 protection (Vella *et al.*, 1996). Interestingly, TNFα (an inflammatory cytokine) is implicated in oxidative stress-related processes, and it has been reported to enhance (Kohno *et al.*, 1996; Zheng *et al.*, 1995) or suppress (Vella *et al.*, 1996) T cell apoptosis. IL-1β was also capable of maintaining cell survival, but together TNFα and IL-1β were ineffective (Vella *et al.*, 1997). Thus, cytokines and oxidative stress can stimulate, block stimulation, induce cell death or block the induction of cell death. The dose of cytokine or oxidant, the characteristics of the cell and/or its metabolic state (e.g. stage in cell cycle), and/or temporal effects of multiple regulatory factors in the complex *in vivo* milieu are all likely to influence cell activation, death or survival.

Differential sensitivity to redox effects

The observation that certain lymphoid subsets may be more prone to redox effects than are others is intriguing and appears to depend on the cell type/system, signalling pathway, or oxidative stress involved. In HIV-infected patients, $CD4^+$ T cells are the most sensitve lymphoid subset, and depletion of $CD4^+$ T cells reflects progression toward clinical disease (AIDS).

Furthermore, viral load does not correlate with $CD4^+$ depletion in individuals infected with HIV (Gougeon *et al.*, 1993). Thus, cell loss may be occurring by some indirect process that appears unrelated to viral cytolysis. It has been suggested that $CD8^+$ cells may be important mediators of $CD4^+$ death, possibly by supplying an oxidative death signal to $CD4^+$ cells (Gmunder and Dröge, 1991). Interestingly, $CD8^+$ cells isolated from lymph tissue of HIV infected patients that express low levels of the antioxidant tumour suppressor protein Bcl-2 more readily undergo apoptosis (Bofill *et al.*, 1995). Bcl-2 has been well documented to obviate oxidant-induced cell death by blocking a common pathway (reviewed by Reed, 1994).

Given the inherent signalling complexity and heterogeneity of the different lymphoid subsets, one may postulate that a given oxidative stress would induce a variety of responses from a mixed cell population. The contention that subsets become skewed by oxidative conditions may be correlated with the absolute number of surface and/or total thiols of a given lymphoid subset. Although not straightforward, as is typical in biology, it is possible that one may predict the fate of a cell undergoing an oxidant challenge by its ability to maintain a redox balance and not tip to either redox extreme. Using the impermeant sulphydryl-reactive fluorescent probe FGT, we quantified surface thiol expression on human lymphoid subsets. The ranking of the lymphocyte populations was $CD19^+$ (B cells) > $CD8^+$ > $CD4^+$ (Lawrence *et al.*, 1996b); this is the same order as lymphocyte sensitivity to γ-radiation (Lawrence, 1981a). Monocytes which are substantially more radioresistant have more surface thiols than B cells (Lawrence, 1981a). The higher thiol content of monocytes correlates with their greater resistance to oxidative stress, a resistance that is not surprising since monocytes/macrophages generate reactive oxygen species (ROS) as a part of their immune defence mechanisms. Differences in subset sensitivity probably lie in the type and duration of oxidative stress examined. However, it must be noted that different agents modify different cellular thiols.

To examine the effects of stimulation and/or chemical oxidation on lymphoid subsets, we cultured human PBMC under physiological conditions in control medium or medium containing a phorbol ester and calcium ionophore $\pm$ *t*-butylhydroquinone (TBHQ; an exogenous oxidant) or TBHQ alone. After 24 h, PBMC were harvested and assayed by flow cytometry to determine what and how subsets were undergoing cell death.

Data in Table 6.2 demonstrate that overall viability decreases when PBMC were stimulated, and TBHQ augmented the death of stimulated cells. TBHQ alone did not significantly alter viability at 25 μm, although higher concentrations enhanced cell loss in particular when added to stimulated cultures. A dramatic loss of $CD4^+$ cells was noted upon PBMC stimulation, and TBHQ did not appear to increase depletion of $CD4^+$ cells at 24 h. This was a true loss of $CD4^+$ cells and not just an internalization of CD4 molecules. Cells were fixed and permeabilized (1% paraformaldehyde followed by 0.1% saponin) to

Table 6.2 Effects of mitogen in the presence or absence of TBHQ on lymphoid subset viability. Total viability and percentage of viable PBMC subsets at 24 h was determined with a propidium iodide (PI) exclusion assay by flow cytometry. Cells were gated as to %CD positive (FITC; FL1) and low PI (FL2) fluorescence (viable cells). Human PBMC in RPMI 1640 with 10% FBS were treated with PMA 25 nm and ionomycin 0.25 μg/ml, TBHQ 25 μm, or PMA 25 nm, ionomycin 0.25 μg/ml, and TBHQ 10 μm and cultured for 24 h under low O_2 conditions (5% O_2, 7% CO_2 and 77% N_2 at 37°C). Results are expressed as the mean for three separate donors (range)

Treatment (% total viability)	% viable $CD4^+$ (range)	% viable $CD8^+$ (range)	% viable $CD19^+$ (range)
Control (97)	36 (31–42)	26 (15–42)	10 (6–12)
PMA/IONO (84)	10 (8–13)	31 (22–48)	11 (7–14)
TBHQ (95)	32 (30–33)	28 (14–46)	11 (7–14)
PMA/IONO/TBHQ (70)	11 (2–18)	33 (15–60)	7 (5–10)

determine if CD4 protein was being internalized during stimulation leading to an artifactual under-representation of this subset. No significant increases in $CD4^+$ cells were observed. TBHQ alone produced a small decrease (10%) in the number of $CD4^+$ cells; however, $CD8^+$ and $CD19^+$ lymphocyte numbers were not altered.

Stimulation by a phorbol ester and calcium ionophore appears to induce apoptosis in $CD4^+$ and $CD8^+$ PBMC subsets (CD4 > CD8), but not $CD19^+$ cells. TBHQ by itself appeared to reduce the number of all cells undergoing apoptosis as compared to control, whereas TBHQ in addition to stimulation enhanced $CD8^+$ apoptosis slightly at 24 h. However, as shown in Table 6.3, TBHQ augmented overall cell death in stimulated cultures, possibly by inducing early apoptosis or cell death by another mechanism such as necrosis. These data suggest that stimulation and chemical oxidation by TBHQ in human PBMC produced a different outcome as compared to γ-radiation

Table 6.3 Effects of mitogen in the presence or absence of TBHQ on apoptosis of lymphoid subsets. Apoptosis of human lymphoid subsets was measured after 24 h. Data indicate the percent of each lymphoid subset undergoing DNA fragmentation as determined by intracellular PI staining and flow cytometry (Nicoletti *et al.*, 1991). Cells are gated as to CD positivity (FITC; FL1) and cell with < 2n amount of DNA (FL2, red fluorescence) were considered to be undergoing apoptosis. Human PBMC were treated and cultured as described in Table 6.2. Results are expressed as the mean for three separate donors ± SD

Treatment	% Apoptotic CD4	% Apoptotic CD8	% Apoptotic CD19
Control	5.3 ± 1.4	4.2 ± 1.1	4.0 ± 2.4
PMA/IONO	9.7 ± 3.4	7.3 ± 4.6	4.7 ± 4.8
TBHQ	3.6 ± 2.8	3.0 ± 2.1	2.8 ± 2.2
PMA/IONO/TBHQ	10.0 ± 2.9	5.8 ± 2.6	3.2 ± 2.3

which preferentially kills $CD19^+$ cells (Duncan and Lawrence, 1991). The effects of oxidants on type-1 versus type-2 reactivities needs to be determined. Preliminary data assayed by determination of intracellular expression of cytokine after stimulation (24 h) and incubation with brefeldin A (for last 5 h) was unreliable for IL4 due to low levels of expression; however, IFNγ expression was measurable. Stimulation (PMA/ionomycin; 24 h) induced expression in 17% (range: 7–32%) of the lymphocytes and TBHQ caused 0–57% inhibition of this expression; non-stimulated lymphocytes did not express IFNγ (0.2–0.3%). There was obvious substantial inter-individual variation with this analysis, but the oxidation inhibited type-1 reactivity. At this time, it cannot be concluded whether type-2 reactivities have a differential sensitivity.

The posited and observed differences in subset sensitivity are probably dependent on the type and duration of oxidative stress experienced by the cells. While both radiation and TBHQ induce generation of reactive oxygen species (ROS) in the cytosol, oxidize GSH to GSSG, and may deplete total thiols, the organic quinone TBHQ specifically affects membrane constituents, such as ATPases and ion channels (Henschke and Elliott, 1995). The inverse correlation between treatments and surface thiol expression is not clear for these two oxidant classes, but may be related to the cellular compartments affected. Moreover, cells of different function, such as Th1 and Th2 lymphocytes, express distinct thiol-bearing proteins with varying redox sensitivities accounting for differential subset skewing by oxidants. It may follow, then, that cells containing a greater proportion of reduced thiols on their surface may be resistant to the effects of lipophilic membrane oxidants, and those with more cytosolic reducing equivalents to radiation. However, more experimentation needs to be done to this end before such generalizations may be stated unequivocally.

The heavy metals lead (Pb) and mercury (Hg) and lymphoid subsets

The immunotoxicity of heavy metals has been recently reviewed (Lawrence, 1995; Lawrence and McCabe, 1995), and the reader is directed to these references for a more complete overview of a variety of metals and their effects on the immune system. This chapter only addresses some select studies with the heavy metals Pb and Hg and limits discussion to immunotoxicological mechanisms to those under recent or current investigation related to T cell subsets.

As our understanding of the consequences of heavy metal exposure becomes more clear, environmental contamination by Pb and Hg continues to be of concern. These ubiquitous metals may be detected in soil, water, and the air we breathe. Most mammals, including humans, have some body burden of Pb and/or Hg. These metals may pose a problem in the free or biologically chelated form. The greatest current source for human Pb and Hg exposures are plumbing, dental amalgam, and waste incineration. Data depicting actual Hg toxicity from dental fillings is controversial and other sources of this toxic metal, such as fish consumption, may be more important.

It has been documented that a major toxicity associated with Hg exposure is the induction of certain autoimmune diseases (e.g. glomerulonephritis; Aten *et al.*, 1992). The overproduction of autoantibodies characteristic of certain types of autoimmune morbidity may be correlated with a relative loss of Th1 function. Interestingly, as might be predicted, Hg has also been shown to skew $CD4^+$ T lymphocytes toward a Th2 predominance in susceptible strains (Goldman *et al.*, 1991; van der Meide *et al.*, 1993; Pelletier *et al.*, 1994; Heo *et al.*, 1996).

Low level human occupational exposures to Pb have been shown to reduce the number of circulating $CD4^+$ cells (Fischbein *et al.*, 1993). Mice subjected to lead in their drinking water, were less able to survive a normally sublethal dose of bacteria (*Listeria monocytogenes*), which is suggestive of a loss of type-1 T cell activity or cell mediated immune response *in vivo* (Lawrence, 1981b). Additional studies have demonstrated that Pb exposure induced a trend toward Th2 phenotype, similar to that described for oxidative stresses, in rodent animal models and human *in vitro* studies (McCabe and Lawrence, 1991; Heo *et al.*, 1996). Other evidence for the type-2 skewing induced by Pb has been obtained from recent experiments demonstrating that cells from Pb-treated mice, or murine splenocytes exposed to Pb *in vitro*, produced more IL-4, a key indicator of Th2 predominance (Heo *et al.*, 1996).

Pb preferentially enhances Th2-like responses, but suppresses Th1-like responses *in vitro, in vivo*, and *ex vivo* (Heo *et al.*, 1996). In contrast to IL-4 induction *in vivo*, IFNγ was downregulated in plasma of BALB/c mice exposed to Pb in the absence of any exogenous antigen exposure. Pb administration also enhanced IL-4 production by a Th2 clone, but suppressed IFNγ production by a Th1 clone. Furthermore, the *in vivo* level of IgE, an immunoglobulin known to be increased following IL-4 upregulation with Hg-induced systemic autoimmunity, was also enhanced in Pb-exposed BALB/c mice. These findings add to the list of the common features between Pb- and Hg-induced immunodysregulation such as enhancement of MHC II expression, induction of T cell proliferation and inhibition of host resistance. As Th subset skewing and/or restricted usage of the TCR repertoire appears to play an important role in the progression of immunopathogenesis, Pb-mediated induction of *in vivo* expansion of T cells with particular specificities (TCR Vβs) and whether any expanded population had a skewed Th1 or Th2 frequency were investigated (Heo *et al.*, 1997). Additionally, *in vivo* effects of Pb on Th1 versus Th2 responses in DO11.10 ovalbumin transgenic (OVAtg) mice were evaluated to examine whether Pb exposure of mice with one predominant TCR Vβ (Vβ8) and antigen (OVA) specificity were skewed toward a particular subset in the presence or absence of antigen. Exposure of BALB/c mice to Pb or Hg induced an oligoclonal response with an increased proportion of Vβ 5^+, Vβ 7^+ and Vβ 13^+ $CD4^+$ peripheral T cells which were predominantly Th2 (IL-4 producers). These T cells are probably responding to Pb-modified/released self antigens. Furthermore, following Pb treatment, OVAtg mice immunized with OVA demonstrated a dominance of Th2 over Th1, observed as increased expression of IgE and a lowered IFNγ level *in vivo*, and increased IgG1 anti-OVA (IL-4 dependent) and lowered IgG2a anti-OVA (IFNγ dependent) levels. This observation suggests that in spite of the predominance of Th with Vβ 8 (a Vβ not significantly altered in the periphery of Pb- or Hg-treated normal BALB/c mice) in OVAtg mice, Pb established an *in vivo* environment which allowed skewing toward type-2 responses in the presence of antigenic stimulation. However, without the antigenic stimulant (OVA), no skewing was apparent. The combination of these results suggests that Pb induces an environment for skewing preferentially toward the type-2 phenotype, but the skewing is dependent on T cell recognition of its appropriate antigenic stimulus. The posited Pb-induced antigens cannot stimulate in the OVA-tg mice, because there are no cells with receptors for an antigen other than OVA. Thus, Pb-treated OVAtg mice only had type-2 like skewing when OVA was present to stimulate the T cells antigenically.

Pb effects on some intracellular signalling pathways of T cells lead to the preferential suppression of Th1 activities. Pb does not appear to inhibit events between phospholipase C-initiation and activation of the PKC/[$Ca2^+$]-dependent pathway in that Pb was still inhibitory when Th1 cells were stimulated by PMA and ionomycin (Heo *et al.*, 1998). The incubation of Th1 cells with Pb induced a transient increase of $[cAMP]_i$, which may contribute to the Pb-induced preferential inhibition of Th1 activation. Reagents elevating intracellular cAMP levels such as PGE_2, PGE_1, cholera toxin, or forskolin are known to suppress Th1 activities but have no effect on or enhance Th2 activities (Betz and Fox, 1991; Hilkens *et al.*, 1995; Katamura *et al.*, 1995; Naito *et al.*, 1996; Novak and Rothenberg, 1990; Tineke *et al.*, 1995). The posited involvement of cAMP in the Pb-induced differential modulation of Th subset activities was further suggested by comparable Pb and cAMP effects on Th1 cells (Heo *et al.*, 1998). A low dose of dibutyryl(db)cAMP (10 μM), a cell-permeable cAMP analogue, inhibited IFNγ production, which, like Pb-induce suppression, was prevented by IL-12. Additionally, the effects of Pb and dbcAMP on the primary antigen (OVA)-driven stimulation of *in vitro* development of naive splenic $CD4^+$ T cells (Thp/Tho cells) from OVA-tg transgenic mice indicated similar outcomes. Pb directed the differentiation of naive Th0 cells toward Th2 by enhancing IL-4 production and suppressing IFNγ production, an imbalance that could be blocked by IL-12 administration. These results correlate well with the *in vivo* findings demonstrating that IL-12 could overcome the inhibitory effects of Pb on mouse resistance to Listeria (Kishikawa and Lawrence, 1997); dbcAMP also appeared to be more inhibitory for IFNγ production than for IL-4 production and IL-12 reversed this inhibition.

The overall concept suggested from the *in vivo* and *in vitro* T cell subset skewing data is as follows: Pb disrupts immune homeostasis, leading to an imbalance between cell-mediated immunity and humoral immunity. Downregulation of effector Th1 activity and Thp development toward the Th1 phenotype is directly involved in the Pb-induced inhibition of cell-mediated immunity, which lowers host resistance to infectious agents. The preferential inhibition of Th1 activity may occur through activation of cAMP-dependent intracellular signaling. Pb-induced enhancement of effector Th2 activity and Thp development toward Th2 may contribute to an immunopathogenesis such as systemic autoimmunity. Expansion of particular TCR $V\beta^+$ Th2 cells by Pb may induce immune responses to self constituents. To test the role of Pb in triggering autoimmune disease, additional epidemiological evidence and animal studies are needed. Further studies on Pb-induced alteration of signalling events are needed to delineate the intracellular mechanism of differential modulation of Th subset activities. As with the influence of inflammatory products on tolerance induction (discussed earlier), Pb and Hg may modify cytokine production and antigen presentation molecules. Posited metal modulation of MHC class II molecules has been reviewed (Lawrence and McCabe, 1995). Recently, human PBMC cultures incubated with Pb have been reported to have altered MHC class II expression (Guo *et al.*, 1996b) and increased TNFα production (Guo *et al.*, 1996a), and TNFα is a cytokine implicated in altering tolerance induction (Vella *et al.*, 1997).

The immunomodulatory effects of Pb have been an expanding area of research over the past few years and a number of mechanisms for heavy metal immunotoxicity and T subset alterations have been proposed. Because heavy metals may not be considered classic oxidants, a distinction needs to be made between heavy metal effects and those of pure oxidants on lymphocyte subpopulations, even though it appears that similar subcellular events may occur in certain instances. As an example, calcium mobilization may be a common signalling pathway affected by some metals and oxidants. Application of organic or inorganic mercury to lymphocytes increases intracellular calcium concentrations (Tan *et al.*, 1993), as is observed with chemical oxidant treatments. Mercury is known to possess a high affinity for reduced thiols. By reacting with free sulphydryls, Hg may be inducing an oxidative stress similar to that caused by classic oxidants via depletion of reducing equivalents, such as GSH. It is also probable that, like certain quinones, Hg may alkylate critical protein sulphydryls leading to altered lymphocyte function and/or cell death. It has been reported that loss of glutathione occurs early in the apoptotic process (Macho *et al.*, 1997). At this time, it is unclear whether Pb can act through similar mechanisms. Preliminary data suggests that there are substantial differences between Hg and Pb in that $HgCl_2$ can induce significant thymocyte apoptosis but $PbCl_2$ has no effect on this process (Li and Lawrence, work in progress). Nevertheless, a number of mechanisms for Pb- and Hg-induced immunotoxicity and Th1/Th2 skewing have been proposed.

Summary

As we obtain a better understanding of the role of environmental contaminants in the distribution of T cell subsets, we shed light on the mechanics of human disease. Indeed, inorganic and organic compounds and radiation have been shown to alter the fidelity of the immune system, thereby provoking disease. The morbidity and mortality associated with immune dysfunction may be mediated through alterations of the Th1/Th2 subset ratio. There is evidence that the stress associated with oxidant and heavy metal exposures may skew $CD4^+$ T cell subsets toward a Th2-like predominance. It may be possible that through manipulation of this immune shift pharmacologically, and thus clinically, we may prevent and/or treat human disease.

References

Abate, C., Patel, L., Rauscher, F.J. and Curran, T. (1990) Redox regulation of Fos and Jun DNA-binding activity *in vitro*. *Science* **249**, 1157–61.

Albright, J.W. and Albright, J.F. (1994) Ageing alters the competence of the immune system to control parasitic infection. *Immunol. Lett.*, **40**, 279–85.

Albright, J.W., Zuniga-Pflucker, J.C. and Albright, J.F. (1995) Transcriptional control of IL-2 and IL-4 in T cells of young and old mice. *Cell. Immunol.*, **164**, 170–5.

Aten, J., Vening, A., Bruijn, J.A., Prins, F.A., Deheer, E. and Weening, J.J. (1992) Antigen specificities of glomerular-bound autoantibodies in membranous glomerulopathy induced by mercuric chloride. *Clin Immunol. Immunopathol.*, **63**, 89–102.

Beiqing, L., Chen, M. and Whisler, R.L. (1996) Sublethal levels of oxidative stress stimulate transcriptional activation of c-jun and suppress IL-2 promoter activation in Jurkat T cells. *J. Immunol.*, **157**, 160–9.

Betz, M. and Fox, B.S. (1991) Prostagladin E2 inhibits production of Th1 lymphokines but not of Th2 lymphokines. *J. Immunol.*, **146**, 108–13.

Bofill, M., Gombert, W., Borthwick, N.J., Akbar, A.N., McLaughlin, J.E., Lee, C.A., Johnson, M.A., Pinching, A.J. and Janossy, G. (1995) Presence of $CD3^+CD8^+$Bcl-2(low) lymphocytes undergoing apoptosis and activated macrophages in lymph nodes of $HIV-1^+$ patients. *Am. J. Pathol.*, **146**, 1542–55.

Broom, J.D. and Jeng, M.W. (1973) Promotion of replication in lymphoid cells by specific thiols and disulfides *in vitro*. *J. Exp. Med.*, **138**, 574–92.

Buttke, T.M. and Sandstrom, P.A. (1995) Oxidative stress as a mediator of apoptosis. *Immunol. Today* **15**, 7–10.

Clerici, M. and Shearer, G.M. (1994) The Th1–Th2 hypothesis of HIV infection: new insights. *Immunol. Today*, **15**, 575–81.

Dai, J., Meij, J.T.A., Padua, R. and Panagia, V. (1992) Depression of cardiac sarcolemmal phospholipase D activity by oxidant-induced thiol modification. **71**, 970–7.

Dröge, W., Mihm, S., Bokstette, M. and Roth, S. (1994) Effects of reactive oxygen intermediates and antioxidants on proliferation and function of T lymphocytes. *Method Enzymol.*, **234**, 135–51.

Duncan, D.D. and Lawrence, D.A. (1988) Four sulfhydryl-modifying compounds cause different structural damage but similar functional damage in murine lymphocytes. *Chem. Biol. Interact.* **68**, 137–52.

Duncan, D.D. and Lawrence, D.A. (1989) Differential lymphocyte growth-modifying effects of oxidants: changes in cytosolic Ca^{+2}. *Toxicol. Appl. Pharmacol.*, **100**, 485–97.

Duncan, D.D. and Lawrence, D.A. (1991) Residual activation events functional after irradiation of mouse splenic lymphocytes. *Radiat. Res.*, **125**, 6–13.
Duvall, E. and Wyllie, A.H. (1986) Death and the cell. *Immunol. Today*, **7**, 115–19.
Fidelus, R.K. (1988) The generation of oxygen radicals: a positive signal for activation. *Cell. Immunol.*, **113**, 175–82.
Fishbein, A., Tsang, P., Luo, J.-C.J., Roboz, J.P., Jiang, J.D. and Bekesi, J.G. (1993) Phenotypic aberrations of $CD3^+$ and $CD4^+$ cells and functional impairments of lymphocytes at low-level occupational exposures to lead. *Clin. Immunol. Immunopathol.*, **66**, 163–8.
Flescher, E., Ledbetter, J.A., Schieven, G.L., Vela-Roch, N., Fossum, D., Dang, H., Ogawa, N. and Talal, N. (1994) Longitudinal exposure of human T lymphocytes to weak oxidative stress suppresses transmembrane and nuclear signal transduction. *J. Immunol.*, **153**, 4880–9.
Gmunder, H. and Dröge, W. (1991) Differential effects of glutathione depletion on T cell subsets. *Cell Immunol.*, **138**, 229–37.
Goldman, M., Druet, P. and Gleichmann, E. (1991) TH2 cells in systemic autoimmunity: insights from allogeneic diseases and chemically-induced autoimmunity. *Immunol. Today*, **12**, 223–7.
Gougeon, M.L., Garcia, S., Heeney, J., Tschopp, R., Lecoeur, H., Guetard, D., Rame, V., Dauguet, C. and Montagnier, L. (1993) Programmed cell death in AIDS-related HIV and SIV infections. *AIDS Res. Human Retroviruses*, **9**, 553–63.
Green, D.R. and Scott, D.W. (1994) Activation-induced apoptosis in lymphocytes. *Curr. Opin. Immunol.*, **6**, 476–87.
Guo, T.L., Mudzinski, S.P. and Lawrence, D.A. (1996a) The heavy metal lead modulates the expression of both tumor necrosis factor α and tumor necrosis factor α receptors in endotoxin-activated human peripheral blood mononuclear cells. *J. Leukocyte Biol.*, **59**, 932–9.
Guo, T.L., Mudzinski, S.P. and Lawrence, D.A. (1996) Regulation of HLA-DR and invariant chain expression by human peripheral blood mononuclear cells with lead, interferon-γ, or interleukin-4. *Cell. Immunol.*, **171**, 1–9.
Handel, M.L., Watts, C.K.W., deFazio, A., Day, R.O. and Sutherland, R.L. (1995) Inhibition of AP-1 binding and transcription by gold and selenium involving conserved cysteine residues in Jun and Fos. *Proc. Natl Acad. Sci. USA*, **92**, 4497–501.
Hecht, D. and Zink, Y. (1992) Inhibition of protein tyrosine phosphatase by hydrogen peroxide. *Biochem. Biophys. Res. Commun.*, **188**, 773–9.
Henschke, P.N. and Elliott, S.J. (1995) Oxidized glutathione decreases luminal Ca^{+2} content of the endothelial cell ins(1,4,5)P3-sensitive Ca^{+2} store. *Biochem. J.*, **312**, 485–9.
Heo, Y., Parsons, P.J. and Lawrence, D.A. (1996) Lead differentially modifies cytokine production *in vitro* and *in vivo*. *Toxicol. Appl. Pharmacol.*, **138**, 149–57.
Heo, Y., Lee, W. and Lawrence, D.A. (1997) *In vivo* the environmental pollutants lead and mercury induce oligoclonal T cell responses skewed toward type-2 reactivities. *Cell. Immunol.*, **179**, 185–95.
Heo, Y., Lee, W. and Lawrence, D.A. (1998) Differential effects of lead and cAMP on development and activities of Th1 and Th2-lymphocytes. *Tox. Sci.* (in press).
Hilkens, C.M.U., Vermeulen, H., van Neerven, R.J.J., Snijdewint, F.G.M., Wierenga, E.A. and Kapsenberg, M.L. (1995) Differential modulation of T helper type 1 (Th1) and T helper type 2 (Th2) cytokine secretion by prostaglandin E_2 critically depends on interleukin-2. *Eur J. Immunol.*, **25**, 59–63.
Hirota, K., Matsui, M., Iwata, S., Nishiyama, A., Mori, K. and Yodoi, J. (1997) AP-1 transcriptional activity is regulated by a direct association between thioredoxin and Ref-1. *Proc. Natl. Acad. Sci. USA*, **94**, 3633–8.

Horvath-Arcidiacono, J.A., Mostowski, H.S. and Bloom, E.T. (1996) IL-12 administered *in vivo* to young and aged mice. Discrepancy between the effects on tumor growth *in vivo* and cytotoxic T lymphocyte generation *ex vivo*: dependence on IFN-gamma. *Int. Immunol.*, **8**, 661–73.

Kanner, S.B., Kavanagh, T.J., Grossmann, A., Shiu-Lok, H., Bolen, J.B., Rabinovitch, P.S. and Ledbetter, J.A. (1992) Sulfhydryl oxidation down-regulated T-cell signaling and inhibits tyrosine phosphorylation of phospholipase Cγ1. *Proc. Natl. Acad. Sci. USA*, **89**, 300–4.

Kass, G.E., Duddy, S.K. and Orrenius, S. (1989) Activation of hepatocyte protein kinase C by redox-cycling quinones. *Biochem. J.*, **260**, 499–507.

Katamura, K., Shintaku, N., Yamauchi, Y., Fukui, T., Ohshima, Y., Mayumi, M. and Furusho, K. (1995) PGE_2 at priming of naive $CD4^+$ T cells inhibits acquisition of ability to produce IFN and IL-2, but not IL-4 and IL-5. *J. Immunol.*, **155**, 4604–12.

Kavanagh, T.J., Grossmann, A., Jinneman, J.C., Kanner, S.B., White, C.C., Eaton, D.L., Ledbetter, J.A. and Rabinovitch, P.S. (1993) The effect of 1-chloro-2,4-dinitrobenzene exposure on antigen receptor (CD3)-stimulated transmembrane signal transduction in purified subsets of human peripheral blood lymphocytes. *Toxicol. Appl. Pharmacol.*, **119**, 91–9.

Kishikawa, H. and Lawrence, D.A. (1997) Interleukin-12 promotes enhanced resistance to *Listeria monocytogenes* infection of lead-exposed mice. *Toxicol. Applied Pharmacol.*, **147**, 180–9.

Kohno, T., Yamada, Y., Hata, T., Mori, H., Yamamura, M., Tomonaga, M., Urata, Y., Gota, S. and Kondo, T. (1996) Relation of oxidative stress and glutathione synthesis to CD95(Fas/APO-1)-mediated apoptosis of adult T cell leukemia cells. *J. Immunol.*, **156**, 4722–8.

Kretzschmar, M. and Muller, D. (1993) Aging, training and exercise. A review of effects on plasma glutathione and lipid peroxides. *Sports Med.*, **15**, 196–209.

Krieger, J.A., Lansiedel, J.C. and Lawrence, D.A. (1997) Differential *in vitro* effects of physiologic and atmospheric oxygen tension on normal peripheral blood mononuclear cell proliferation, cytokine and immunoglobulin production. *Int. J. Immunopharmacol.*, **18**, 545–52.

Lander, H.M., Ogiste, J.S., Teng, K.K. and Novogrodsky, A. (1995) p21ras as a common signaling target of reactive free radicals and cellular redox stress. *J. Biol Chem.*, **270**, 21195–8.

Lawrence, D.A. (1981a) Antigen activation of T-cells in *Handbook of Cancer Immunology*, vol. 8 (ed. H. Waters), Garland Press, NY, pp. 257–320.

Lawrence, D.A. (1981b) *In vivo* and *in vitro* effects of lead on humoral and cell mediated immunity. *Infect Immun.*, **31**, 136–43.

Lawrence, D.A. (1995). Metal ions in autoimmune and infectious diseases, in *Handbook on Metal–Ligand Interactions in Biological Fluids*, Vol. 2 (ed. G. Berthon), Marcel Dekker, NY, pp. 904–13.

Lawrence, D.A. and McCabe, M.J. (1995). Immune modulation by toxic metals, in *Organ–Specific Metal Toxicology* (eds M. Waalkes, R. Goyer, and C. Klaassen), Academic Press, NY, pp. 305–37.

Lawrence, D.A., Colinas, R.J. and Walsh, A.C. (1996a) Influence of oxygen partial pressure on human and mouse myeloid cell line characteristics. *Fundam. Appl. Toxicol.*, **29**, 287–93.

Lawrence, D.A., Song, R. and Weber, P.B. (1996b) Surface thiols of human lymphocytes and their changes after *in vitro* and *in vivo* activation. *J. Leukocyte Biol.*, **60**, 611–18.

Levy, E.M., Wu, J., Salibian, M. and Black, P. (1992) The effect of changes in thiol subcompartments on T-cell colony formation and cell cycle progression: relevance to AIDS. *Cell. Immunol.*, **140**, 370–80.

Lin, K.I., Lee, S.H., Narayanan, R., Baraban, J.M., Hardwick, J.M. and Ratan, R.R. (1995) Thiol agents and Bcl-2 identify an alphavirus-induced apoptotic pathway that requires activation of the transcription factor NF-kappa B. *J. Cell Biol.*, **131**, 1149–61.

Liu, B., Hackshaw, K.V. and Whisler, R.L. (1996) Calcium signals and protein tyrosine kinases are required for induction of c-jun in Jurkat cells stimulated by the T cell receptor complex and oxidative signals. *J. Interferon Cytokine Res.*, **16**, 77–90.

Los, M., Dröge, W., Stricker, K., Baeuerle, P., and Schulze-Osthoff, K. (1995) Hydrogen peroxide as a potent activator of T lymphocyte functions. *Eur. J. Immunol.*, **25**, 159–65.

Mahon, T.M. and O'Neill, L.A. (1995) Studies into the effect of the tyrosine kinase inhibitor herbimycin A on NF-κB activation in T lymphocytes. *J. Biol. Chem.*, **270**, 28557–64.

Macho, A., Hirsch, T., Marzo, I., Marchetti, P., Dallaporta, B., Susin, S.A., Zamzami, N. and Kroemer, G. (1997) Glutathione depletion is an early and calcium elevation is a late event of thymocyte apoptosis. *J Immunol.*, **158**, 4612–19.

Marchetti, P., Decaudin, D., Macho, A., Zamzami, N., Hirsch, T., Susin, S.A. and Kroemer, G. (1997). Redox regulation of apoptosis: impact of thiol oxidation status on mitochondrial function. *Eur. J. Immunol.*, **27**, 289–96.

Masoro, E. (1990) The role of free radicals in senescence. *Annu. Rev. Gerontol. Geriatr.*, **10**, 183–97.

Mbawuike, I.N., Acuna, C., Caballero, D., Pham-Nguyen, K., Gilbert, B., Petribon, P. and Harmon, M. (1996) Reversal of age-related deficient influenza virus-specific CTL responses and IFN-gamma production by monophosphoryl lipid A. *Cell. Immunol.*, **173**, 64–78.

McCabe, M.J. and Lawrence, D.A. (1991) Lead, a major environmental pollutant, is immunomodulatory by its differential effects on $CD4^+$ T cell subsets. *Toxicol. Appl. Pharmacol.*, **111**, 13–23.

Messina, J.P. and Lawrence, D.L. (1989) Cell cycle progression of glutathione-depleted human peripheral blood mononuclear cells is inhibited at S phase. *J. Immunol.*, **143**, 1974–81.

Meydani, S.N., Wu, D., Santos, M.S. and Hayek, M.G. (1995) Antioxidants and immune response in aged persons: overview of present evidence. *Am. J. Clin. Nutr.*, **62**, 1462s–1476s.

Meyer, M., Pahl, H.L. and Baeuerle, P.A. (1994) Regulation of the transcription factors NF-κ B and AP-1 by redox changes. *Chem.-Biol. Interact.*, **91**, 91–100.

Meyer, M., Schreck, R. and Baeuerle, P.A. (1993) H_2O_2 and antioxidants have opposite effects on activation of NF-κB and AP-1 in intact cells:AP-1 as secondary anti-oxidant-responsive factor. *EMBO J.*, **12**, 2005–15.

Mishell, R.I. and Dutton, R.W. (1967) Immunization of dissociated spleen cell cultures from normal mice. *J. Exp. Med.*, **126**, 423.

Naito, Y., Endo, H., Arai, K.-I., Coffman, R.L. and Arai, N. (1996) Signal transduction in Th clones: target of differential modulation by PGE_2 may reside downstream of the PKC-dependent pathway. *Cytokine*, **8**, 346–56.

Natarajan, V., Taher, M.M., Roehm, B., Parinandi, N.L., Schmid, H.H.O., Kiss, Z. and Garcia, J.G.N. (1993) Activation of endothelial cell phospholipase D by hydrogen peroxide and fatty acid hydroperoxide. *J. Biol. Chem.*, **268**, 930–7.

Nicoletti, I., Migliorati, G., Pagliacci, M.C., Grignani, F. and Riccardi, C. (1991) A rapid and simple method for measuring thymocyte apoptosis by propidium iodide staining and flow cytometry. *J. Immunol. Meth.*, **139**, 271–9.

Noelle, R.J. and Lawrence, D.A. (1981a) Modulation of T-cell functions. II. Chemical basis for the involvement of cell surface thiol-reactive sites in control of T-cell proliferation. *Cell. Immunol.*, **60**, 453–69.

Noelle, R.J. and Lawrence, D.A. (1981b) Determination of glutathione in lymphocytes and possible association of redox state and proliferative capacity of lymphocytes. *Biochem. J.*, **198**, 571–9.

Novak, T.J. and Rothenberg, E.V. (1990) cAMP inhibits induction of interleukin 2 but not of interleukin 4 in T cells. *Proc. Natl Acad. Sci. USA*, **87**, 9353–7.

Novogrodsky, A., Ravid, A., Rubin, A.L. and Stenzel, K.H. (1982) Hydroxyl radical scavengers inhibit lymphocyte mitogenesis. *Proc. Natl Acad. Sci. USA*, **79**, 1171–4.

Orrenius, S. (1993) In *Free Radicals: From Basic Science to Medicine* (eds G. Poli, E. Albano and M. Dianzani), Birkhauser Verlag, Basel, pp. 15–36.

Paganelli, R., Scala, E., Rosso, R., Cossarizza, A., Bertollo, L., Barbieri, D., Fabrizi, A., Lusi, E.A., Fagiolo, U. and Franceschi, C. (1996) A shift to Th0 cytokine production by $CD4^+$ cells in human longevity: studies on two healthy centenarians. *Eur. J. Immunol.*, **26**, 2030–4.

Pelletier, L., Castedo, M., Bellon, B. and Druet, P. (1994) Mercury and autoimmunity. in *Immunotoxicology and Immunopharmacology* (eds J.H. Dean, M.I. Luster, A.E. Munson and H. Amos), Raven Press, New York, pp. 539–52.

Qin, J., Clore, G.M., Kennedy, W.P., Kuszewski, J. and Gronenborn, A.M. (1996) The solution structure of human thioredoxin complexed with its target from Ref-1 reveals peptide chain reversal. *Structure*, **4**, 613–20.

Ramsdell, F., Seaman, M.S., Miller, R.E., Picha, K.S., Kennedy, M.K. and Lynch, D. (1994) Differential ability of Th1 and Th2 cells to express Fas ligand and to undergo activation-induced cell death. *Int. Immunol.*, **6**, 1545–53.

Reed, J.C. (1994) Bcl-2 and the regulation of programmed cell death. *J. Cell. Biol.*, **124**, 1–6.

Richie, J.P. (1992). The role of glutathione in aging and cancer. *Exp. Gerontol.*, **27**, 615–26.

Ristow, S.S., Starkey, J.R., Stanford, D.R., Davis, W.C. and Brooks, C.G. (1985) Cell surface thiols, but not glutathione, are essential for cytolysis by a cloned murine natural killer cell line. *Immunol. Invest.*, **14**, 401–14.

Röcken, M., Urban, J.F. and Shevach, E.M. (1992) Infection breaks T-cell tolerance. *Nature*, **359**, 79–82.

Röcken, M., Urban, J.F. and Shevach, E.M. (1994) Antigen-specific activation, tolerization and reactivation of the interleukin 4 pathway *in vivo*. *J. Exp. Med.*, **179**, 1885–93.

Schieven, G.L., Kirihara, J.M., Burg, D.L., Geahlin, R.L. and Ledbetter, J.A. (1993) $P72^{syk}$ tyrosine kinase is activated by oxidizing conditions that induce lymphocyte tyrosine phosphorylation and Ca^{++} signals. *J. Biol. Chem.*, **268**, 16688–92.

Schieven, G.L., Kirihara, J.M., Burg, D.L., Geahlen, R.L. and Ledbetter, J.A. (1994) ZAP-70 tyrosine kinase, CD45, and T cell receptor involvement in UV- and H_2O_2-induced T cell signal transduction. *J. Biol. Chem.*, **269**, 20718–26.

Sekkat, C., Dornand, J. and Gerber, M. (1988) Oxidative phenomena are implicated in human T-cell stimulation *Immunology*, **63**, 431–7.

Staal, F.J.T., Roederer, M., Herzenberg, L.A. and Herzenberg, L.A. (1990) Intracellular thiols regulate activation of nuclear factor κB and transcription of human immunodeficiency virus. *Proc. Natl. Acad. Sci. USA*, **87**, 9943–7.

Staal, F.J.T., Ela, S.W., Roederer, M., Anderson, M.T., Herzenberg, L.A. and Herzenberg, L.A. (1992) Glutathione deficiency and human immunodeficiency virus infection. *Lancet*, **339**, 909–12.

Taher, M.M. and Abd-Elfattah, A.S.A. (1996) Redox regulation of signal transduction in smooth muscle cells: protein kinase C inhibition by thiol agents. *Biochem. Mol. Biol. Int.*, **40**, 159–71.

Tan, X.X., Tang, C., Castoldi, A.F., Manzo, L. and Costa, L.G. (1993) Effects of inorganic and organic mercury on intracellular calcium levels in rat T lymphocytes. *J. Toxicol. Environ. Health*, **38**, 159–70.

Tineke, C.T.M., Kraan, P., Boeije, L.C.M., Smeenk, R.J.T., Wijdenes, J. and Aarden, L.A. (1995) Prostaglandin-E_2 is a potent inhibitor of human interleukin 12 production. *J. Exp. Med.*, **181**, 775–9.

van der Meide, P.H., de Labie, M.C., Botman, C.A., van Bennekom, W.P., Olsson, T., Aten, J. and Weening, J.J. (1993) Mercuric chloride down-regulates T cell interferon-gamma production in brown Norway, but not in Lewis rats: role of glutathione. *Eur. J. Immunol.*, **23**, 675–81.

Vella, A.T., McCormack, J.E., Linsley, P.S., Kappler, J.W. and Marrack, P. (1995) Lipopolysaccharide interferes with the induction of peripheral T cell death. *Immunity*, **2**, 261–70.

Vella, A.T., Mitchell, T., Groth, B., Linsley, P.S., Green, J.M., Thompson, C.B., Kappler, J.W. and Marrack, P. (1997) CD28 engagement and proinflammatory cytokines contribute to T cell expansion and long-term survival *in vivo*. *J. Immunol.*, **158**, 4714–20.

Weksler, M.E. (1993) Immune senescence and adrenal steroids: immune dysregulation and the action of dehydroepiandrosterone (DHEA) in old animals. *Eur. J. Clin. Pharmacol.*, **45**, S21–23.

Xanthoudakis, S., Miao, G. and Curran, T. (1994) The redox and DNA-repair activities of Ref-1 are encoded by nonoverlapping domains. *Proc. Natl. Acad. Sci. USA* **91**, 23–7.

Zheng, L., Fisher, G., Miller, R.E., Peschon, J., Lynch, D.H. and Lenardo, M.J. (1995) Induction of apoptosis in mature T cells by tumour necrosis factor. *Nature*, **377**, 348–51.

7 The Effect of UV Exposure on the Activation of T-Cell Subsets

Stephen E. Ullrich

Introduction

The ultraviolet (UV) radiation found in sunlight is a major environmental toxin. Aside from being the primary cause of non-melanoma skin cancer, the most prevalent form of human cancer, exposure to UV radiation also causes erythema and sunburn, premature ageing of the skin, ocular damage, such as cataract formation, and the induction of immune suppression. Because of the link between the immunosuppressive effects of UV radiation and its ability to induce skin cancer, intensive studies into the mechanisms responsible for the induction of immune suppression by UV radiation have been carried out by a number of laboratories. Due to these efforts, the mechanisms underlying the immunotoxic effects of UV radiation are perhaps the best studied of any environmental immunotoxin. This chapter reviews what is known of the induction of immune suppression by UV radiation, and will focus principally on the differential activation of T cell subsets following exposure to the UV radiation in sunlight.

Photoimmunology and photocarcinogenesis

Photoimmunology is the study of the effects of photons on the immune system. Although photoimmunology can trace its origins to antiquity, where sunlight was used to treat skin diseases, its origins as a scientific discipline stem primarily from studies on UV radiation-induced skin tumours in mice and from investigations on the effects of sunlight and sunlamp exposure on cutaneous immunity. The photons of interest are those wavelengths in the UV region of the solar spectrum, particularly the mid-range UVB (290–320 nm) radiation. Many of the biological effects of sunlight are attributed to UVB radiation including, tanning, sunburn, vitamin D formation, cataract formation, premature ageing of the

---

T Lymphocyte Subpopulations in Immunotoxicology. Edited by Ian Kimber and MaryJane K. Selgrade.
 ISBN 0 471 97194 4

skin, skin cancer and alterations in immune responsiveness. Short wavelength UVC (200–290 nm) is also biologically active. Nevertheless, because UVC radiation is not normally present in our environment (all of the UVC radiation is removed by the atmospheric ozone layer), the effects of UVC on immunity have received much less attention. Long wave UVA (320–400 nm) triggers photoallergic reactions in humans and is used in combination with photosensitizing drugs (i.e. psoralens) in the treatment of a variety of skin diseases, including psoriasis, vitiligo, and cutaneous T cell lymphoma. UVA contributes to cataract formation and to skin ageing. In addition, recent studies have suggested that UVA may cooperate with UVB in the induction of skin cancers in mice (Berg *et al.*, 1993), and studies with platy fish suggest that UVA may be involved in the induction of malignant melanoma (Setlow *et al.*, 1993).

The association between photocarcinogenesis and photoimmune suppression was first made by Kripke in a set of experiments performed in the mid-1970s. Contrary to the behaviour of the vast majority of chemically induced murine skin tumours, tumours induced by UV radiation failed to grow progressively when transferred into normal, immunocompetent, syngeneic hosts. Progressive growth occurred only when the UV-induced skin tumours were transferred into immune suppressed recipients (Kripke, 1974). From these observations, Kripke concluded that the tumours induced by UV radiation exhibit a highly antigenic phenotype, which is recognized as foreign by the immune system of the syngeneic host, thus leading to immune rejection. How then did these ‘regressor’ tumours grow progressively and develop in the primary irradiated host in the first place? The answer to this question came with the realization that in addition to being carcinogenic, UV radiation was also immune suppressive. Exposing mice to subcarcinogenic doses of UV radiation suppressed tumour rejection and permitted the progressive growth of the UV-induced tumours. Subsequent studies demonstrated that one consequence of UV-irradiation was the induction of antigen-specific suppressor T cells ($CD3^+$, $CD4^+$, $CD8^-$), that controlled the development of the primary tumours in the UV-irradiated animals, and transferred the immunosuppressed state to syngeneic recipients (Daynes and Spellman, 1977; Fisher and Kripke, 1982; Ullrich and Kripke, 1984). It should be noted, however, that the immune suppression induced by UV-irradiation is very selective. Aside from tumour rejection, the suppression of delayed-in-time hypersensitivity reactions and a systemic impairment of antigen-presenting cell function, most other immune responses in the UV-irradiated animals are normal (Table 7.1). Thus, these early studies in experimental animals established a link between the immunosuppressive effects of UV radiation and its ability to induce skin cancer.

A similar conclusion was reached concerning the effects of UV radiation on skin cancer induction in humans. UV exposure suppresses cell-mediated immune reactions in mice and in UV-irradiated human volunteers (Cooper *et al.*, 1992; Czarnecki *et al.*, 1995; Hersey *et al.*, 1983). UV exposure, however, only suppresses the induction of contact hypersensitivity (CHS) in a proportion

Table 7.1 Immunological profile of UV-irradiated mice

Suppressed	Normal
Rejection of UV-induced skin cancers	IgM antibody production
Contact hypersensitivity	Skin allograft rejection
Delayed type hypersensitivity	Tumour allograft rejection
Systemic antigen presenting cell function	Macrophage phagocytic and tumoricidal activity
	Mitogen-induced lymphoproliferation
	Antigen-induced lymphoproliferation
	Generation of cytotoxic T lymphocytes

(40–50%) of normal human volunteers. When, however, biopsy-proven skin cancer patients were exposed to UV radiation, and their CHS reaction measured, almost all of the patients were susceptible to the suppressive effects of UV radiation. While these findings support the idea that there is a genetic component to the activation of immune suppression by UV in humans, similar to that which is seen with mice (Streilein and Bergstresser, 1988), they also indicate that the immune suppression induced by UV exposure is a major risk factor for skin cancer in humans (Yoshikawa *et al.*, 1990). This clinical study, therefore, supports the conclusion arrived at using experimental animals, and shows that the immune suppression induced by UV radiation serves a co-factor for skin cancer induction.

As mentioned above, the association between the immune suppressive effects of UV radiation, and its carcinogenic potential became the driving force behind studies to determine the mechanisms through which UV exposure interferes with immune responsiveness. The skin and the elements of the cutaneous immune system are the targets of the UV radiation present in sunlight. UV radiation is relatively weak, non-ionizing radiation, and all the electromagnetic energy is absorbed by the uppermost layers (epidermis and dermis) of the skin. All the effects of the photons contained within sunlight on the immune system, therefore, are initiated within the skin, either by directly influencing immune components that reside within the skin, or by inducing epidermal cells to secrete soluble mediators that secondarily affect the cells of the immune system.

Effects of UV exposure on cutaneous immunity

Most attention has focused on the epidermal Langerhans cell and the keratinocyte as the cellular targets of UV radiation within the skin. In the local model of immune suppression, where the contact allergen is applied directly to the UV-irradiated skin, it is clear that the effect of UV on Langerhans cell function plays a major role in the induction of immune suppression. Exposure to UV alters the morphology of Langerhans cells, down-regulates the expression of major histocompatibility complex class II antigens on their surface

(Toews *et al.*, 1980), alters cell surface expression of costimulatory molecules, such as CD80, CD86 (Weiss *et al.*, 1995), and intracellular adhesion molecule-1 (ICAM-1/CD54) (Tang and Udey, 1991). These changes are thought to alter the functional ability of Langerhans cells. For example, UV-irradiated human Langerhans cells fail to stimulate allogeneic T cells in the epidermal mixed lymphocyte reaction (Hurks *et al.*, 1995). Nor can sensitization through UV-irradiated, Langerhans cell depleted skin induce a CHS reaction *in vivo* (Cooper *et al.*, 1992; Cruz *et al.*, 1990). It appears that the migration of UV-damaged, cyclobutane pyrimidine dimer positive Langerhans cells to the draining lymph nodes plays an essential role in unresponsiveness (Bucana *et al.*, 1994; Vink *et al.*, 1996).

In addition to its effect on Langerhans cells, UV-irradiation can also affect other antigen-presenting cells in the skin, and the modulation of their function may contribute further to immune suppression. Both dermal dendritic cells (Kurimoto *et al.*, 1994) and infiltrating macrophages (Baadsgaard *et al.*, 1990; Hammerberg *et al.*, 1994; Stevens *et al.*, 1995) have been shown to play a role in the induction of immune suppression and tolerance following UV-irradiation. Moreover, susceptibility to UV-induced immune suppression in the local model is controlled by genetic influences (Kurimoto and Streilein, 1994) that appear to affect the ability of Langerhans cells (Dai and Streilein, 1996) and dermal dendritic cells (Kurimoto *et al.*, 1994) to present antigen.

The immune modulators released by UV-irradiated keratinocytes can also affect antigen-presenting cell function in the skin. Activated keratinocytes are a rich source of cytokines, and other immune modulators and some of the factors released by UV-irradiated keratinocytes, such as, interleukin (IL)-10 (Enk *et al.*, 1993), *cis*-urocanic acid (Kurimoto and Streilein, 1992), and tumour necrosis factor-alpha (TNF-α) (Vermeer and Streilein, 1990) block antigen-presentation by Langerhans cells. It has also been suggested that the IL-10 released by infiltrating macrophages (Kang *et al.*, 1994) and the TNF-α secreted by dermal cells following UV exposure (de Kossodo *et al.*, 1995) may contribute to suppression of cutaneous immunity.

UV radiation may also promote immune suppression by blocking the production of certain keratinocyte-derived cytokines. The secretion of IL-7 and colony stimulating factor-1 (CSF-1 or macrophage-CSF) is suppressed following UV exposure (Schuhmachers *et al.*, 1996; Takashima *et al.*, 1995). Because these two cytokines are growth factors for immune competent cells within the skin (IL-7 promotes the growth of dendritic epidermal T cells, and CSF-1 is a Langerhans cell growth factor), it is possible that the removal of these cytokines from the epidermal milieu promotes the death of these cells. In addition, IL-15, a cytokine that binds to the α and β chains of the IL-2 receptor, and promotes the activation of T cells, is constitutively released by epidermal cells (Sorel *et al.*, 1996), and its secretion is reported to be suppressed following UV exposure (Blauvelt *et al.*, 1996). Blauvelt *et al.* suggest that removing IL-15 from the epidermis may contribute to the suppressed T cell responses [i.e. CHS and delayed type hypersensitivity (DTH)] that are seen following UV exposure. Thus, it is now becoming clear that UV exposure can induce immune suppression by either promoting the secretion of immune regulatory factors, such as IL-10, prostaglandin-E_2 (PGE_2) and TNF-α, or by interfering with the elaboration of growth factors for critical immunocompetent cells within the epidermis, which may induce apoptotic cell death within these populations.

Effects of UV exposure on systemic immunity

UV radiation also has systemic effects on the immune system. For example, splenic adherent cells from mice exposed to UV radiation do not efficiently present antigens to T cells (Greene *et al.*, 1979). In addition, CHS is suppressed and antigen-specific suppressor T cells are found in the spleens of mice that are irradiated at one site, and the sensitizing hapten is applied at a distant non-irradiated site (Noonan *et al.*, 1981). Clearly, the weak, non-ionizing, UV radiation does not penetrate directly to the spleen to induce these defects. Instead it has become apparent that cytokines released by UV-irradiated keratinocytes play an essential role in the induction of systemic immune suppression. A large number of immunoregulatory soluble mediators, cytokines, and hormones are produced by UV-irradiated keratinocytes and many have been implicated in the induction of systemic immune suppression following UV-irradiation (for a partial listing see Table 7.2). These include, PGE_2 (Chung *et al.*, 1986), TNF-α (Rivas and Ullrich, 1994; Yoshikawa and Streilein, 1990), cis-urocanic acid (Norval *et al.*, 1995), IL-1 (Robertson *et al.*, 1987), alpha-melanocyte stimulating hormone (α-MSH) (Grabbe *et al.*, 1996; Schauer *et al.*,

Table 7.2 Keratinocyte-derived cytokines

Interleukins		*Growth and chemoctatic factors*	
IL-1*	IL-8*	PDGF	TGF-α*
IL-1 RA*	IL-10*	BFGF	a-MSH*
IL-4*	IL-12*	NGF	hu GRO
IL-6*	IL-15*	NKGF	MIP-2
IL-7*		TGF-β*	IP-10*
Interferons		*Colony-stimulating factors*	
IFN-α		GM-CSF*	
IFN-β		M-CSF*	
		G-CSF*	
Tumour necrosis factors		*Suppresive factors*	
TNF-α*		Contra-IL-1*	
TNF-β*		*cis*-Urocanic acid*	
		Prostaglandin-E_2*	

*Expression (mRNA and/or protein secretion) modulated by UV-irradiation.

1994), and IL-10 (Brown *et al.*, 1995; Rivas and Ullrich, 1992, 1994; Ullrich, 1994). These cytokines are probably acting in concert to induce immune suppression. For instance, *cis*-urocanic acid has been shown to up-regulate the production of PGE_2 (Hart *et al.*, 1993), which has been shown to up-regulate IL-4 and IL-10 production and down-regulate IL-12 secretion by monocytes (Ayala *et al.*, 1994; Strassmann *et al.*, 1994; Tineke *et al.*, 1995). Moreover, α-MSH up-regulates PGE_2 synthesis (Davidson *et al.*, 1992) which can then inhibit IL-12 production and promote the secretion of IL-4 and 10. Similarly, Krutmann and colleagues have demonstrated that IL-1 modulates TNF-α production by keratinocytes, which can then suppress the expression of ICAM-1, an essential co-stimulatory molecule involved in T cell activation (Krutmann, 1994). In addition, we have recently reported that injecting UV-irradiated mice with antibodies to *cis*-urocanic acid blocks the induction of immune suppression. Because injecting this antibody alters the kinetics of IL-10 secretion in the skin of UV-irradiated mice, it appears that *cis*-urocanic acid affects IL-10 production *in vivo* (Moodycliffe *et al.*, 1996). Regardless of the exact mechanism involved, one consequence of cytokine production by UV-irradiated keratinocytes appears to be suppression of T helper 1 (Th1) cell function, coupled with the activation of T helper 2 (Th2) cells, which may explain the suppression of inflammatory immune reactions (CHS, DTH, and tumour rejection) observed following UV exposure.

Effects of UV exposure on the activation of Th1 and Th2 cells

Human and mouse $CD4^+$ and $CD8^+$ T cells can be segregated into different subsets based on the cytokines they secrete and the immune responses they participate in (Mosmann *et al.*, 1991; Romagnani *et al.*, 1993) (also see Chapters 2 and 3). Type 1 cells (Th1) secrete interleukin IL-2, interferon (IFN)-γ, provide help for cell-mediated immune reactions, and help B cells to secrete complement fixing antibodies. Type 2 cells (Th2), secrete IL-4, IL-5, IL-6, and IL-10, help B cells produce non-complement fixing antibodies and are involved with immune reactions generally associated with allergy, IgE production and the activation of mast cells and eosinophils. The cytokines produced by each subset of cells can serve also to regulate the activity of the other. The IFN-γ produced by Th1 cells interferes with the proliferation of Th2 cells, and the IL-10 secreted by Th2 cells blocks the secretion of cytokines by Th1 cells (Mosmann, 1991).

Perhaps the most important signal that affects Th1 or Th2 cell activation is the cytokine microenvironment present during the initial activation of the cells. The presence of IL-12 drives the activation and differentiation of Th1 cells (Hsieh *et al.*, 1993; Magram *et al.*, 1996), whereas Th2 cells are activated when IL-4 is present (Hsieh *et al.*, 1992; Kanagawa *et al.*, 1993) (see Chapter 2).

One of the first indications that UV exposure influences the balance of Th1 and Th2 cells came from the studies of Araneo *et al.* (1989). T cells were isolated from mice exposed to UV and then immunized with antigen. These cells were then cultured *in vitro* with antigen, and IL-4, IL-2 and IFN-γ secretion was measured. Compared with normal control cultures, T cells from mice exposed to UV radiation were found to secrete decreased amounts of IL-2 and IFN-γ, whereas the secretion of the Th2 cytokine IL-4 was increased following UV exposure.

The modulation of Th1 or Th2 cell activation appears to result from an alteration of antigen-presenting cell function. This UV-induced alteration in antigen-presenting cell activity occurs after direct UV-irradiation of the antigen-presenting cell population (epidermal Langerhans cells), or can occur indirectly via the action of UV-induced keratinocyte-derived cytokines on antigen-presenting cells. Simon and colleagues found that while UV-irradiated Langerhans cells were quite capable of presenting antigen to Th2 clones, they were unable to stimulate Th1 cells. Furthermore, not only was antigen-presentation to Th1 cells blocked, presentation of antigen by UV-irradiated Langerhans cells tolerized the Th1 cells (Simon *et al.* 1990, 1991). Almost identical results were reported by Enk *et al.* (1993) when normal Langerhans cells were treated with IL-10 and then used to present antigen to Th1 and Th2 cells. Here also, presentation to Th1, but not Th2 cells was blocked, and the Th1 cells cultured with the IL-10-treated Langerhans cells were tolerized.

Following UV exposure IL-10 is produced by keratinocytes and IL-10 is found in the serum of UV-irradiated mice (Rivas and Ullrich, 1992, 1994). It has been shown that UV-induced IL-10 can affect the activation of Th1 and Th2 cells. When adherent spleen cells from UV-irradiated mice were used to present nominal antigen to T cell clones, the presentation to Th1 cells was significantly suppressed. On the other hand, increased antigen presentation to Th2 cells was observed when the antigen-presenting cells were isolated from UV-irradiated mice. Both the suppression of Th1 cell activation, and the enhancement of Th2 cell function was reversed by injecting the UV-irradiated mice with antibodies to IL-10 (Ullrich, 1994). Thus, it appears that IL-10 works in both the local and systemic models to alter antigen presenting cell function that results in suppressed Th1 cell function while permitting the presentation of antigen to Th2 cells.

Additional data support the hypothesis that UV exposure causes a shift to a Th2-like immune reaction. Antibodies to IL-4 and/or IL-10 were found to block the activity of the UV-induced antigen-specific $CD4^+$ suppressor T cells, suggesting that these cells are Th2-like (Rivas and Ullrich, 1994). A similar conclusion was drawn by Yagi and colleagues, who cloned antigen-specific, Th2-like, $CD3^+$, $CD4^+$, $CD8^-$ suppressor T cells from the spleens of UV-irradiated animals. The suppressive activity of these cells, *in vivo* and *in vitro*, was blocked by monoclonal anti-IL-10 antibodies, indicating that the suppressor cells function by releasing Th2 cytokines (Yagi *et al.*, 1996).

As mentioned above, Th1 cells are essential for cell mediated immune reactions, such as DTH but also provide help to B cells for the secretion of complement fixing antibodies (IgG2a and IgG2b). Th2 cells, on the other hand, provide B cell help for the production of non-complement fixing antibodies (IgG1). The effects of UV exposure on both DTH and antibody formation have been determined. In these studies, the effect of UV exposure on the immune response to *Borrelia burgdorferi*, the causative agent of Lyme disease, was measured. Although antibody formation and DTH are often mutually exclusive, Parish and colleagues demonstrated that exclusivity was primarily a function of antigen dose. In these studies, immunization at lower doses of antigen resulted in DTH but not antibody formation, whereas high dose immunization resulted in the production of antibody with no discernible DTH. At intermediate doses, however, both antibody formation and DTH could be achieved (Parish, 1972). Therefore, the dose of antigen was first titrated and an inoculum of *B. burgdorferi* that generated both a DTH response and antibody formation in normal mice was used. The effect that UV-irradiation had on either immune reaction was measured. As expected, prior UV exposure suppressed the generation of a DTH reaction to *B. burgdorferi*, and injecting the UV-irradiated animals with monoclonal anti-IL-10 completely restored the ability to generate DTH. When the effect of UV exposure on the formation of *B. burgdorferi*-specific antibodies was examined, it was found that UV suppressed the generation of complement fixing IgG2a and IgG2b, in a dose-dependent manner. Moreover, when the UV-irradiated mice were injected with anti-IL-10, the levels of *B. burgdorferi*-specific IgG2a and IgG2b in the serum of the treated animals returned to control levels. A different situation was found when *B. burgdorferi*-specific IgG1 was measured. In this case, no suppression of IgG1 levels was noted; in fact, a slight enhancement of IgG1 serum concentration was observed at later time points in UV-irradiated mice (Brown *et al.*, 1995).

To test further the hypothesis that UV-irradiation interferes with Th1 cell activation, while allowing for the activation of Th2 cells, the effects of recombinant IL-12 (rIL-12) on UV-induced immune suppression was examined. Interleukin-12 is the primary cytokine responsible for driving precursor Th0 cells along the Th1 cell pathway. Macrophage-derived IL-12 promotes the differentiation of naive $CD4^+$ T cells into Th1 cells during primary antigenic stimulation (Hsieh *et al.*, 1993), and studies with IL-12-deficient animals indicate that mature Th1 cells are completely lacking *in vivo* in mice where the IL-12 gene was 'knocked-out' (Magram *et al.*, 1996). Furthermore, rIL-12 has been reported to block the activation/differentiation of Th2 cells *in vivo* (McKnight *et al.*, 1994). Thus, administration of IL-12 serves to enhance Th1 cell activity, while blocking the induction of Th2 cells. If indeed, UV-irradiation alters the activation of Th1 cells, then administration of IL-12 should restore the immune response in UV-irradiated mice. This is exactly what was found in studies reported by Schmitt *et al.* (1995). When mice were injected with recombinant IL-12 (0.5–1 μg per mouse) 4 h after UV exposure, the suppression of CHS and/or DTH was completely reversed. Doses of IL-12 lower than, or equal to 0.1 μg/mouse were ineffective at restoring immune function. Associated with this reversal of immune suppression was an abrogation of suppressor cell generation. No immune suppression was noted when spleen cells from the UV-irradiated, IL-12-treated mice were transferred into syngeneic recipients, whereas suppressor cells were generated in mice exposed to UV and treated with the vehicle. We suggest that suppressor cell induction was blocked because the injected IL-12 interfered with the differentiation of Th2 cells in the UV-irradiated mice.

The effect that IL-12 had on the activity of pre-formed suppressor cells was examined next. In these experiments the suppressor cells were generated in a set of animals exposed to UV radiation, and then transferred into normal recipient mice. The recipient mice were then injected with either rIL-12 or the vehicle control. As expected from previous experiments, transfer of spleen cells from UV-irradiated mice into normal recipients suppressed the generation of DTH or CHS in the recipient mice. When, however, the recipient mice that received the suppressor cells were injected with rIL-12, the transferred suppressor cells failed to down-regulate the immune reaction. Injecting rIL-12 into mice that received spleen cells from non-UV-irradiated donors did not potentate the immune response, suggesting that the IL-12 did not simply accelerate the immune reaction in the recipients, but actually blocked the activity of the suppressor cells. Because IL-12 counteracts the biological activities of IL-4 and IL-10, it is proposed that one consequence of IL-12 administration in the recipient mice is to suppress the release of these anti-inflammatory cytokines by the antigen-activated UV-induced Th2 cells.

The finding that IL-12 can overcome UV-induced immune suppression was recently confirmed by Schwarz *et al.* (1996). In this study immune suppression and the induction of hapten-specific suppressor cells by UV exposure were blocked by a single injection of rIL-12. Of particular interest in this study was an experiment that examined the effect of rIL-12 on tolerance induction in UV-irradiated mice. Sensitization of mice through UV-irradiated skin results in suppression of CHS and the induction of hapten-specific tolerance. Tolerance is measured by the inability of the mice to respond to re-sensitization with the same hapten, and can be observed 14–21 days after the initial sensitization. Schwarz and colleagues found that a single injection of rIL-12, prior to UV exposure, totally abrogated the induction of tolerance. Because of the essential role that IL-12 plays in the induction of Th1 cells, and since it has been demonstrated that IL-12 interferes with the activation of Th2 cells *in vivo*, the results published in these two studies provide further support for the hypothesis that UV exposure down-regulates Th1 cell function while allowing the activation and expansion of Th2 cells (Table 7.3).

Although most of the above mentioned studies have focused on the role of UV-induced IL-10 in modulating immune responsiveness, other immune

Table 7.3 Evidence for a shift towards a Th-2-like immune reaction following UV-irradiation

• Lymphocytes isolated from UV-irradiated mice secrete primarily Th2 cytokines when stimulated *in vitro*	Araneo *et al.* (1989)
• UV-irradiated Langerhans cells tolerize Th1 cells but do present antigen to Th2 cells	Simon *et al.* (1990, 1991)
• IL-10 containing supernatants suppress Th1-cell driven immune reactions (DTH and IFN-γ production)	Rivas and Ullrich (1992)
• IL-10-treated Langerhans cells tolerize Th1 cells but do present antigen to Th2 cells	Enk *et al.* (1994)
• UV-irradiation blocks the ability of splenic adherent cells to present antigen to Th1 cells, while enhancing presentation to Th2 cells. Injecting the UV-irradiated mice with anti-IL-10 reverses these effects	Ullrich (1994)
• The suppressor T cells induced following UV-irradiation mediate their immune suppressive effects by releasing IL-4 and IL-10	Rivas and Ullrich (1994)
• UV-irradiation suppresses the production of antibody subclasses helped by Th1 cells (IgG2 and IgG2b) but not the production of subclasses helped by Th2 cells. Injecting the UV-irradiated mice with anti-IL-10 reverses these effects	Brown *et al.* (1995)
• Injecting UV-irradiated mice with IL-12 blocks the induction of immune suppression, induction of suppressor cells, and interferes with the activity of UV-induced suppressor cells	Schmitt *et al.* (1995)
• Injecting UV-irradiated mice with IL-12 blocks the induction of immune suppression, induction of suppressor cells, and blocks the induction of immune tolerance	Schwarz *et al.* (1996)
• Cloned suppressor cells from UV-irradiated mice secrete IL-10 and behave like Th2 cells	Yagi *et al.* (1996)

modulators released by keratinocytes following UV exposure may also play a pivotal role in shifting the immune response to one favoured by Th2 cells. Prostaglandin E_2 and IL-1 promote the activation of Th2 cells (Araneo *et al.*, 1989; Betz and Fox, 1991; Hilkens *et al.*, 1995). Moreover, PGE_2 suppresses the secretion of IL-12 by blood monocytes (Tineke *et al.*, 1995) and up-regulates the secretion of IL-10 and IL-4 (Ayala *et al.*, 1994; Strassmann *et al.*, 1994), which would favour the suppression of Th1 cell function and the activation of Th2 cells. Similarly, *cis*-urocanic acid, which is found in the skin and serum of UV-irradiated mice (Gibbs *et al.*, 1993; Moodycliffe *et al.*, 1993), up-regulates PGE_2 secretion by blood monocytes (Hart *et al.*, 1993), which through its effects on IL-4 and IL-10, may promote the activation of Th2 cells.

Effect of UV exposure on the immune response to pathogenic microorganisms

Infectious disease remains a major public health problem world-wide and as mentioned above, UV-irradiation suppresses the immune response to micro-organisms (Brown *et al.*, 1995). Indeed, studies using mice and artificial sun-lamps, have demonstrated clearly that UV exposure can suppress the immune responses to bacteria, fungi, intracellular parasites and virus (reviewed by Jeevan and Kripke (1993)). This has prompted some to question whether the UVB radiation in sunlight can exacerbate infectious disease in humans. Unfortunately, very little information is available in answer to this question. Sunlight triggers viral recrudescence of herpes simplex virus and some evidence suggests a role for UV-induced immune suppression in reactivating the virus (Miura *et al.*, 1994). UV-exposure also reduced the granulomatous reaction and decreased the number of $CD4^+$ T cells infiltrating the site of elicitation of a DTH reaction to leprosy antigens in healthy lepromin-positive individuals, suggesting that UV can down-modulate immune reactions to infectious agents in human beings (Cestari *et al.*, 1995). In addition, some have suggested that a shift away from a Th1-like reaction towards a Th-2-predominated response precipitates the progression from AIDS related complex to full-blown AIDS in human immunodeficiency virus (HIV) infected individuals (Clerici *et al.*, 1993; Maggi *et al.*, 1994). Interleukin-10 has been shown to play a role in the T cell dysregulation seen in HIV^+ individuals (Clerici *et al.*, 1994), and more recent findings have suggested that HIV gp120 induces lymphocytes to secrete IL-10, which activates T cell anergy, thus contributing to the immune dysfunction seen in AIDS (Schols and De Clercq, 1996). Because UV exposure induces human keratinocytes to produce IL-10 (Grewe *et al.*, 1995), and data in mice suggest that UV-irradiation causes a shift towards a Th2-like reaction, the question is whether exposure to sunlight will adversely affect the immune status of HIV^+ individuals. Clearly, data from experimental models demonstrate that UV exposure has the potential to modify immune reactions to infectious agents, but at the present time it is not clear if increased exposure to UV radiation will translate into increased incidence of infectious disease in the human population.

DNA damage as a trigger for the induction of immune suppression by UV-irradiation

The exact identity of the molecular target within the skin that triggers the induction of immune suppression following UV-irradiation is not clear. Some have suggested urocanic acid (Noonan and De Fabo, 1992), while others have proposed that UV-induced oxidative stress initiates cytokine production and the induction of immune suppression (Devary *et al.*, 1993; Simon *et al.*, 1994). UV-irradiation also damages DNA, and compelling evidence provided by Kripke and colleagues suggest that DNA damage can initiate the cascade of events that activates UV-induced immune suppression.

In a series of experiments done with marsupials (*Monodelphis domestica*), it was found that the induction of immune suppression was reversed when the animals were exposed to visible light immediately after UV-irradiation. *Monodelphis domestica* were used in these experiments because these animals possess an enzyme, the photoreactivating enzyme, that is activated by visible light to repair pyrimidine dimers (the primary lesion found in UV-irradiated DNA) *in situ*, and return the DNA to its native conformation. Dimer repair occurs when visible light is applied after the UV light, but not when the photoreactivating light is given before the UV radiation. Similarly, restoration of immune function was only seen when the photoreactivating light was applied immediately after, but not before, the UV-irradiation. Because visible light activates the marsupial pyrimidine dimer repair pathway, these findings suggest that the formation of pyrimidine dimers in irradiated DNA is required for immune suppression (Applegate *et al.*, 1989).

Another series of experiments attempted to repair pyrimidine dimer formation in the mouse by introducing the repair enzyme to the nucleus of irradiated cells, and determining what effect this procedure has on UV-induced immune suppression. Unlike the marsupial, mouse cells do not have a photoreactivation mechanism to repair pyrimidine dimer formation, so liposomes were constructed containing the bacteriophage excision repair enzyme T4 endonuclease V (T4N5) to introduce the excision repair enzyme to the target cells. Immunohistochemical analysis demonstrated that these liposomes can penetrate through the stratum corneum, migrate to the level of the basal keratinocytes, fuse with the cell membrane and deposit the repair enzyme into the cells. By immunoelectron microscopic analysis, the enzyme can be seen clearly in the cytoplasm and in the nucleus of the treated cells (Yarosh *et al.*, 1994). Moreover, application of the T4N5-containing liposomes to UV-irradiated mouse skin repairs pyrimidine dimer formation, as fewer dimers were present in the DNA of epidermal cells treated with UV and the T4N5-containing liposomes than found in controls treated with UV only, or exposed to UV and treated with liposomes containing a heat-inactivated enzyme preparation (Kripke *et al.*, 1992).

To determine whether dimer repair overcame UV-induced immune suppression, mice were exposed to UV, and the T4N5-containing liposomes, or liposomes containing heat-inactivated enzyme, were applied to the UV-irradiated skin. The mice were then immunized and the development of an immune response was measured. Two different antigens were used, the mice were injected with *C. albicans* and the development of DTH was measured, or the mice were painted with 2,4-dinitrofluorobenzene and CHS was measured. Regardless of the antigen used, application of the liposomes to the skin of UV-irradiated mice totally overcame the induction of immune suppression. When, however, the mice were treated with liposomes containing the heat-inactivated enzyme, the suppression observed was identical to that found in the UV-irradiated controls. Because of the unique specificity of the enzyme for pyrimidine dimers, these data suggest that the formation of cyclobutane pyrimidine dimers in the DNA is the initiating event in UV-induced immune suppression, since repairing this lesion totally reversed the suppressive effects (Kripke *et al.*, 1992).

Further support for this hypothesis comes from the experiments recently reported by Vink *et al.* (1996). It was known from previous work that contact sensitization promotes the migration of Langerhans cells from the epidermis to the draining lymph nodes. Presumably, the Langerhans cell takes up the hapten in the skin, and migrates to the lymph node to present the hapten to T cells (Kripke *et al.*, 1990). In addition, it is known that prior UV-irradiation alters the antigen-presenting cell function of the migrating Langerhans cells, so that when they arrive at the draining lymph node, rather than activating an effector response, they induce immune suppression and activate suppressor T cells (Okamoto and Kripke, 1987). Vink *et al.* tested the hypothesis that UV-induced DNA damage plays a role in the alteration of antigen presenting cell function. They took advantage of the availability of a monoclonal antibody that recognizes pyrimidine dimers to identify the UV-damaged cells in the draining lymph nodes. Mice were exposed to UV and painted with the fluorescent contact sensitizer, fluorescein isothiocyanate (FITC), which allowed the investigators to follow the fate of the migrating antigen-presenting cells. Dimer positive dendritic cells were initially seen in the epidermis, then in the dermis, and within 24 h, were found in the draining lymph nodes of UV-irradiated mice. Cells containing pyrimidine dimers persisted for at least 4 days after UV exposure. The hapten positive dendritic cells (Ia^+, $FITC^+$) recovered from the draining lymph nodes of UV-irradiated mice were ineffective as antigen-presenting cells. Staining these cells with the monoclonal antibody specific for pyrimidine dimers indicated that the DNA of these cells was damaged. Application of the T4N5-containing liposomes immediately following UV exposure not only restored antigen-presenting cell function to the Ia^+, $FITC^+$ cells, but significantly reduced the number of dimer positive cells in the dendritic cell population. Here again, treatment of the UV-irradiated mice with liposomes containing the heat-inactivated enzyme did not restore antigen-presenting cell function to the Ia^+, $FITC^+$ dendritic cells, nor did the heat-inactivated enzyme reduce the numbers of pyrimidine dimer positive cells recovered from the draining lymph nodes. From these observations, it was concluded that dimer positive dendritic cells migrate from the skin to the draining lymph node, where their ability to present antigen is severely depressed. Repair of the lesion, through the use of the T4N5-containing liposomes, not only decreased the numbers of dimer positive cells, but increased the ability of the hapten-positive, Ia^+, dendritic cells to present antigen and activate the immune response. These data therefore further support the hypothesis that UV-induced DNA damage is the initiating event in the activation of immune suppression.

As mentioned above, the cytokines released from UV-irradiated keratinocytes play an essential role in the development of immune suppression. Does UV-induced DNA damage activate keratinocytes to release these immune modulatory cytokines? Two independent approaches were used to test this hypothesis. In the first, an attempt was made to induce DNA damage, in a relatively non-specific manner, and measure what effect this would have on cytokine release and immune suppression. This was accomplished by applying liposomes containing the DNA restriction enzyme *Hin*dIII to the skin of mice. The restriction enzyme *Hin*dIII cuts at palindromic AAGCTT sequences and results in DNA double strand breaks. The activity of the liposomal delivery system was measured by the Comet assay. DNA damage was induced by liposomes containing *Hin*dIII, but not by liposomes containing the heat inactivated enzyme. Application of the *Hin*dIII-containing liposomes to mouse skin, mimicked the effect of UV exposure and resulted in suppression of CHS, both locally and systemically. In addition, the *Hin*dIII-containing liposomes induced keratinocytes (both *in situ* and *in vitro*) to produce IL-10 and TNF-α (O'Conner *et al.*, 1996). Thus, these data show that non-specific DNA damage will activate immune suppression.

The second approach again relied on the use of the T4N5-containing liposomes. Will repairing pyrimidine dimer formation in UV-irradiated keratinocytes block the release of immune suppressive cytokines? Advantage was taken of the fact that UV-irradiation of keratinocyte cultures *in vitro* causes the release of IL-10, and the IL-10 present in the supernatant fluid suppresses the induction of an immune response when injected into mice (Rivas and Ullrich, 1992; Ullrich, 1994). Keratinocyte cultures were first exposed to UV radiation and then treated with the T4N5-containing liposomes. The supernatant fluid was collected 24 h later, a portion was injected into mice, while some was retained and the IL-10 content was measured by ELISA (*E*nzyme-*L*inked *I*mmunoab*S*orbant *A*ssay). As expected, injecting the supernatants from UV-irradiated keratinocytes into mice suppressed the induction of DTH in these animals. Treating the UV-irradiated keratinocyte cultures with T4N5-containing liposomes reversed the induction of immune suppression. No reversal was seen when the UV-irradiated keratinocytes were treated with liposomes containing the heat inactivated enzyme. The inhibition of immune suppression by the T4N5-containing liposomes was associated with a suppression of IL-10 production by the irradiated keratinocytes, as IL-10 levels in the supernatant fluid of these cultures were reduced to almost background levels. Because application of the T4N5-containing liposomes totally reversed the induction of immune suppression when applied to the skins of UV-irradiated mice (Kripke *et al.*, 1992), it was determined if UV-induced IL-10 production *in vivo* was affected by repairing pyrimidine dimer formation. The application of the T4N5-containing liposomes to UV-irradiated mouse skin blocked the production of IL-10 by keratinocytes *in situ*, and blocked the appearance of IL-10 in the serum of the UV-irradiated animals (Nishigori *et al.*, 1996).

In further support of the hypothesis that DNA damage is the trigger for induction of UV-induced immune suppression is the recent observation made by Hurks *et al.* (1995) that the action spectrum for UV-induced suppression of antigen presenting cell function of UV-irradiated human epidermal cells is identical to the published action spectrums for the induction of pyrimidine dimers and 6-4 photoproducts in UV-irradiated DNA. These findings, therefore, suggest that UV-induced DNA damage, specifically the formation of pyrimidine dimers, is the trigger that leads to an impairment of the antigen-presenting cell function of epidermal Langerhans cells, and activates keratinocytes to secrete immune modulatory cytokines, resulting in the initiation of a cascade of events that leads to immune suppression.

Summary and conclusions

The focus of photoimmunological investigations is to determine how the photons contained primarily in the UVB region of the solar spectrum activate immune suppression. The association between immune suppression and photocarcinogenesis has been clearly established in animal models and data generated with biopsy proven skin cancer patients suggest that photoimmune suppression acts as a risk factor for skin cancer induction in humans. In the course of these studies, however, important contributions to our basic understanding of immunological function have been made, in particular the role of antigen presenting cells in the activation of immune suppressive T cells, and the identity of T cells with immune suppressive activity.

Our working model for the effects of UV on the immune system is presented in Figure 7.1. The molecular target of UVB-radiation is the DNA, and the primary lesion formed in the DNA is the pyrimidine dimer. Dimer positive cells Langerhans cells take up antigens (or haptens), migrate to the draining lymph nodes where they efficiently present antigen to Th2 cells, but tolerize Th1 cells. The result is a suppression of Th1 cell-mediated functions, such as CHS, DTH, and perhaps tumour rejection. The formation of pyrimidine dimers in keratinocytes results in the activation of cytokine genes, and the production of immune modulatory cytokines, such as IL-10. These cytokines can then work locally to affect the antigen presenting properties of epidermal Langerhans cells, which then migrate to the draining lymph node and efficiently present antigen to Th2 cells, but at the same time tolerize Th1 cell function. Alternatively, the cytokines produced by UV-irradiated, dimer positive keratinocytes can work systemically on tissue antigen presenting cells to alter their function in such a way so that Th1 cell activation is suppressed, whereas presentation to Th2 cells is intact or even enhanced. The addition of exogenous IL-12 may be blocking immune suppression at this step by providing the necessary signal (usually provided by antigen-presenting cells, whose

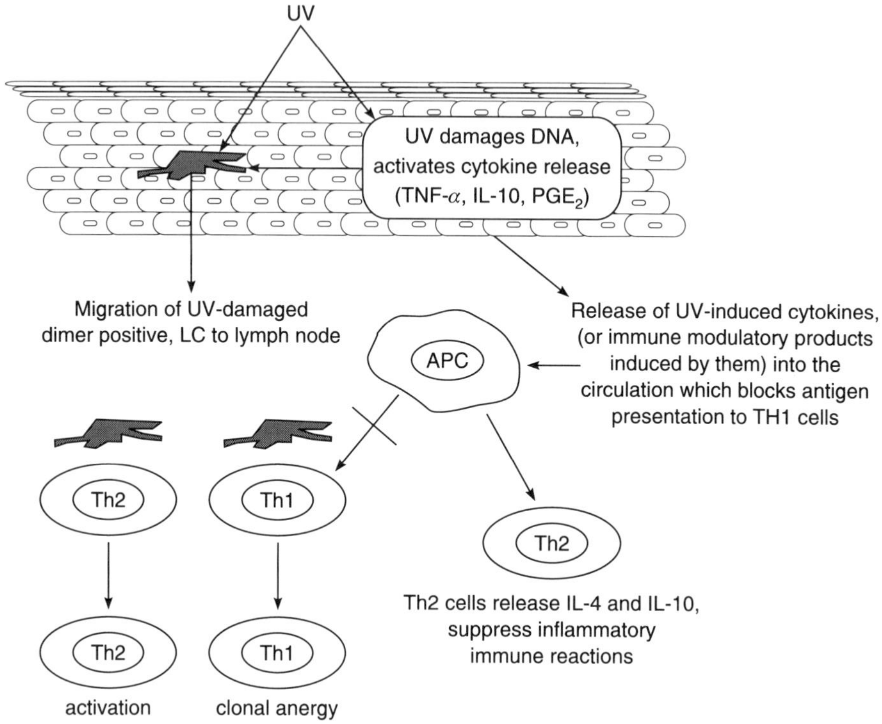

Figure 7.1 Effects of UV exposure on T cell activation.

normal function is blocked by UV-induced IL-10) to drive the differentiation of Th1 cells. Furthermore, it is proposed that the antigen specific UV-induced suppressor T cells ($CD3^+$, $CD4^+$, $CD8^-$) are simply antigen activated Th2 cells, which when stimulated by the appropriate antigen, secrete IL-4 and IL-10, thus suppressing inflammatory immune reactions. It is suggested that exogenous IL-12 blocks suppressor cell function by counteracting the biological activity of the IL-4 and IL-10 secreted by the UV-induced suppressor cells. What is not clear is whether the IL-1, IL-6, IL-10, etc. found in the serum of UV-irradiated individuals is derived directly from the UV-damaged keratinocytes, or whether some of the products released by UV-irradiated keratinocytes (perhaps PGE_2, *cis*-urocanic acid, or α-MSH) stimulate blood monocytes to secrete these cytokines, which accounts for the increased serum levels observed in UV-irradiated individuals. Nevertheless, the net effect is suppressed Th1 cell function, and an activation of Th2 cells.

In addition, it is suggested that the mechanisms involved in the induction of immune suppression by UV radiation may serve as a model system to study the effects of other environmental toxins on the immune system, especially those that interact with skin. For example, in preliminary studies, we have noted that treating keratinocyte cultures with toxins that have been shown to be immune suppressive, such as benz[a]pyrene, 7,12-dimethylbenz[a]anthracene and 12-*O*-tetradecanoly-phorbol-13-acetate, causes the release of IL-10 (Ullrich, unpublished observations). It is interesting to note that many of the known immunotoxic compounds (Luster *et al.*, 1992) are also carcinogens. Perhaps DNA damage is the initiating event in the induction of immunotoxicity by these compounds. It is tempting to speculate that the application of an immunotoxin to the skin triggers DNA damage, which results in cytokine production and the induction of immune suppression. Moreover, it is possible to use the reagents described above (T4N5 and HindIII-containing liposomes) to test this hypothesis and determine if a common mechanism for the induction of immune suppression is involved.

Acknowledgements

The work performed in my laboratory, was supported by grants from the National Institutes of Health (AR 40824 and ES 07327). That support, and the efforts of Angus Moodycliffe, David Schmitt, and Todd Giese in performing the experiments described here, are acknowledged with thanks.

References

Applegate, L.A., Ley, R.D., Alcalay, J. and Kripke, M.L. (1989) Identification of the molecular target for the suppression of contact hypersensitivity by UV radiation. *J. Exp. Med.*, **170**, 1117–31.

Araneo, B., Dowell, T., Moon, H.B. and Daynes, R.A. (1989) Regulation of murine lymphocyte production *in vivo.* UV radiation exposure depresses IL-[illegible], and enhances IL-4 production by T cells through an IL-1 dependent mechanism. *J. Immunol.*, **143**, 1737–44.

Ayala, A., Lehman, D.L., Herdon, C.D. and Chaudry, I.H. (1994) Mechanism of enhanced susceptibility to sepsis following hemorrhage: IL-10 suppression of T cell response is mediated by eicosanoid-induced IL-4 release. *Arch. Surg.*, **129**, 1172–8.

Baadsgaard, O., Salvo, B., Mannie, A., Dass, B., Fox, D. and Cooper, K.C. (1990) *In vivo* ultraviolet-exposed human epidermal cells activate T suppressor cell pathways that involve $CD4^{+}CD45RA^{+}$ suppressor-inducer T cells. *J. Immunol.*, **145**, 2854–61.

Berg, R.J.W., de Gruijl, F.R. and van der Leun, J.C. (1993) Interaction between UVA and UVB radiation in skin cancer induction in hairless mice. *Cancer Res.*, **53**, 4212–17.

Betz, M. and Fox, B. (1991) Prostaglandin E_2 inhibits production of Th1 but not Th2 lymphokines. *J. Immunol.*, **146**, 108–13.

Blauvelt, A., Asada, H., Klaus-Kovtum, V., Altman, D.J., Lucey, D.R. and Katz, S.I. (1996) Interleukin-15 mRNA is expressed by human keratinocytes, Langerhans cells, and blood-derived dendritic cells and is downregulated by ultraviolet B radiation. *J. Invest. Dermatol.*, **106**, 1047–52.

Brown, E.L., Rivas, J.M., Ullrich, S.E. Young, C.R., Norris, S.J. and Kripke, M.L. (1995) Modulation of immunity to *Borrelia burgdorferi* by ultraviolet irradiation: Differential effect on Th1- and Th2-type immune responses. *Eur. J. Immunol.*, **25**, 3017–22.

Bucana, C., Tang, J.-M., Dunner, K., Strickland, F.M. and Kripke, M.L. (1994) Phenotypic and ultrastructural properties of antigen presenting cells involved in contact sensitization of normal and UV-irradiated mice. *J. Invest. Dermatol.*, **102**, 928–33.

Cestari, T.F., Kripke, M.L., Baptista, P.L., Bakos, L. and Bucana, C.D. (1995) Ultraviolet radiation decreases the granulomatous response to Lepromin in humans. *J. Invest. Dermatol.*, **105**, 8–13.

Chung, H.-T., Burnham, D.K., Robertson, B., Roberts, L.K. and Daynes, R.A. (1986) Involvement of prostaglandins in the immune alterations caused by the exposure of mice to ultraviolet radiation. *J. Immunol.*, **137**, 2478–84.

Clerici, M., Lucey, D.R., Berzofsky, J.A., Pinto, L.A., Wynn, T.A., Blatt, S.P., Dolan, M.J., Hendrix, C.W., Wolf, S.F. and Shearer, G.M. (1993) Restoration of HIV-specific cell-mediated immune responses by interleukin-12. *Science*, **262**, 1721–4.

Clerici, M., Wynn, T.A., Berzofsky, J.A., Blatt, S.P., Hendrix, C.W., Sher, A., Coffman, R.L. and Shearer, G.M. (1994) Role of interleukin-10 in T helper cell dysfunction in asymptomatic individuals infected with the human immunodeficiency virus. *J. Clin. Invest.*, **93**, 768–75.

Cooper, K.D., Oberhelman, L., Hamilton, T.A., Baadsgaard, O., Terhune, M., Levee, G., Anderson, T. and Koren, H. (1992) UV exposure reduces immunization rates and promotes tolerance to epicutaneous antigens in humans: relationship to dose, CD1a-DR$^+$ epidermal macrophage induction, and Langerhans cell depletion. *Proc. Natl Acad. Sci. USA*, **89**, 8497–501.

Cruz, P.D., Tigelaar, R.E. and Bergstresser, P.R. (1990) Langerhans cells that migrate to skin after intravenous infusion regulate the induction of contact hypersensitivity. *J. Immunol.*, **144**, 2486–92.

Czarnecki, D., Zalcberg, J., Kulinskaya, E. and Kay, T. (1995) Impaired cell-mediated immunity of apparently normal patients who had multiple skin cancers. *Cancer*, **76**, 228–31.

Dai, R. and Streilein, J.W. (1996) Sensitizing capacity of Langerhans cells obtained from ultraviolet-B-exposed murine skin. *Immunology*, **86**, 661–7.

Davidson, J., Milton, A.S. and Rotondo, D. (1992) Alpha-melanocyte-stimulating hormone suppresses fever and increases plasma levels of prostaglandin E2 in the rabbit. *J. Physiol. (Lond.)*, **451**, 491–502.

Daynes, R.A. and Spellman, C.W. (1977) Evidence for the generation of suppressor cells by UV radiation. *Cell. Immunol.*, **31**, 182–7.

de Kossodo, S., Cruz, P.D., Dougherty, I., Thompson, P., Silva-Valdez, M. and Beutler, B. (1995) Expression of the tumor necrosis factor gene by dermal fibroblasts in response to ultraviolet irradiation or lipopolysaccharide. *J. Invest. Dermatol.*, **104**, 318–22.

Devary, Y., Rosette, C., DiDonato, J.A. and Karin, M. (1993) NF-κB activation by ultraviolet light is not dependent on a nuclear signal. *Science*, **261**, 1442–5.

Enk, A.H., Angeloni, V., Udey, M.C. and Katz, S.I. (1993) Inhibition of Langerhans cell antigen-presenting function by IL-10: a role for IL-10 in induction of tolerance. *J. Immunol.*, **151**, 2390–8.

Fisher, M.S. and Kripke, M.L. (1982) Suppressor T lymphocytes control the development of primary skin cancers in UV-irradiated mice. *Science*, **216**, 1133–4.

Gibbs, N.K., Norval, M., Traynor, N.A., Wolf, M., Johnson, B.E. and Crosby, J. (1993) Action spectra for the *trans* to *cis* photoisomerisation of urocanic acid *in vitro* and in mouse skin. *Photochem. Photobiol.*, **57**, 584–90.

Grabbe, S., Bhardwaj, R.S., Mahnke, K., Simon, M.M., Schwarz, T. and Luger, T.A. (1996) α-Melanocyte-stimulating hormone induces hapten-specific tolerance in mice. *J. Immunol.*, **156**, 473–8.

Greene, M.I., Sy, M.S., Kripke, M.L. and Benacerraf, B. (1979) Impairment of antigen-presenting cell function by UV radiation. *Proc. Natl Acad. Sci. USA*, **76**, 6591–5.

Grewe, M., Gyufko, K. and Krutmann, J. (1995) Interleukin-10 production by cultured human keratinocytes: regulation by UVB and UVA1 radiation. *J. Invest. Dermatol.*, **104**, 3–6.

Hammerberg, C., Duraiswamy, N. and Cooper, K.D. (1994) Active induction of unresponsiveness (tolerance) to DNFB by *in vivo* ultraviolet-exposed epidermal cells is dependent upon infiltrating class II MHC^+ $CD11b^{bright}$ monocytic/macrophagic cells. *J. Immunol.*, **153**, 4914–24.

Hart, P.H., Jones, C.A., Jones, K.L., Watson, C.J., Santucci, I., Spencer, L.K. and Finlay-Jones, J.J. (1993) *cis*-Urocanic acid stimulates human peripheral blood monocyte prostaglandin E-2 production and suppresses indirectly tumour necrosis factor-alpha levels. *J. Immunol.*, **150**, 4514–23.

Hersey, P., Haran, G., Hansic, E. and Edwards, A. (1983) Alteration of T cell subsets and induction of suppressor T cell activity in normal subjects after exposure to sunlight. *J. Immunol.*, **131**, 171–4.

Hilkens, C.M.U., Vermeulen, H., van Neerven, R.J.J., Snijdewint, F.G.M., Wierenga, E.A. and Kapsenberg, M.L. (1995) Differential modulation of T helper type 1 and T helper type 2 cytokine secretion by prostaglandin E_2 critically depends on interleukin-2. *Eur. J. Immunol.*, **25**, 59–63.

Hsieh, C.-S., Heimberger, A.B., Gold, J.S., O'Garra, A. and Murphy, K.M. (1992) Differential regulation of T helper phenotype development by interleukins 4 and 10 in an $\alpha\beta$ T-cell-receptor transgenic system. *Proc. Natl Acad. Sci. USA*, **89**, 6065–9.

Hsieh, C.-S., Macatonia, S.E., Tripp, C.S., Wolf, S.F., O'Garra, A. and Murphy, K.M. (1993) Development of Th1 CD4+ T cells through IL-12 produced by Listeria-induced macrophages. *Science*, **260**, 547–9.

Hurks, H.M.H., Out-Luiting, C., Vermeer, B.J., Claas, F.H.J. and Mommaas, A.M. (1995) The action spectra for UV-induced suppression of MLR and MECLR show that immunosuppression is mediated by DNA damage. *Photochem. Photobiol.* **62**, 449–53.

Jeevan, A. and Kripke, M.L. (1993) Impact of ozone depletion on immune function. *World Resource Review*, **5**, 141–55.

Kanagawa, O., Vaupel, B.A., Gayama, S., Koehler, G. and Kopf, M. (1993) Resistance of mice deficient in IL-4 to retrovirus-induced immunodeficiency syndrome (MAIDS). *Science*, **262**, 240–2.

Kang, K., Hammerberg, C., Meunier, L. and Cooper, K.D. (1994) CD11b+ macrophages that infiltrate human epidermis after *in vivo* ultraviolet exposure potently produce IL-10 and represent the major secretory source of epidermal IL-10. *J. Immunol.*, **153**, 5256–64.

Kripke, M.L. (1974) Antigenicity of murine skin tumors induced by UV light. *J. Natl Cancer Inst.*, **53**, 1333–6.

Kripke, M.L., Munn, C.G., Jeevan, A., Tang, J.-M. and Bucana, C. (1990) Evidence that cutaneous antigen-presenting cells migrate to regional lymph nodes during contact sensitization. *J. Immunol.*, **145**, 2833–8.

Kripke, M.L., Cox, P.A., Alas, L.G. and Yarosh, D.B. (1992) Pyrimidine dimers in DNA initiate systemic immunosuppression in UV-irradiated mice. *Proc. Natl Acad. Sci. USA*, **89**, 7516–20.

Krutmann, J. (1994). Regulatory interactions between epidermal cell adhesion molecules and cytokines, in *Epidermal Growth Factors and Cytokines* (eds T. Luger, and T. Schwarz), Marcel Dekker, New York, pp. 415–31.

Kurimoto, I. and Streilein, J.W. (1992) *cis*-Urocanic acid suppression of contact hypersensitivity induction is mediated via tumor necrosis factor-α. *J. Immunol.*, **148**, 3072–8.

Kurimoto, I. and Streilein, J.W. (1994) Characterization of the immunologic basis of ultraviolet-B light effects on contact hypersensitivity induction. *Immunology*, **81**, 353–8.

Kurimoto, I., Arana, M. and Streilein, J.W. (1994) Role of dermal cells from normal and ultraviolet B-damaged skin in induction of contact hypersensitivity and tolerance. *J. Immunol.*, **152**, 3317–23.

Luster, M.I., Portier, C., Pait, D.G., White, K.I., Gennings, C., Munson, A.E. and Rosenthal, G.J. (1992) Risk assessment in immunotoxicity I. Sensitivity and predictability of immune tests. *Fundam. Appl. Toxicol.*, **18**, 200–10.

Maggi, E., Giudizi, M.G., Biagiotti, R., Annunziato, F., Manetti, R., Piccinni, M.-P., Parronchi, P., Sampognaro, S., Giannarini, L., Zuccati, G. and Romagnani, S. (1994) Th2 like CD8+ T cells showing B cell helper function and reduced cytolytic activity in human immunodeficiency virus type 1 infection. *J. Exp. Med.*, **180**, 489–95.

Magram, J., Connaughton, S.E., Warrier, R.R., Carvajal, D.M., Wu, C.-Y., Ferrante, J., Stewart, C., Sarmiento, U., Faharty, D.A. and Gately, M.K. (1996) IL-12-deficient mice are defective in IFN-γ production and type 1 cytokine responses. *Immunity* **4**, 471–81.

McKnight, A.J., Zimmer, G.J., Fogelman, I., Wolf, S.F. and Abbas, A.K. (1994) Effects of IL-12 on helper cell-dependent immune responses *in vivo*. *J. Immunol.*, **152**, 2172–9.

Miura, S., Kulka, M., Smith, C.C., Imafuku, S., Burnett, J.W. and Aurelian, L. (1994) Cutaneous UV radiation inhibits herpes simplex virus-induced lymphoproliferation in latently infected subjects. *Clin. Immunol. Immunopathol.*, **72**, 62–9.

Moodycliffe, A.M., Norval, M., Kimber, I. and Simpson, T.J. (1993) Characterization of a monoclonal antibody to *cis*-urocanic acid: detection of *cis*-urocanic acid in the serum of irradiated mice by immunoassay. *Immunology*, **79**, 667–72.

Moodycliffe, A.M., Bucana, C.D., Kripke, M.L., Norval, M. and Ullrich, S.E. (1996) Differential effects of a monoclonal antibody to *cis*-urocanic acid on the suppression of delayed and contact hypersensitivity following ultraviolet irradiation. *J. Immunol.*, **157**, 2891–9.

Mosmann, T.R. (1991) Regulation of immune responses by T cells with different cytokine secretion phenotypes: role of a new cytokine, cytokine synthesis inhibitory factor (IL10). *Int. Arch. Allergy Appl. Immunol.*, **94**, 110–15.

Mosmann, T.R., Schumacher, J.H., Street, N.F., Buss, R., O'Garra, A., Fong, T.A.T., Bond, M.W., Moore, K.W.M., Sher, A. and Fiorentino, D.F. (1991) Diversity of cytokine synthesis and function of mouse $CD4^+$ T cells. *Immunol. Rev.*, **123**, 209–29.

Nishigori, C., Yarosh, D.B., Ullrich, S.E., Vink, A.A., Bucana, C.D., Roza, L. and Kripke, M.L. (1996) Evidence that DNA damage triggers interleukin 10 cytokine production in UV-irradiated murine keratinocytes. *Proc. Natl Acad. Sci. USA*, **93**, 10354–9.

Noonan, F.P. and De Fabo, E.C. (1992) Immunosuppression by ultraviolet B radiation: initiation by urocanic acid. *Immunol. Today*, **13**, 250–7.

Noonan, F.P., Kripke, M.L., Pedersen, G.M. and Greene, M.I. (1981) Suppression of contact hypersensitivity by UV radiation is associated with defective antigen presentation. *Immunology*, **43**, 527–33.

Norval, M., Gibbs, N.K. and Gilmour, J. (1995) The role of urocanic acid in UV-induced immunosuppression: Recent advances (1992–1994). *Photochem. Photobiol.*, **62**, 209–17.

O'Conner, A., Nishigori, C., Yarosh, D., Alas, L., Kibitel, J., Burley, L., Cox, P., Bucana, C., Ullrich, S. and Kripke, M. (1996) DNA double strand breaks in epidermal cells cause immune suppression *in vivo* and cytokine production *in vitro*. *J. Immunol.*, **157**, 271–8.

Okamoto, H. and Kripke, M.L. (1987) Effector and suppressor circuits of the immune response are affected *in vivo* by different mechanisms. *Proc. Natl Acad. Sci. USA*, **84**, 3841–5.

Parish, C.R. (1972) The relationship between humoral and cell-mediated immunity. *Transplant. Rev.*, **13**, 35–66.

Rivas, J.M. and Ullrich, S.E. (1992) Systemic suppression of DTH by supernatants from UV-irradiated keratinocytes: an essential role for interleukin 10. *J. Immunol.*, **149**, 3865–71.

Rivas, J.M. and Ullrich, S.E. (1994) The role of IL-4, IL-10, and TNF-α in the immune suppression induced by ultraviolet radiation. *J. Leukocyte. Biol.*, **56**, 769–75.

Robertson, B., Gahring, L., Newton, R. and Daynes, R.A. (1987) *In vivo* administration of IL-1 to normal mice decreases their capacity to elicit contact hyper-sensitivity responses: prostaglandins are involved in this modification of the immune response. *J. Invest. Dermatol.*, **88**, 380–7.

Romagnani, S., Del Prente, G., Parronchi, P., De Carli, M., Macchia, D., Sampognaro, S., Piccinni, M.-P., Giudizi, M.-G., Biagiotti, R. and Almerigogna, F. (1993) Human Th1 and Th2 subsets. *Int. Arch. Allerg. Immunol.*, **99**, 242–5.

Schauer, E., Trautinger, F., Kock, A., Schwarz, A., Bhardwaj, R., Simon, M., Ansel, J.C., Schwarz, T. and Luger, T.A. (1994) Proopiomelanocortin-derived peptides are synthesized and released by human keratinocytes. *J. Clin. Invest.*, **93**, 2258–62.

Schmitt, D.A., Owen-Schaub, L. and Ullrich, S.E. (1995) Effect of IL-12 on immune suppression and suppressor cell induction by ultraviolet radiation. *J. Immunol.*, **154**, 5114–420.

Schols, D. and De Clercq, E. (1996) Human immunodeficiency virus type 1 gp120 induces anergy in human peripheral blood lymphocytes by inducing interleukin-10 production. *J. Virol.*, **70**, 4953–60.

Schuhmachers, G., Ariizumi, K., Kitajima, T., Edelbaum, D., Xu, S., Shadduck, R.K., Gilmour, G.L., Bergstresser, P.R. and Takashima, A. (1996) UVB radiation interrupts cytokine-mediated support of an epidermal-derived dendritic cells line (XS52) by a dual mechanism. *J. Invest. Dermatol.*, **106**, 1023–9.

Schwarz, A., Grabbe, S., Aragane, Y., Sandkuhl, K., Riemann, H., Luger, T.A., Kubin, M., Trinchieri, G. and Schwarz, T. (1996) Interleukin-12 prevents ultraviolet B-induced local immunosuppression and overcomes UVB-induced tolerance. *J. Invest. Dermatol.*, **106**, 1187–91.

Setlow, R.B., Grist, E., Thompson, K. and Woodhead, A.D. (1993) Wavelengths effective in induction of malignant melanoma. *Proc. Natl Acad. Sci. USA*, **90**, 6666–70.

Simon, J.C., Cruz, P.C., Bergstresser, P.R. and Tigelaar, R.E. (1990) Low dose ultraviolet B-irradiated Langerhans cells preferentially activate $CD4^+$ cells of the T helper 2 subset. *J. Immunol.*, **145**, 2087–91.

Simon, J.C., Tigelaar, R.E., Bergstresser, P.R., Edelbaum, D. and Cruz, P.D. (1991) Ultraviolet B radiation converts Langerhans cells from immunogenic to tolerogenic antigen-presenting cells: induction of specific clonal anergy in $CD4^+$ T helper 1 cells. *J. Immunol.*, **146**, 485–91.

Simon, M.M., Aragane, Y., Schwarz, A., Luger, T.A. and Schwarz, T. (1994) UVB light induces a nuclear factor κB (NFκB) activity independently from chromosomal DNA damage in cell-free cytosolic extracts. *J. Invest. Dermatol.*, **102**, 422–7.

Sorel, M., Cherel, M., Dremo, B., Bouyce, I., Guilbert, J., Dubois, S., Lebeau, B., Raher, S., Minvielle, S. and Jacques, Y. (1996) Production of interleukin-15 by human keratinocytes. *Eur. J. Dermatol.*, **6**, 209–12.

Stevens, S.R., Shibaki, A., Meunier, L. and Cooper, K.D. (1995) Suppressor T cell-activating macrophages in ultraviolet irradiated human skin induce a novel, TGF-β-dependent form of T cell activation characterized by deficient IL-2rα expression. *J. Immunol.*, **155**, 5601–7.

Strassmann, G., Patil-Koota, V., Finkelman, F., Fong, M. and Kambayashi, T. (1994) Evidence for the involvement of IL-10 in the differential deactivation of murine peritoneal macrophages by prostaglandin E_2. *J. Exp. Med.*, **180**, 2365–70.

Streilein, J.W. and Bergstresser, P.R. (1988) Genetic bases of UVB effects on contact hypersensitivity. *Immunogenetics*, **27**, 252–8.

Takashima, A., Matsue, H., Bergstresser, P.R. and Ariizumi, K. (1995) Interleukin-7-dependent interaction of dendritic epidermal T cells with keratinocytes. *J. Invest. Dermatol.*, **105**(1 Suppl), 50S–53S.

Tang, A. and Udey, M.C. (1991) Inhibition of epidermal Langerhans cell function by low dose ultraviolet radiation: UVB radiation selectively modulates ICAM-1 expression by murine Langerhans cells. *J. Immunol.*, **146**, 3347–56.

Tineke. C.T.M., van der Pouw, K., Boeije, L.C.M., Smeenk, R.J.T., Wijdenes, J. and Aarden, L.A. (1995) Prostaglandin-E2 is a potent inhibitor of human interleukin-12 production. *J. Exp. Med.*, **181**, 775–9.

Toews, G.B., Bergstresser, P.R. and Streilein, J.W. (1980) Epidermal Langerhans cell density determines whether contact hypersensitivity or unresponsiveness follows skin painting with DNFB. *J. Immunol.*, **124**, 445–9.

Ullrich, S.E. (1994) Mechanism involved in the systemic suppression of antigen-presenting cell function by UV irradiation: keratinocyte-derived IL-10 modulates antigen-presenting cell function of splenic adherent cells. *J. Immunol.*, **152**, 3410–16.

Ullrich, S.E. and Kripke, M.L. (1984) Mechanisms in the suppression of tumor rejection produced in mice by repeated UV irradiation. *J. Immunol.*, **133**, 2786–90.

Vermeer, M. and Streilein, J.W. (1990) Ultraviolet-B-light induced alterations in epidermal Langerhans cells are mediated by tumor necrosis-alpha. *Photodermatol. Photoimmunol. Photomed.*, **7**, 258–65.

Vink, A.A., Strickland, F.M., Bucana, C., Cox, P.A., Roza, L., Yarosh, D.B. and Kripke, M.L. (1996) Localization of DNA damage and its role in altered antigen-presenting cell function in ultraviolet-irradiated mice. *J. Exp. Med.*, **183**, 1491–1500.

Weiss, J.C., Renkl, A.C., Denfeld, R.W., de Roche, R., Spitzlei, M., Schöpf, E. and Simon, J.C. (1995) Low-dose UVB radiation perturbs the functional expression of B7.1 and B7.2 co-stimulatory molecules on human Langerhans cells. *Eur. J. Immunol.*, **25**, 2858–62.

Yagi, H., Tokura, Y., Wakita, H., Furukawa, F. and Takigawa, M. (1996) TCRV$\beta 7^+$ Th2 cells mediate UVB-induced suppression of murine contact photosensitivity by releasing IL-10. *J. Immunol.*, **156**, 1824–31.

Yarosh, D., Bucana, C., Cox, P., Kibitel, J. and Kripke, M.L. (1994) Localization of liposomes containing a DNA repair enzyme in murine skin. *J. Invest. Dermatol.*, **103**, 461–8.

Yoshikawa, T. and Streilein, J.W. (1990) Tumor necrosis factor-alpha and ultraviolet light have similar effects on contact hypersensitivity in mice. *Regional Immunol.*, **3**, 139–44.

Yoshikawa, T., Rae, V., Bruins-Slot, W., van-den-Berg, J.W., Taylor, J.R. and Streilein, J.W. (1990) Susceptibility to effects of UVB radiation on induction of contact hypersensitivity as a risk factor for skin cancer in humans. *J. Invest. Dermatol.*, **95**, 530–6.

8 T Helper Cell Clones and *In Vitro* Assessment of Immunotoxicity

B. Paige Lawrence and Nancy I. Kerkvliet

Introduction

$CD4^+$ T lymphocytes play a pivotal role in the development and regulation of cell-mediated and humoral immune responses. As summarized in Chapter 2, distinct subsets of $CD4^+$ T cells are defined by a particular profile of secreted cytokines. Briefly, T helper 1 (Th1) cells produce interleukin(IL)-2 and IL-3, interferon(IFN)-γ and tumour necrosis factor (TNF), and orchestrate cellular immune responses against viral pathogens, tumour cells and transplanted tissue. T helper 2 (Th2) cells secrete IL-3, IL-4, IL-5, IL-6 and IL-13, and drive antibody-mediated responses against bacterial pathogens, mucosal immunity and the induction of mast cell and eosinophil responses (Romagnani, 1996).

Current immunological theory is that undifferentiated $CD4^+$ T (Th0) cells secrete a less polarized cytokine profile and can be driven to differentiate into either Th1 or Th2 effector cells. However, the specific factors that determine the direction of a Th0 cell response are still unclear. Evidence from several laboratories indicates that the dose and route of antigen administration, the type of antigen presenting cell (APC) involved, the presence of particular hormones, costimulatory molecule expression, cytokines and peptide affinity for major histocompatibility complex (MHC) determinants collectively influence the direction of a response (Constant and Bottomly, 1997; O'Garra and Murphy, 1996; Romagnani, 1996). Accumulating evidence from the study of pathogenic diseases and immune disorders indicates that this Th1/Th2 dichotomy is an important immunoregulatory mechanism. The cytokines produced by each Th subset downregulate cytokine production and effector function of the other subset, and in this manner regulate an immune response.

Certain natural and synthetic chemicals are known or suspected to affect the immune system. Broadly speaking, xenobiotics can cause two different types of immune dysfunction: immune suppression, possibly leading to an increased

T Lymphocyte Subpopulations in Immunotoxicology. Edited by Ian Kimber and MaryJane K. Selgrade.
 ISBN 0 471 97194 4

susceptibility to infectious diseases and cancer; or immune activation, leading to either autoimmune disease or hypersensitivity disorders. In most cases, the mechanisms by which chemicals cause immune dysfunction are unknown. Exposure to an enormous variety of natural and synthetic chemicals has been linked to the development of hypersensitivity or allergy, and one possible mechanism is via aberrations in T cell-dependent functions and divergent cytokine production (see Chapters 11–14). Likewise, immunosuppression following *in vivo* exposure of mice or *in vitro* exposure of murine and human lymphocytes to a variety of chemicals has been associated with altered T cell functions, including T cell-dependent B cell responses, cytotoxic T lymphocyte responses, delayed-type hypersensitivity, cytokine production, altered T cell proliferation and phenotypic alterations in T cell subpopulations.

For example, exposure to polyaromatic hydrocarbons (PAHs) such as benzo[a]pyrene (B[a]P) and dimethylbenzanthracene (DMBA) has been shown to alter signalling through the T cell receptor (Archuleta *et al.*, 1993), and to suppress mitogen-stimulated T cell proliferation (Burchiel *et al.*, 1990; Lee *et al.*, 1996), IL-2 production (Myers *et al.*, 1988), and the antibody response to sheep red blood cells (SRBC, Ladics *et al.*, 1991). Other environmental contaminants, including halogenated aromatic hydrocarbons such as 2,3,7,8-tetrachlorodibenzo-*p*-dioxin (TCDD) and polychlorinated biphenyls (PCB) (Kerkvliet *et al.*, 1996), mycotoxins (Marin *et al.*, 1996; Pahl *et al.*, 1996), octylphenols (Nair-Menon *et al.*, 1996) and benzene metabolites (Li *et al.*, 1996) have also been shown to alter T cell function.

One hypothesis for the mechanism of these alterations in T cell-dependent immune functions is that exposure to chemicals leads to the preferential activation of one subset of Th cells, which then drives the dysfunction of the immune system, leading to either enhanced immune reactions or immunosuppression. Reports of altered T cell function associated with chemically induced immune modulation support this theory. For example, the immunosuppressive effects of cocaine on T cell-dependent functions have been described in both humans and experimental animal models. Examination of T cell cytokine production in the presence of physiologic concentrations of cocaine indicates that cocaine alters cytokine production from both mitogen- and antigen-activated T cells (Klein *et al.*, 1993; Falchetti *et al.*, 1995), suggesting that part of cocaine's immunomodulatory effects may be mediated by alteration in Th cell cytokine production.

In addition to the suppression of immune responses, exposure to chemicals can also lead to enhanced immune function, in the form of either hypersensitivity or autoimmune disease. Exposure to an enormous variety of natural and synthetic chemicals has been linked to the development of hypersensitivity or allergy. For example, Dearman *et al.* (1995, 1996) and Vento *et al.* (1996) have demonstrated that chemical allergens which provoke a Th1-type cytokine response go on to elicit a cell-mediated type of allergic reaction, whereas chemicals which induce the production of Th2 cytokines are associated with

IgE-mediated immediate hypersensitivity. Likewise, exposure of certain rodent strains to the environmental contaminant mercury ($HgCl_2$) provides a model for chemically induced autoimmune disease and experimental evidence suggests that exposure to $HgCl_2$ preferentially induces Th2-type responses (Pelletier, 1994).

Collectively, these observations provide the framework for the hypothesis that exposure to natural and synthetic chemicals preferentially influences one population of Th cells, thereby perturbing the Th1/Th2 balance, leading to disease. In this manner, the Th1/Th2 model provides a useful paradigm for examining the mechanisms by which chemicals disrupt Th cell function. Understanding how xenobiotics disrupt this balance may ultimately lead to a greater understanding of how the immune system maintains a healthy equilibrium between two apparently opposing subsets of T cells. This chapter describes the utility of T clones in identifying and elucidating mechanisms of immunotoxicity associated with differential effects on Th1 and Th2 cells.

Antigen-specific Th clones as an *in vitro* model

In an attempt to elucidate the cellular and biochemical mechanisms of action of a wide variety of compounds which impact T cell function *in vivo*, many investigators have used *ex vivo* lymphocyte populations and cultured T cell tumour lines. To date, most *in vitro* studies of the effects of chemicals have used polyclonal activators or phorbol esters and calcium ionophores to stimulate T cells in culture (Chiappelli *et al.*, 1995; Marin *et al.*, 1996). While mitogens and phorbol esters are potent polyclonal activators, they do not necessarily mimic antigen-specific T cell activation or antigen-specific immune responses. Nor do they model T cell-APC interactions, including the need for cognate interactions between T cells and APC.

Antigen-specific $CD4^+$ Th cell lines, or clones, offer an alternative to tumour cell lines and polyclonal activators. Unlike tumour cell lines, Th cell clones require regular restimulation with antigen presented in the context of MHC. To this end, Th cell clones are typically restimulated every 10–14 days with antigen and irradiated syngeneic APC. Unlike mitogen activated lymphocytes or tumour cell lines, Th cells more accurately model *in vivo* antigen-specific responses. A complete review of the literature on Th clones is not within the scope of this chapter; however, a superficial overview of Th cell regulation by exogenous soluble molecules is worthwhile to demonstrate the value of Th clones as an *in vitro* model. As mentioned previously, Th1 and Th2 cells appear to be differentially sensitive to regulation by soluble immunoregulatory molecules such as cytokines. The theory that Th1 and Th2 cells cross-regulate each other's function via the cytokines they secrete was first postulated following work with Th clones (Fernandez-Botran *et al.*, 1988; Fiorentino *et al.*, 1989; Gajewski and Fitch, 1988). Subsequent experimentation provided evidence for this mechanism *in vivo*. Th clones have also been used to demonstrate the differential sensitivity of Th1 and Th2 cells to the effects of prostaglandin (PG)E_2, a potent immunomodulatory arachidonic acid metabolite. PGE_2 selectively inhibits Th1, but not Th2, cytokine production (Betz and Fox, 1991). Using antigen-specific murine Th clones, the differential mechanism of PGE_2 effects on Th subsets has been elucidated (Li and Fox, 1993).

Antigen-specific Th clones have also been widely used to elucidate the mechanism of action of immunomodulatory pharmaceuticals. For example, psoriasis vulgaris is an autoimmune disease associated with Th1 cytokine production. One effective therapy is the use of dimethylfumarate (DMF); however, until recently the mechanism of action of this drug remained unclear. Monomethylfumarate (MMF) is a biometabolite of DMF, and was proposed to be one of the active agents. Using antigen-specific $CD4^+$ murine T cell clones and isolated human lymphocytes, De Jong *et al.* (1996) demonstrated that *in vitro* exposure to MMF leads to direct and selective up-regulation of IL-4 and IL-5 production. Interestingly, no decrease in IFN-γ production was observed, suggesting that in the case of MMF the mechanism may be the preferential activation of Th2 cells, rather than a down-modulation of Th1 cells.

Application of Th clones to immunotoxicity studies

MECHANISTIC STUDIES

While antigen-specific Th clones have been widely used to examine the normal function of Th cells and the effects of pharmaceuticals on T cell function, they have not been widely used in immunotoxicological studies. Based on their ability to address antigen-specific T cell activation, T cell-APC interactions and the preferential action of xenobiotics on Th1 or Th2 cells, Th clones provide a potentially powerful tool for investigating the cellular and molecular mechanisms of action of chemically induced alterations in T cell function. Recent investigations using two different xenobiotics, lead and TCDD, illustrate the use of antigen-specific Th clones for the examination of the effects of xenobiotics on Th1 and Th2 cytokine production. Exposure of experimental animals to lead (Pb) has been shown to be immunomodulatory, and epidemiological evidence suggests that Pb may also cause deleterious effects on the human immune system (McCabe, 1994; Chapter 6). With specific reference to T cell function, exposure to Pb has been associated with both suppressed and enhanced T cell-dependent immune responses (McCabe, 1994). In the case of enhanced responses, one downstream effect is increased numbers of antigen-specific immunoglobulin-secreting B cells. When exposed to $PbCl_2$, BALB/c mice showed increased plasma IL-4 and IgE, and decreased IFN-γ production; again suggesting that exposure to Pb may alter the Th1/Th2 cytokine balance. This hypothesis is supported by a recent study with antigen-specific murine Th1 and Th2 clones derived from BALB/c mice (Heo *et al.*, 1996). In these experiments, rabbit gamma globulin (RGG)-specific Th1 and Th2 clones were activated with antigen and irradiated syngeneic splenocytes as APC in the presence of increasing concentrations of $PbCl_2$. Exposure to Pb had a differential effect on Th1 and Th2 cells. $PbCl_2$ decreased proliferation of Th1 clones while enhancing the proliferation of Th2 clones. As an indicator of Th1 and Th2 cytokine production, IFN-γ and IL-4 levels in culture supernatants were measured 48 h following activation. Exposure to $PbCl_2$ decreased IFN-γ and increased IL-4 production by RGG-specific Th1 and Th2 clones, respectively. A parallel, *in vivo*, examination of Th subsets in Pb-treated BALB/c mice demonstrated a similar trend: increased plasma IL-4 levels and decreased IFN-γ, compared with saline-treated control animals. Thus, both *in vivo* and *in vitro* exposure to Pb preferentially increased Th2 cytokine production and the proliferation of Th2 cells, suggesting that one mechanism of Pb-induced immune disease is via perturbations of the balance between Th cell subsets.

T clones have also been used to study the mechanism of immunotoxicity associated with exposure to TCDD. Many studies have demonstrated that T cell function is particularly sensitive to the *in vivo* immunosuppressive action of TCDD (Kerkvliet and Burleson, 1994; Chapter 4). For example, TCDD and structurally related PCBs suppress the induction of the cell-mediated response to P815 tumour cells (Clark *et al.*, 1981; Kerkvliet and Steppan, 1988; De Krey and Kerkvliet, 1995). Underlying the suppressed CTL response is the diminished production of IL-2, IFN-γ and TNF-α (Kerkvliet *et al.*, 1996). In other studies, TCDD enhanced T cell proliferation and altered serum IFN-γ, IL-6 and granulocyte/macrophage-colony stimulating factor (GM-CSF) levels in response to the polyclonal T cell activator anti-CD3 (Neumann *et al.*, 1993; Prell *et al.*, 1995). In addition, exposure to TCDD suppressed antigen-specific IL-2 production and T cell proliferation in response to ovalbumin (Lundberg *et al.*, 1992). However, despite this descriptive work of TCDD-induced alterations in T cell proliferation and cytokine production, little is known about the underlying mechanisms that mediate these *in vivo* effects of TCDD.

To determine the mechanism of TCDD-induced changes in cytokine production, the direct effect of TCDD on $CD4^+$ T cells was examined using antigen-specific murine Th clones. Four different murine Th clone cell lines were activated, with their specific antigen and irradiated spleen cells in the presence or absence of TCDD, and cytokine release over a 72-h time course was measured. In contrast to experiments in the whole animal, treatment with TCDD led to no consistent suppression or enhancement of the production of either Th1 cytokines (IFN-γ, IL-2 and TNF) or Th2 cytokines (IL-4 and IL-6) (Lawrence *et al.*, 1996). It has been demonstrated that serum can influence the *in vitro* responsiveness of cells to TCDD (Harper *et al.*, 1995; Morris *et al.*, 1991); therefore, experiments were also conducted under serum-free conditions or in the presence of mouse serum instead of fetal bovine serum. However, no consistent effect of TCDD on cytokine production was observed when T cell clones were cultured with 2.5% mouse serum, and serum-free conditions failed to support normal cytokine production. As a positive control, treatment of these Th clones with the immunosuppressive drug cyclosporin A (0.1 μg/ml) consistently and completely blocked all cytokine production.

Several other investigators have reported discrepancies between the *in vivo* and *in vitro* effects of TCDD on T cell function; namely, that it is difficult to reproduce the *in vivo* effects of TCDD on T cells *in vitro*. In light of this, use of these Th clones was redirected toward an in-depth investigation of the capability of TCDD to directly act on Th cells. The immunotoxic effects of TCDD are mediated by a ligand-activated transcription factor called the aryl hydrocarbon (Ah) receptor (Harper *et al.*, 1993; Kerkvliet *et al.*, 1990a,b; Silkworth *et al.*, 1984). Therefore, an extensive analysis of the T cell Ah receptor was conducted using Th clones and purified murine splenic T cells. This investigation led to a description of the intracellular distribution of the AhR in resting and activated T lymphocytes, and to the discovery that, although AhR is present in the nucleus of TCDD-treated T cells, it does not behave in the same manner as observed in hepatocytes or other cells (i.e. the T cell Ah receptor does not appear to bind to its consensus regulatory element of DNA, suggesting that TCDD may not be able to act directly on T cells) (Lawrence *et al.*, 1996). This result explains the inconsistent effects of TCDD on T cells in culture. Clearly, T clones can only be used to demonstrate preferential effects on Th1 and Th2 responses when the compound acts directly on T cells.

The results of these studies demonstrate the usefulness of Th clones as an *in vitro* model for the examination of the effects of chemicals on antigen-dependent T cell functions. In the case of Pb, T clones provide an excellent model to examine the biochemical mechanism which underlies the direct action of Pb on cytokine production, and the results of the study by Heo *et al.* (1996) support the concept that some chemicals have differential effects on Th subsets. In the case of TCDD, T clones provide insight into the indirect nature of the effects of TCDD on T cell function.

SCREENING STUDIES

In addition to their usefulness in mechanistic investigations, another potential use of antigen-specific Th clones is as a screen, to probe the potential effects of other xenobiotics on Th1 and Th2 cell function. Because Th clones can be activated with antigen and APC or via crosslinking of the T cell receptor with an antibody, they provide a context in which the ability of a particular xenobiotic to act directly on T cells, or indirectly via alterations in APC function, can be addressed. With this in mind, a preliminary study of the effects two mechanistically distinct chemicals, 17β-oestradiol (17β-Es) and 1,2-dichloroethane (DCE), on cytokine production by antigen-specific Th1 and Th2 clones has been conducted. 1,2-Dihaloethanes are widely used industrial solvents and insecticidal fumigants, and the toxicity of DCE and related dihaloethanes is relevant to human health concerns, as they have been shown to be mutagenic, carcinogenic, hepatotoxic and nephrotoxic (US Department of Health and Human Services, 1992, 1994). A proposed mechanism of dihaloethane toxicity is through the formation of a conjugate with glutathione, leading to the production of a reactive species which forms conjugates with cellular macromolecules (Jean and Reed, 1992). Little is known about the immunotoxic effects of dihaloethanes; however, it has been reported that exposure to DCE at 1/100th of the LD_{50} (for 14 days by gavage) suppressed both humoral and cell-mediated immune responses in CD-1 mice (Munson *et al.*, 1982).

17β-Es is a hormone as well as the prototype for a class of oestrogen-like chemicals found in the environment. Most, if not all, of the effects of oestrogen are mediated by the nuclear oestrogen receptor, which acts as a ligand-dependent transcription factor to modulate gene expression (Gorski *et al.*, 1993). Like other sex hormones, 17β-Es has been shown to modulate the immune system (Grossman, 1985). Although exposure to exogenous 17β-Es has been shown to cause thymic atrophy and to depress cell-mediated immunity and natural killer cell activity, its effects on cytokine production have not been fully elucidated.

Using both T_H1 and T_H2 cell lines, we examined the dose-dependent effects of 17β-Es and DCE on cytokine production. As with TCDD, Th clones were activated with antigen and irradiated syngeneic splenocytes in the presence of either DCE or 17β-Es and cytokine production was measured over a 72 h period. DCE caused a dose-dependent increase in the amount of IFN-γ detected in culture supernatants from Th1 cells (F1.A.2 cells, Figure 8.1A). Treatment with DCE led to a 50% increase in IL-4 levels produced by Th2 cells (10.5.17 cells, Figure 8.2A). Interestingly, DCE had a similar effect on cytokine production following the activation of Th clones with anti-CD3 (i.e. in the absence of accessory cells), suggesting that DCE may be able to affect T cells directly (Figures 8.1B and 8.2B). Because many types of cells (including spleen cells) can make IL-6, the effects of DCE on Th2 cell IL-6 production were examined following activation with anti-CD3. Since there are no APCs in this experiment, the IL-6 produced was coming directly from the Th2 clones. Similar to IL-4 production, DCE dose-dependently increased IL-6 levels in culture supernatants (Figure 8.2C).

In similar time course experiments, the effect of 17β-Es on Th1 and Th2 cytokine production was examined. When Th clones were activated with antigen and irradiated splenocytes in the presence of 1–100 nM 17β-Es, IFN-γ production by the Th1 cell line F1.A.2 was reduced in a dose-dependent manner to about 50% of the control (Figure 8.3A). In a Th2 cell line, 17β-Es decreased IL-4 production by approximately 60% (Figure 3B), but had no effect on IL-6 production (data not shown). Similar effects of 17β-Es on IL-4 and IL-6 levels were observed using another KLH-specific Th2 cell line, F4 (data not shown). In contrast to DCE, comparable changes in cytokine levels were not observed when the Th clones were activated with anti-CD3,

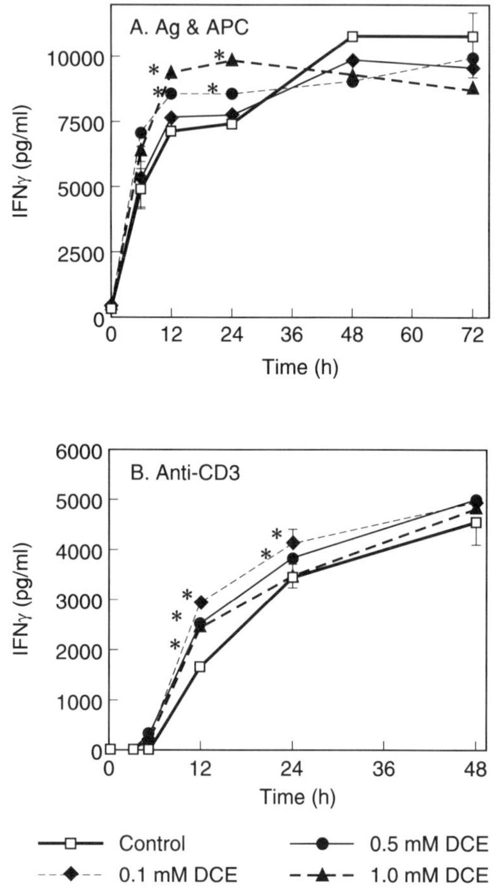

Figure 8.1 The effects of *in vitro* exposure to DCE on T_H1 cell IFN-γ production. 1×10^5 F1.A.2 cells/ml were stimulated with either 2×10^6/ml irradiated syngeneic spleen cells and 50 μg/ml cytochrome c (A) or 5 μg/ml immobilized anti-CD3 (B) in the presence of either 0.02% DMSO (control) or 0.1, 0.5 or 1.0 nM DCE. Release of IFN-γ into culture supernatants was measured using an IFN-γ specific ELISA (Pharmingen, San Diego, CA, USA). Data represent the average of two separate groups of cells and error bars indicate the SEM of these replicate cultures. An * denotes samples which were significantly different from controls ($p < 0.05$).

which implies that 17β-Es may act on the APC, leading to an indirect alteration of Th cell cytokine production. These data suggest that exposure to DCE may enhance both Th1 and Th2 cell responses, while exposure to 17β-Es decreases cytokine production by Th1 and Th2 cells. The functional significance of these perturbations remains to be determined; however, these preliminary findings suggest that in addition to their usefulness in mechanistic studies, antigen-

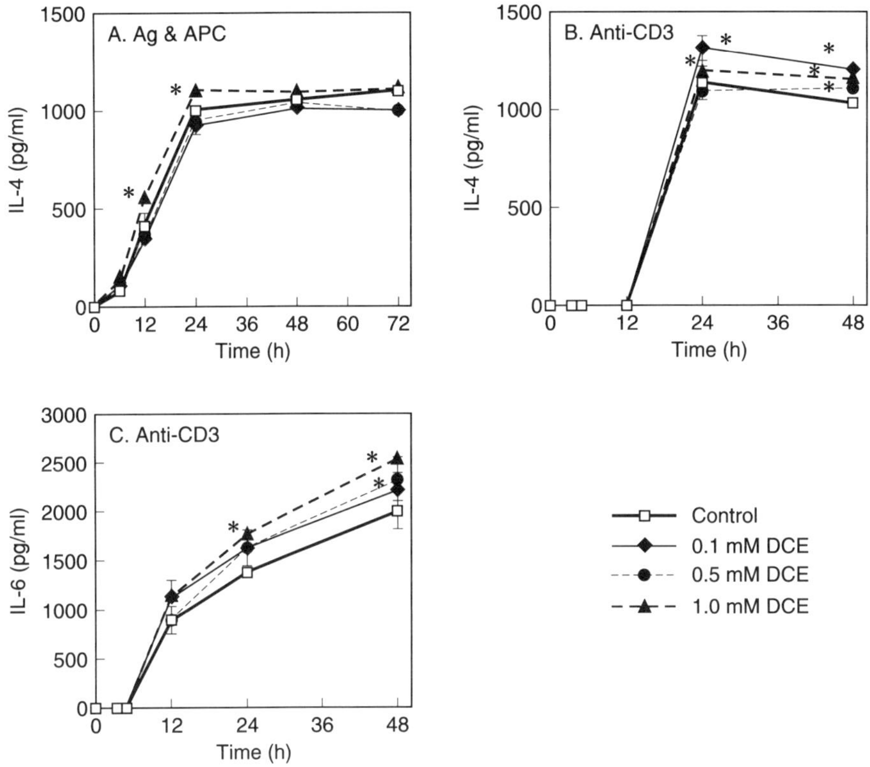

Figure 8.2 The effects of *in vitro* exposure to DCE on T_H2 cell cytokine production. 1×10^5 10.5.17 cells/ml were stimulated with either 2×10^6/ml irradiated syngeneic spleen cells and 10 μg/ml KLH (A) or 5 μg/ml immobilized anti-CD3 (B) in the presence of either 0.02% DMSO (control) or 0.1, 0.5 or 1.0 nM DCE. Release of IL-4 (A and B) and IL-6 (C) into culture supernatants was measured using cytokine-specific ELISAs (Pharmingen). Data represent the average of two separate groups of cells and error bars indicate the SEM of these replicate cultures. An * denotes samples which were significantly different from controls ($p < 0.05$).

specific Th clones may serve as a practical system for immunotoxicological screening. In this manner, the toxicity of putative immunomodulatory compounds could be probed specifically for an impact on Th cell function.

Summary

Although several advantages of this system over *ex vivo* cells or tumour cell lines have already been discussed, a critical evaluation of the use of Th clones as an immunotoxicological model requires the enumeration of potential pitfalls. First, this system does not account for the metabolism of chemicals, which

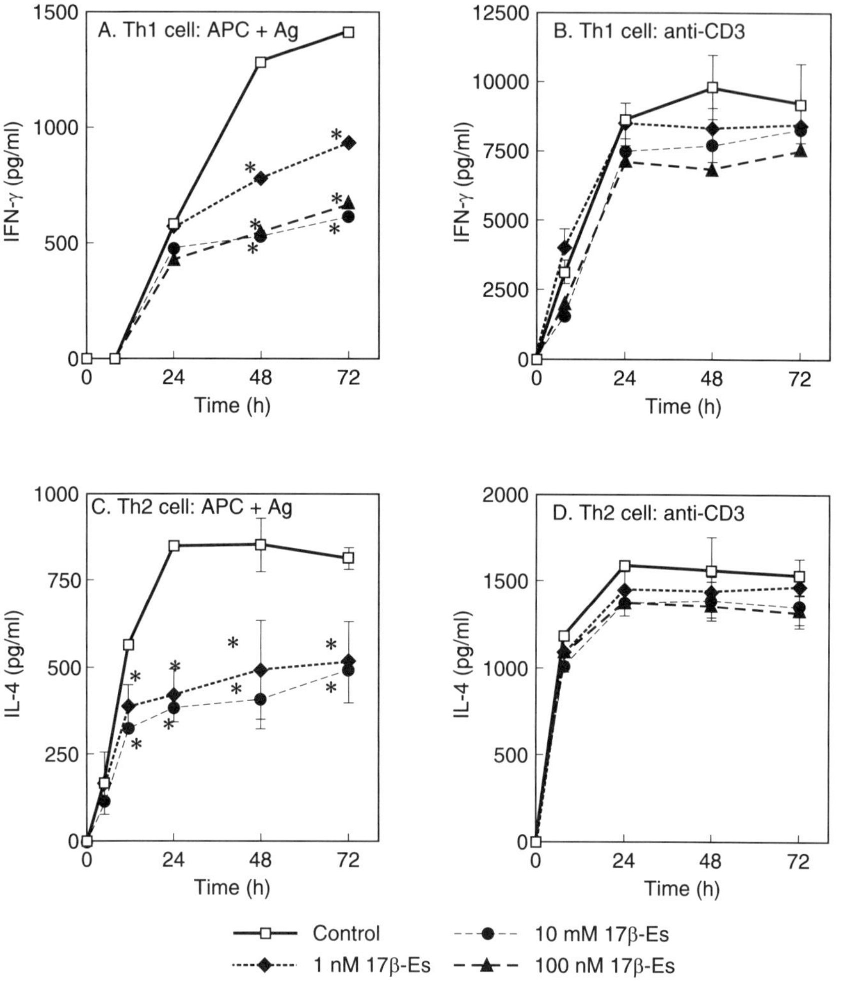

Figure 8.3 The effects of *in vitro* exposure to 17β-Es on T_H1 cell IFN-γ production and T_H2 cell IL-4 production. 1×10^5 T_H clones (F1.A.2 or 10.5.17)/ml were stimulated with either 2×10^6/ml irradiated syngeneic spleen cells and antigen (50 μg/ml cytochrome c (A), or 10 μg/ml KLH (C)) or 5 μg/ml immobilized anti-CD3 (C and D) in the presence of either 0.02% DMSO (control) or 1.0, 10 or 100 nM 17β-Es. Release of IFN-γ and IL-4 into culture supernatants was measured by ELISA. Data represent the average of two separate groups of cells and error bars indicate the SEM of these replicate cultures. An * denotes samples which were significantly different from controls ($p < 0.05$).

may lead to the production of species that are either more or less immunotoxic than the parent compound. Second, maintenance of Th clones is labour-intensive, and without meticulous and consistent care these cells may yield highly variable data. Finally, Th clones have been previously activated and thus do not model a primary immune response. Therefore, activation of Th1 and Th2 clones is essentially a memory response, and established Th clone cell lines can not address what drives Th1/Th2 functional divergence.

It has recently been discovered that $CD8^+$ lymphocytes have a homologous counterpart to $CD4^+$ Th cells. These so-called Tc1 and Tc2 cells secrete the same profile of cytokines as their $CD4^+$ counterparts (Sad *et al.*, 1995). To date, there is very little information regarding the specific influence of exogenous compounds on Tc cell cytokine production. Thus, the use of antigen-specific $CD8^+$ T cell clones is another area for future growth of this type of *in vitro* system. Given the expanding use of T clones and the desire for a more mechanistic understanding of the effects of xenobiotics, there is likely to be an expansion of the use of antigen-specific T clones as an *in vitro* immunotoxicological model.

Conclusions

Antigen-specific Th clones mimic the *in vivo* activation of T cells by processed antigen presented in the context of MHC. $CD4^+$ Th cell clones have been extensively used in studies of the basic function of T cells, including T cell signal transduction, regulation of gene expression and T cell interactions with other cells. Th clones have also been used to study the effects of both natural and synthetic compounds on T cell functions. While most of these studies have focused on physiological mediators of T cell function, Th clones have also been used to examine the mechanism of action of pharmaceuticals and several environmental contaminants. Antigen-specific T clones provide a powerful immunotoxicological model for the determination of a compound's mechanism of action, including the preferential activation of one T cell subset and the molecular and biochemical effects on Th function.

References

Archuleta, M., Schieven, G., Ledbetter, J., Deanin, G. and Burchiel, S. (1993) 7,12-Dimethylbenz[a]anthracene activates protein tyrosine kinases fyn and lck in the HPB-ALL human T cell line and increases tyrosine phosphorylation of phospholipase C-gamma-1, formation of inositol 1,4,5-trisphosphate, and mobilization of intracellular calcium. *Proc. Natl Acad. Sci. USA*, **90**, 6105–9.

Betz, M. and Fox, B. (1991) Prostaglandin E_2 inhibits production of Th1 lymphokines but not of Th2 lymphokines. *J. Immunol.*, **146**, 108–13.

Burchiel, S.W., Davis, D.A.P., Gomez, M.P., Montano, R.M., Barton, S.L. and Seamer, L.C. (1990) Inhibition of lymphocyte activation in splenic and gut-associated lymphoid tissue following oral exposure of mice to 7,12-dimethylbenz[a]anthracene. *Toxicol. Appl. Pharmacol.* **105**, 434–42.

Chiappelli, F., Kung, K.A., Villanueva, P., Lee, P., Frost, P. and Prieto, N. (1995) Immunotoxicity of cocaethylene. *Immunotox. Immunopharm.*, **17**, 399–417.

Clark, D.A., Gauldie, J., Szewczuk, M.R. and Sweeney, G. (1981) Enhanced suppressor cell activity as a mechanism of immunosuppression by 2,3,7,8-tetrachlorodibenzo-*p*-dioxin. *Proc. Exp. Biol. Med.*, **168**, 290–9.

Constant, S.L. and Bottomly, K. (1997) Induction of Th1 and Th2 CD4+ T cell responses: The alternative approaches. *Annu. Rev. Immunol.*, **15**, 297–322.

De Jong, R., Bezemer, A., Zomerdijk, T., van de Pouw-Kraan, T., Ottenhoff, T. and Nibbering, P. (1996) Selective stimulation of T helper 2 cytokine responses by the anti-psoriasis agent monomethylfumarate. *Eur. J. Immunol.*, **26**, 2067–74.

De Krey, G., and Kerkvliet, N. (1995) Suppression of cytotoxic T lymphocyte activity by 2,3,7,8-tetrachlorodibenzo-*p*-dioxin occurs *in vivo*, but not *in vitro*, and is independent of corticosterone elevation. *Toxicology*, **97**, 105–12.

Dearman, R., Basketter, D. and Kimber, I. (1995) Differential cytokine production following chronic exposure of mice to chemical respiratory and contact allergens. *Immunology*, **86**, 545–50.

Dearman, R., Basketter, D. and Kimber, I. (1996) Characterization of chemical allergens as a function of divergent cytokine secretion profiles induced in mice. *Toxicol. Appl. Pharmacol.* **138**, 308–16.

Department of Health and Human Services; Agency for Toxic Substances and Disease Registry (1992) *Toxicological Profile for 1,2-Dibromoethane*. Public Health Service, US Department of Health and Human Services, Washington, DC.

Department of Health and Human Services; Agency for Toxic Substances and Disease Registry (1994) *Toxicological Profile for 1,2-Dichloroethane*. Public Health Service, US Department of Health and Human Services, Washington, DC.

Falchetti, R., Di Francesco, P., Lanzilli, G., Gaziano, R., Casalinuovo, I., Ravagnan, G. and Garaci, E. (1995) *In vitro* effects of cocaine on cytokine secretion induced in murine splenic CD4(+) T cells by antigen specific stimulation. *Cell. Immunol.*, **164**, 57–64.

Fernandez-Botran, R., Sanders, V.M., Mossman, T.R. and Vitetta, E.S. (1988) Lymphokine-mediated regulation of the proliferative response of clones to T helper 1 and T helper 2 cells. *J. Exp. Med.*, **168**, 543–52.

Fiorentino, D.F., Bond, M.W. and Mosmann, T.R. (1989) Two types of mouse helper T cell. IV. Th2 clones secrete a factor that inhibits cytokine production by Th1 clones. *J. Exp. Med.*, **170**, 2081–9.

Gajewski, T.F. and Fitch, F.W. (1988) Anti-proliferative effect of IFN-γ in immune regulation. I. IFN-γ inhibits the proliferation of Th2 but not Th1 murine helper T lymphocyte clones. *J. Immunol.*, **140**, 4245–52.

Gorski, J., Furlow, J.D., Murdoch, F.E., Fritsch, M., Kaneko, K., Ying, C. and Malayer, J.R. (1993) Perturbations in the model of estrogen receptor regulation of gene expression. *Biol. Reprod.*, **48**, 8–14.

Grossman, C.J. (1985) Interactions between gonadal steroids and the immune system. *Science*, **227**, 257–61.

Harper, N., Connor, K. and Safe, S. (1993) Immunotoxic potencies of polychlorinated biphenyl (PCB), dibenzofuran (PCDF) and dibenzo-*p*-dioxin (PCDD) congeners in C57Bl/6 and DBA/2 mice. *Toxicology*, **80**, 217–27.

Harper, N., Steinberg, M., Thomsen, J. and Safe, S. (1995) Halogenated aromatic hydrocarbon-induced suppression of the plaque-forming cell response in B6C3F1 splenocytes cultured with allogenic mouse serum: Ah receptor structure activity relationships. *Toxicology*, **99**, 199–206.

Heo, Y., Parsons, P.J. and Lawrence, D.A. (1996) Lead differentially modifies cytokine production *in vitro* and *in vivo*. *Toxicol. Appl. Pharmacol.*, **138**, 149–57.

Jean, P.A. and Reed, D.J. (1992) Utilization of glutathione during 1,2-dihaloethane metabolism in rat hepatocytes. *Chem. Res. Toxicol.*, **5**, 386–91.

Kerkvliet, N.I. and Burleson, G. (1994) Immunotoxicity of TCDD and related halogenated aromatic hydrocarbons, in *Immunotoxicology and Immunopharmacology* (eds J.H. Dean, M.I. Luster, A.E. Munson and I. Kimber), Raven Press, New York, pp. 97–121.

Kerkvliet, N.I. and Steppan, L. (1988) Suppression of allograft immunity by 3,4,5,3′,4′,5′-hexachlorobiphenyl. I. effects of exposure of tumour rejection and cytotoxic T cell activity *in vivo*. *Immunopharmacology*, **16**, 1–12.

Kerkvliet, N.I., Steppan, L.B, Brauner, J.A., Deyo, J.A., Henderson, M.C., Tomar, R.S. and Buhler, D.R. (1990a) Influence of the Ah locus on the humoral immunotoxicity of 2,3,7,8-tetrachlorodibenzo-*p*-dioxin: evidence of Ah receptor-dependent and Ah receptor-independent mechanisms of immunosuppression. *Toxicol. Appl. Pharmacol.*, **105**, 26–36.

Kerkvliet, N.I., Steppan, L.B., Smith, B.B., Youngberg, J.A., Henderson, M.C. and Buhler, D.R. (1990b) Role of the Ah locus in suppression of cytotoxic T lymphocyte activity by halogenated aromatic hydrocarbons (PCB and TCDD): structure-activity relationships and effects in C57Bl/6 mice congenic at the Ah locus. *Fundam. Appl. Toxicol.*, **14**, 532–41.

Kerkvliet, N.I., Baecher-Steppan, L., Shepherd, D.M., Oughton, J., Vorderstrasse, B. and DeKrey, G. (1996) Inhibition of TC-1 cytokine production, effector cytotoxic T lymphocyte development and alloantibody production by 2,3,7,8-tetrachlorodibenzo-*p*-dioxin. *J. Immunol.*, **157**, 2310–19.

Klein, T.W., Katsuhiko, M., Newton, C.A., Young, J., Widen, R.E. and Friedman, H. (1993) Cocaine suppresses proliferation of phytohemagglutinin-activated human peripheral blood T cells. *Int. J. Immunopharmacol.*, **15**, 77–86.

Ladics, G., Kawabata, T. and White Jr, K. (1991) Suppression of the in vitro humoral immune response of mouse splenocytes by 7,12-dimethylbenz[a]anthracene metabolites and inhibition of immunosuppression by alpha-naphthoflavone. *Toxicol. Appl. Pharmacol.*, **110**, 31–44.

Lawrence, B.P., Leid, M. and Kerkvliet, N.I. (1996) Distribution and behavior of the Ah receptor from murine T lymphocytes. *Toxicol. Appl. Pharmacol.*, **138**, 275–84.

Lee, M., Kirlin, W., Sturrup, M., Rodriguez, J. and Urso, P. (1996) Variation in the response of T cells to concanavalin A after *in vitro* exposure to benzo[a]pyrene and 2-aminofluorene. *Immunopharm. Immunotox.*, **18**, 309–21.

Li, Q., Geiselhart, L., Mittler, J., Mudzinski, S., Lawrence, D. and Freed, B. (1996) Inhibition of human T lymphoblast proliferation by hydroquinone. *Toxicol. Appl. Pharmacol.*, **139**, 317–23.

Li, T. and Fox, B. (1993) Effect of prostaglandin E_2 (PGE_2) on IL-3/granulocyte-macrophage colony-stimulation factor production by T helper cells. *J. Immunol.*, **150**, 1680–90.

Lundberg, K., Dencker, L. and Gronvik, K. (1992) 2,3,7,8-Tetrachlorodibenzo-*p*-dioxin (TCDD) inhibits the activation of antigen specific T cells in mice. *Int. J. Immunopharmacol.*, **14**, 699–705.

Marin, M., Murtha, J., Dong, W. and Pestka, J. (1996) Effects of mycotoxins on cytokine production and proliferation in EL-4 thymoma cells. *J. Toxicol. Environ. Health*, **48**, 379–96.

McCabe, M.J. (1994) Mechanisms and consequences of immunomodulation by lead, in *Immunotoxicology and Immunopharmacology* (eds J.H. Dean, M.I. Luster, A.E. Munson, and I. Kimber), Raven Press, New York, pp. 143–62.

Morris, D.L., Jordan, S.D. and Holsapple, M.P. (1991) Effects of 2,3,7,8-tetrachlorodibenzo-*p*-dioxin (TCDD) on humoral immunity: I. Similarities to *Staphylococcus aureus* Cowan strain I (SAC) in the in vitro T-dependent antibody response. *Immunopharmacology*, **21**, 159–70.

Munson, A.E., Sanders, V.M., Douglas, K.A., Sain, L.E., Kauffmann, B.M. and White, K.J. (1982) *In vivo* assessment of immunotoxicity. *Environ. Health Perspect.*, **43**, 41–52.

Myers, M., Blanton, R. and Bick, P. (1988) Inhibition of IL-2 responsiveness following exposure to benzo[a]pyrene is due to alterations in accessory cell function. *Int. J. Immunopharmacol.*, **10**, 177–86.

Nair-Menon, J., Campbell, G. and Blake, C. (1996) Toxic effects of octylphenol on cultured rat and murine splenocytes. *Toxicol. Appl. Pharmacol.*, **139**, 437–44.

Neumann, C., Oughton, J. and Kerkvliet, N. (1993) Anti-CD3 induced T cell activation-II. effect of 2,3,7,8-tetrachlorodibenzo-*p*-dioxin (TCDD). *Int. J. Immunopharmacol.*, **15**, 543–50.

O'Garra, A. and Murphy, K. (1996) Role of cytokines in development of Th1 and Th2 cells, in *Chemical Immunology: Th1 and Th2 Cells in Health and Disease* (ed. S. Ramangnani), Karger, Basel, pp. 1–13.

Pahl, H., Krauss, B., Schulze-Osthoff, K., Decker, T., Traenckner, E., Vogt, M., Myers, C., Parks, T., Warring, P., Muhlbacher, A., Czernilofsky, A. and Baeuerle, P. (1996) The immunosuppressive fungal metabolite gliotoxin specifically inhibits transcription factor NFκB. *J. Exp. Med.*, **183**, 1829–40.

Pelletier, L., Castedo, M., Bellon, B. and Druet, P. (1994) Mercury and autoimmunity, in *Immunotoxicology and Immunopharmacology* (eds J.H. Dean, M.I. Luster, A.E. Munson and I. Kimber), Raven Press, New York, pp. 539–52.

Prell, R.A., Oughton, J.A. and Kerkvliet, N.I. (1995) Effect of 2,3,7,8-tetrachlorodibenzo-*p*-dioxin on anti-CD3-induced changes in T cell subsets and cytokine production. *Int. J. Immunopharmacol.*, **17**, 951–61.

Romagnani, S. (1996) Th1 and Th2 in human disease. *Clin. Immunol. Immunopathol.*, **80**, 225–35.

Sad, S., Marcotte, R. and Mosmann, T.R. (1995) Cytokine-induced differentiation of precursor mouse $CD8^+$ T cells into cytotoxic $CD8^+$ T cells secreting Th1 or Th2 cytokines. *Immunity*, **2**, 271–7.

Silkworth, J.B., Antrim, L. and Kaminsky, L.S. (1984) Correlations between polychlorinated biphenyl immunotoxicity, the aromatic hydrocarbon locus and liver microsomal enzyme induction in C57Bl/6 and DBA/2 mice. *Toxicol. Appl. Pharmacol.*, **75**, 156–65.

Vento, K., Dearman, R., Kimber, I., Basketter, D. and Coleman, J. (1996) Selectivity of IgE responses, mast cell sensitization and cytokine expression in the immune response of Brown Norway rats to chemical allergens. *Cell. Immunol.*, **172**, 246–53.

9 Influence of the Sympathetic Nervous System on T Helper Cell Function

Virginia M. Sanders, Deborah J. Kasprowicz and D. Susanne Ramer-Quinn

Introduction

For many years, the nervous and immune systems were researched as separate disciplines. This approach has contributed much to the understanding of each system individually. However, for future mechanistic studies *in vivo*, an understanding of the routes through which the two systems relate to each other following exposure to a xenobiotic that targets either one or both systems (Figure 9.1) will help us to understand more fully the mechanisms responsible for xenobiotic-induced immune toxicities. If the nervous system is able to modulate Th cell function, then it is possible that a xenobiotic could affect Th cell function indirectly by inducing a change in nervous system function. This possibility emphasizes the need for a better understanding of not only the mechanisms by which the nervous system influences the function of $CD4^+$ Th cells, but also the mechanisms through which xenobiotics may disrupt this neuroimmune relationship. This chapter reviews the evidence supporting the existence of a functional link between the nervous and immune systems, particularly with regard to the role of the sympathetic neurotransmitter, norepinephrine (NE), in $CD4^+$ Th cell function. Mechanisms by which a xenobiotic, such as the heavy metal lead, may disturb the neuroimmune relationship and contribute to immune toxicity are also discussed.

Anatomical and physiological evidence for a role of the sympathetic nervous system in modulating T helper cell function

An important function of the autonomic nervous system is to effect *modest* adjustments in many diverse cellular activities to maintain the constancy of the body's internal environment by regulating the activities of organs that are not

---

T Lymphocyte Subpopulations in Immunotoxicology. Edited by Ian Kimber and MaryJane K. Selgrade.
 ISBN 0 471 97194 4

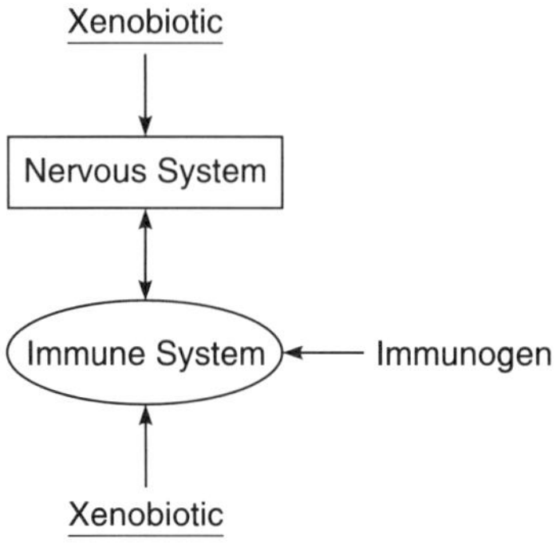

Figure 9.1 The direct effect of a xenobiotic on either the nervous system or the immune system may indirectly affect the functional relationship that exists between these two systems.

under voluntary, conscious control. The autonomic nervous system is divided into the parasympathetic and sympathetic systems that are characterized by the neurotransmitter released by post-ganglionic nerve fibres. The neurotransmitter synthesized and released by nerve terminals associated with the sympathetic division of the autonomic nervous system is noradrenaline (NA), while the neurotransmitter associated with the parasympathetic division is acetylcholine. Few reports exist for either parasympathetic innervation of lymphoid tissue or receptors for acetylcholine on lymphocytes. However, four key discoveries illustrate the existence of a link between the immune and sympathetic nervous systems and they have been reviewed extensively elsewhere (Madden *et al.*, 1995; Sanders, 1995). These discoveries are 1) the identification of sympathetic innervation of lymphoid organs, 2) the finding that the sympathetic neurotransmitter, NA, is released from sympathetic nerve terminals residing within lymphoid organs, 3) the identification of adrenergic receptors on the surface of lymphoid cells, and 4) the identification of a role for neurotransmitters in modulating T and B cell functions.

The discovery that is credited with establishing this area of research known as neuroimmunology was the identification of sympathetic innervation of lymphoid organs. Both primary lymphoid organs, such as the bone marrow and thymus (Bulloch and Pomerantz, 1984; Calvo, 1968; Felten *et al.*, 1985; Livnat *et al.*, 1985; Williams and Felten, 1981; Williams *et al.*, 1981), and secondary organs, such as the spleen and lymph nodes (Dahlstrom and Zetterstrom, 1965; Felten *et al.*, 1984, 1987a; Reilly *et al.*, 1976, 1979), were found to be innervated by sympathetic fibres. In 1929, Riegele first identified splenic innervation by silver-staining spleen tissue sections (Riegele, 1929). More recent innervation studies have used either neurohistochemical staining for tyrosine hydroxylase (TH), the rate limiting enzyme in NA synthesis, or ultrastructural examination of the splenic microenvironment (McCuskey and McCuskey, 1977; Reilly *et al.*, 1976).

Some regions of the spleen are more highly innervated than others (Felten DL *et al.*, 1987a,b; Felten SY *et al.*, 1988). Sympathetic nerve fibres travel along the periphery of the marginal sinus, a region of T and B cell entry and antigen capture, with some fibres in direct apposition with macrophages and B cells. The T cell-rich periarteriolar lymphatic sheath (PALS) contains nerve fibres in direct apposition to T cells. In contrast, follicular regions of the spleen that are rich in B cells are sparsely innervated. In a manner similar to the spleen, lymph node areas of T cell entry and T cell activation are densely innervated (Felten *et al.*, 1984; Williams and Felten, 1981) and the thymus contains sympathetic nerve fibres that localize primarily to the cortex, but not the medulla (Williams and Felten, 1981).

The reversible destruction of peripheral NA-containing sympathetic nerve endings is induced by treatment with the neurotoxin, 6-hydroxydopamine (6-OHDA) (Kostrzewa and Jacobwitz, 1974), and up to 95% of NA in the spleen of an adult animal is depleted following 6-OHDA treatment (Felten *et al.*, 1984, 1985; Williams *et al.*, 1981). A recent study has examined NA levels in the spleen using a microdialysis technique (Shimizu *et al.*, 1994). NA levels were measured before and after the administration of interleukin-1β (IL-1β), an immune response-associated cytokine that increases the level of corticotrophin-releasing factor (CRF) to activate directly the sympathetic nervous system (Irwin *et al.*, 1990). Using the microdialysis technique, basal levels of NA in the spleen were found to be $\sim 1.6 \times 10^{-6}$ M. This level increased to $\sim 3 \times 10^{-6}$ M following the administration of IL-1β. In addition, following the injection of CRF, splenic NA levels increased to 5×10^{-6} M (Shimizu *et al.*, 1994). Taken together, these findings were important for establishing that NA is released directly from sympathetic nerve terminals in the spleen and that the level of NA increases following antigen exposure. Thus, the abundance of sympathetic innervation in T cell areas of lymphoid organs suggests the potential for a direct effect of NA on T cells and suggests that a xenobiotic that disrupts the level of innervation to these T cell-rich areas may ultimately affect Th cell function.

Adrenergic receptor expression on immune cells

The ability of NA directly to modulate immune cell function depends upon the expression of adrenergic receptors on the surface of immune cells. Lymphocytes express the β-adrenergic receptor (βAR), particularly of the β_2-subtype (Bishopric *et al.*, 1980; Conolly and Greenacre, 1977; Loveland *et al.*, 1981; Meurs *et al.*, 1982; Pochet *et al.*, 1979; Williams *et al.*, 1976). However, the density of beta-adrenergic receptor expression varies among the functional subsets of immune cells. Human $CD4^+$ and $CD8^+$ cells differ in the number of the βAR per cell, but show similar βAR affinity. $CD8^+$ T cytotoxic (Tc) cells express a higher density of the βAR than CD4$^+$ Th cells (Van Tits *et al.*, 1990; Khan *et al.*, 1986). Furthermore, resting Th1 cells, but not resting Th2 cells, express a detectable level of the β_2AR ligand binding site on their cell surface (Sanders *et al.*, 1997). Thus, CD4$^+$ Th cell populations differentially express the receptor that binds NA and this may provide a mechanism by which NA selectively modulates effector Th cell subset function.

CD4$^+$ Th cells are activated by antigen-presenting cells, such as macrophages and B cells. Both B cells and macrophages express the βAR and the level of expression is higher than the level expressed by CD4$^+$ Th cells, but lower than the level expressed by CD8$^+$ Tc cells (Cremaschi *et al.*, 1991; Fuchs *et al.*, 1988a; Krawietz *et al.*, 1982; Radojcic *et al.*, 1991). Thus, effective Th cell activation *in vivo* may be influenced by the stimulation of the βAR on antigen-presenting cells to induce changes in their level of expression of costimulatory molecules that are necessary for Th cell activation. Such molecules may include the B7-2 molecule on B cells that can be upregulated by β_2AR stimulation (D.J. Kasprowicz and V.M. Sanders, unpublished findings). However, it should be noted that, *in vivo*, the difference in the level of βAR expression between Th cells and B cells may be related to the level of NA released within the microenvironment of these cells. As reviewed above, B cell-enriched follicles are sparsely innervated while T cell-enriched areas are highly innervated with sympathetic nerve fibres. Therefore, the B cells contained within the follicular microenvironment, as compared with the T cells in the PALS, may express a higher level of the β_2AR, since the concentration of NA is low in the follicle as compared with the PALS.

Additionally, mitogen or pharmacological activation of immune cells either increases or decreases the level of expression, but not the affinity, of the βAR on unfractionated populations of T cells or thymomas (Cazaux *et al.*, 1995; Radojcic *et al.*, 1991; Westly and Kelley, 1987). A recent radioligand binding study showed that the level of expression of the β_2AR on Th1 cells increases as early as 4 h after activation with anti-CD3 monoclonal antibody (mAb) in comparison with the level of expression on resting cells (Ramer-Quinn *et al.*, 1997). In contrast, Th2 cell activation did not induce a detectable level of expression of the β_2AR at any time point examined after activation. Thus, if the level of expression of the β_2AR is important for NA to modulate effector Th cell function, then a xenobiotic that influences the level of expression of the β_2AR on CD4$^+$ Th cells during the course of an immune response may affect the ability of these cells to respond to signals delivered by NA.

Role of NA and beta-2-adrenergic receptor stimulation in modulating Th cell-mediated functions

As described in Chapter 2, CD4$^+$ T cells are divided into functionally distinct subsets based on their cytokine secretion profile (Mosmann *et al.*, 1991). Th1 cells produce interleukin-2 (IL-2) and interferon-γ (IFN-γ) and primarily function in cell-mediated immune responses, such as delayed-type hypersensitivity (DTH) reactions and in the generation of $CD8^+$ Tc cells. Th1 cells also help B cells to make certain antibody isotypes. Th2 cells produce interleukin-4 (IL-4), interleukin-5 (IL-5), interleukin-6 (IL-6), interleukin-10 (IL-10) and interleukin-13 (IL-13) and primarily function in humoral immune responses by helping B cells to produce antibodies. IFN-γ and IL-4 are opposing cytokines that promote the secretion of either IgG_{2a} or IgG_1 and IgE, respectively, by B cells (Finkelman *et al.*, 1990; Mosmann *et al.*, 1991; Stevens *et al.*, 1988) and influence the development of precursor Th cells into effector Th1 and Th2 cells, respectively (Seder and Paul, 1994).

Early studies show that activation of the β_2AR by NA or βAR agonists induced an elevation in the intracellular concentration of cAMP that mediates an inhibition of T cell proliferation induced by mitogen- or anti-CD3 mAbs (DeRubertis *et al.*, 1974; Diamantstein and Ulmer, 1975; Estes *et al.*, 1971; Hadden *et al.*, 1970;Johnson and Gordon, 1981; Johnson *et al.*, 1981; Ledbetter *et al.*, 1986; Smith *et al.*, 1971a,b; Vischer, 1976; Weinstein *et al.*, 1975). This inhibition involves a mechanism by which cAMP induces an inhibition of early biochemical events (Alava *et al.*, 1992; Park *et al.*, 1992; Van Tits *et al.*, 1991) that lead to a decrease in IL-2 secretion (Mary *et al.*, 1989; Wacholtz *et al.*, 1991) and IL-2 receptor expression (Feldman *et al.*, 1987; Krause and Deutsch, 1991). A recent study reported that the stimulation of the βAR on human peripheral blood T cells interferes with the assembly of F actin and subsequent pseudopodia formation that normally occurs following activation of the T cell with anti-CD3 monoclonal antibody (Selliah *et al.*, 1995). This finding suggests that βAR stimulation disrupts normal biochemical, morphological, and functional changes that must occur in the T cell following activation through the T cell receptor. Thus, the modulatory effect of βAR stimulation appears to be inhibitory, but this inhibition may be dependent on the mode of cell activation, the cytokine milieu at the time of activation, or the specific immune parameter being evaluated.

Chemical sympathectomy with 6-OHDA reduces cell-mediated immune responses, such as DTH reactions and the generation of $CD8^+$ Tc cells. Administration of 6-OHDA before or after contact sensitization with the trinitrophenyl antigen decreases the DTH ear swelling reaction, demonstrating that NA may have a role in altering T cell-mediated events (Madden *et al.*, 1989). Normally, after contact sensitization, antigen-specific T cells are activated to proliferate in the lymph node during the sensitization phase of the immune response. After challenge, antigen specific T cells migrate to the site of antigen challenge to participate in a local inflammatory response during the effector phase of the immune response. The DTH reaction is reduced when 6-OHDA is administered on the day of antigen sensitization or challenge, indicating that both the sensitization and effector phases of the response are altered (Madden *et al.*, 1989). In cell transfer experiments, normal lymphocytes showed enhanced migration into the lymph nodes of mice that were treated with 6-OHDA, while lymphocytes obtained from 6-OHDA-treated mice showed decreased migration into lymph nodes of normal mice (Madden *et al.*, 1994a), suggesting that the migration of T cells to the site of challenge during the effector phase of the response may be affected by NA. However, a recent study shows that NA binds to the βAR to decrease the number of T cell–endothelial cell interactions that must occur to mediate T cell migration involved in either homing to lymphoid tissues or recruitment to inflammatory sites (Carlson *et al.*, 1996). The mechanisms responsible for this NA-induced decrease in the number of T cell–endothelial cell interactions do not appear to involve changes in the level of expression of adhesion molecules on either the T cell or endothelial cell, but rather appear to involve a change in the kinetics of the T cell/endothelial cell adhesion process. Thus, both *in vivo* and *in vitro* data suggest that T cell migration may be influenced by the sympathetic nervous system.

The sensitization phase of the DTH reaction is also affected by 6-OHDA treatment. Antigen specific Th cell cytokine production is decreased *in vitro* when draining lymph node cells from sensitized 6-OHDA-treated mice are cultured with TNP-modified spleen cells (Madden *et al.*, 1989). In these cultures, IL-2 production from $CD4^+$ Th cells was reduced and this decrease in IL-2 was probably responsible for the decrease in the generation of hapten-specific $CD8^+$ Tc cell. Also, CTL responses from cells obtained from 6-OHDA-treated mice were reduced *in vivo* and *in vitro* in response to alloantigen (Madden and Livnat, 1991). These findings suggest that effector Th cell function requires NA for a 'normal' $CD8^+$ Tc cell response, since the removal of NA *in vivo* decreases the responsiveness of T cells involved in DTH reactions.

In contrast to the previously described studies on lymphocyte function in immunized sympathectomized mice, a recent investigation has examined lymphocyte function in unimmunized sympathectomized mice. T cells in lymph nodes of denervated mice exhibit reduced proliferation and increased IFN-γ production in response to the mitogen, Con A, as compared with normal mice, suggesting that when NA is present, it exerts a mixed enhancing/inhibitory effect on lymph node Th cell function *in vivo* (Madden *et al.*, 1994b). In contrast, spleen cells from denervated mice showed enhanced proliferation and enhanced IL-2 and IL-4 production in response to antigen challenge *in vitro* (Kruszewska *et al.*, 1995), suggesting that when NA is present, it exerts an inhibitory effect on some splenic Th cell functions *in vivo*. Therefore, the effect of NA on $CD4^+$ Th cell function appears to be dependent on the microenvironment in which the cells reside, the immune parameter being measured, and the type of immune cell involved in the functional response.

A number of recent studies has addressed the role of NA and βAR stimulation in modulating the functions of specific T cell populations, particularly effector Th1 and Th2 cells. These studies show that exposure of Th1 and Th2 cells to the β_2AR agonist terbutaline *prior to* Th cell activation by an antigen-presenting B cell results in the inhibition of IL-2 and IFN-γ production by Th1 cells, but causes no change in IL-4 production by Th2 cells (Sanders *et al.*, 1997). Furthermore, exposure of Th1 and Th2 cells to terbutaline *following* activation by anti-CD3 mAb also differentially effects cytokine production by these Th cell subsets (Ramer-Quinn *et al.*, 1997). Stimulation of the β_2AR on Th1 cells following activation inhibits IL-2 and does not change IFN-γ production. Exposure of activated Th2 cells to a β_2AR agonist did not alter the level of cytokine production. The primary mechanism for this differential modulation of Th1 and Th2 cell function by terbutaline was found to involve a differential expression of the β_2AR on these effector Th cell subsets. As mentioned in the previous section of this chapter, both resting and activated Th1 cells express detectable levels of the β_2AR as determined by radioligand binding, while Th2 cells do not (Ramer-Quinn *et al.*, 1997; Sanders *et al.*, 1997). Thus, the effect of NA on effector Th cell function may depend on the specific effector Th cell subset residing in lymphoid tissue, the level of β_2AR expression on Th1 cells, and the activation state of the cell.

Exposure of lymphocytes *in vitro* to either NA, a β_2AR agonist, or cAMP-elevating agents during the early phases of the Th cell-dependent antibody response increases antibody production (Burchiel and Melmon, 1979a,b; Kouassi *et al.*, 1988; Roper *et al.*, 1990; Sanders and Powell-Oliver, 1992; Sanders and Munson, 1984a,b; Shreurs *et al.*, 1982; Stein and Phipps, 1991; Teh and Paetkau, 1976; Watson *et al.*, 1973). This finding suggests that NA exerts an enhancing effect on the Th cell-dependent antibody response *in vitro*. To determine whether NA influences the Th cell-dependent antibody response *in vivo*, 6-OHDA-treated animals were evaluated for their ability to generate an antibody response following immunization with a Th cell-dependent antigen. Results showed that antibody production is decreased in the absence of NA as compared with controls (Ackerman *et al.*, 1991b; Cross *et al.*, 1986; Fuchs *et al.*, 1988b; Hall *et al.*, 1982; Kasahara *et al.*, 1977; Livnat *et al.*, 1985; Madden *et al.*, 1989, 1991), suggesting that NA also exerts an enhancing effect on the Th cell-dependent antibody response *in vivo*. However, this enhancing effect of NA on the Th cell-dependent antibody response may be targeted to either the B cell, the Th cell, or both. Therefore, the role of NA in modulating the Th cell-dependent antibody response may be difficult to interpret in light of the potential for NA to affect both Th and B cell function directly.

Although findings differ when using sympathectomized mice, it is clear that NA most likely affects the level of a Th cell-dependent antibody response at the level of both Th and B cell function. Thus, a xenobiotic that affects either the level of NA in a lymphoid organ or the level of expression of the β_2AR on lymphocytes during the course of an immune response may also affect the outcome of that immune response.

Role of NA in Th cell-related disease states

Cell-mediated immune responses are diminished in aged individuals who suffer from an increased incidence of infectious disease, cancer and autoimmunity. A decline in sympathetic innervation of lymphoid organs, including the spleen and lymph nodes occurs in normal aged rodents (Ackerman *et al.*, 1991a; Bellinger *et al.*, 1992a,b; Felten *et al.*, 1987), and may be related to the diminished ability of T cells from aged rodents to proliferate, produce IL-2, and generate CTL responses (Madden and Livnat, 1991; Madden *et al.*, 1992, 1994b; Miller, 1996; Thoman and Weigle, 1989). Thus, NA may be necessary to maintain an optimal level of immune cell function during the ageing process.

The sympathetic nervous system has been shown to influence the severity of experimental autoimmune encephalomyelitis (EAE), an inflammatory autoimmune disease of the central nervous system in which Th1 cells specific for myelin basic protein (MBP) are involved in the initiation of the disease. In striking contrast, Th2 cells are implicated in amelioration of the disease (Liblau *et al.*, 1995). Chemical sympathectomy with 6-OHDA increases the severity of clinical EAE and increases mortality, even though the severity of the lesions did not differ from those in non-sympathectomized EAE controls (Chelmicka-Schorr *et al.*, 1988). In addition, EAE was more severe when donor lymph node cells from MBP-immunized mice were transferred to sympathectomized recipients (Chelmicka-Schorr *et al.*, 1992). Thus, NA may play a down-modulatory role in the development and expression of EAE.

To confirm further the down-modulatory role of NA in the EAE model, the βAR antagonist propranolol was administered peripherally to MBP-immunized animals, resulting in exacerbated disease and increased susceptibility of animals to EAE (Chelmicka-Schorr *et al.*, 1989). In contrast, treatment with βAR agonists decreased the susceptibility of animals to EAE (Chelmicka-Schorr *et al.*, 1989) and suppressed the frequency of relapse (Wiegmann *et al.*, 1995). Interestingly, the level of expression of the β_2AR increases on splenocytes from rats with EAE as compared with control rats (Wiegmann *et al.*, 1995) and on lymphocytes from patients with multiple sclerosis (Karaszewski *et al.*, 1990), suggesting that a lower basal level of NA may be present in the microenvironment of these cells from individuals with the disease. Collectively, these data suggest that NA is normally released by the sympathetic nervous system either to suppress the immune response involved in this autoimmune disease or to enhance some aspect of the immune system that maintains immune homeostasis. Therefore, a diminution in lymphoid organ innervation may precipitate the development of an immune dysfunction that contributes to the development and progression of autoimmune diseases, such as EAE.

The role of Th cell-derived cytokines in the development and progression of disease states is well documented. For example, the balance *in vivo* that exists between the level of Th1 and Th2 cytokines is important in the development and expression of rheumatoid arthritis. While Th1 cytokines enhance the development of this disease, Th2 cytokines reduce the severity (Miossec *et al.*, 1996). Sympathetic nervous system function is also linked to the modulation of rheumatoid arthritis. Selective denervation of popliteal and inguinal lymph nodes exacerbates experimental rheumatoid arthritis in Lewis rats via an earlier onset and enhanced severity of arthritic symptoms (Felten *et al.*, 1992). $CD4^+$ Th cell-derived cytokines also play an important role in the aetiology of systemic lupus erythematosus (SLE). Sympathetic innervation is diminished in secondary lymphoid organs before the onset of symptoms in autoimmune MRL-*lpr*/*lpr* mice, an animal model of SLE (Breneman *et al.*, 1993). Thus, these findings suggest that the sympathetic nervous system is required for modulating effector $CD4^+$ Th cell function for either the prevention or progression of autoimmune diseases in which effector Th cells play a prominent role. These findings suggest also that a xenobiotic-induced effect on some parameter associated with normal sympathetic nervous system function may indirectly affect Th cell function and contribute to either the development or progression of disease states.

Possible xenobiotic-induced changes in sympathetic nervous system function

The previous sections of this chapter discuss data that support the existence of a functional relationship between the nervous and immune systems. These data have strong implications for immunotoxicological research. It is possible that a xenobiotic-induced immune system dysfunction *in vivo* may reflect either a direct action of the xenobiotic on immune cells or a direct action of the xenobiotic on a nervous system parameter that influences immune system function (Figure 9.2). The latter possibility may be important when a xenobiotic targets the nervous system at much lower concentrations than it targets the immune system. In this manner, the xenobiotic-induced change in nervous system function may contribute to the development of an immune system dysfunction without any direct effect of the xenobiotic being exerted on the immune system itself.

The indirect mechanism by which a xenobiotic induces immune toxicity *in vivo* through a direct effect of the xenobiotic on the nervous system may explain why considerably higher concentrations of some xenobiotics must be used with immune cells cultured *in vitro* to induce an immune toxicity comparable with that induced *in vivo*. Conversely, it is also possible that a xenobiotic-induced change in nervous system function may counteract a direct xenobiotic-induced effect on the immune system by modulating another immune cell function to compensate for the xenobiotic-induced immune effect. Of course, none of the above-mentioned possibilities have been tested to date, but hopefully, they will be investigated by immunotoxicologists in the future. The following section includes some examples of xenobiotics that directly affect nervous system function. These neuro-targeted effects may result in a change in either the delivery or transduction of NA signals to immune cells during the

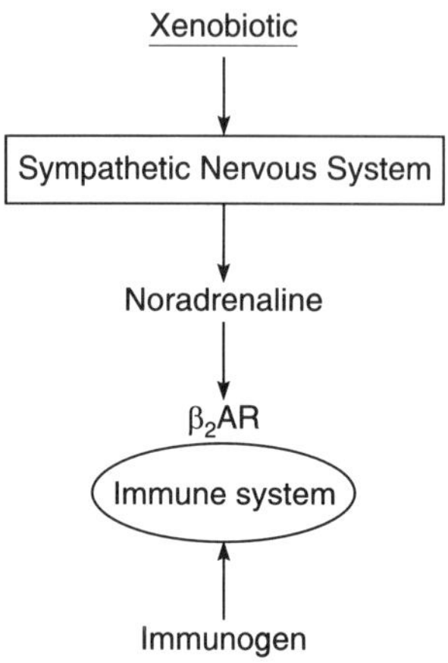

Mechanism of Toxicity	Xenobiotic
Denervation	6-Hydroxydopamine
Inhibit NE synthesis	Carbon disulphide
Inhibit NE uptake	Trimethyltin
Inhibit termination of NE release	Chlordimeform
Increase level of neurotransmitter receptor expression	Cessation of smoking
Decrease level of neurotransmitter receptor expression	Lead
Increase peripheral sympathetic tone	Lead

Figure 9.2 A parameter associated with the nervous system may be directly affected by a xenobiotic. This effect on the nervous system may indirectly influence immune system function. The nervous system parameters that may be directly affected by a xenobiotic are listed.

course of an immune response to antigen. In light of the findings presented above, showing a role for NA in modulating immune system function, a direct xenobiotic-induced change in neuronal function may indirectly result in notable changes in immune system function. In this section, we will also discuss the possible mechanisms by which the heavy metal, lead, may induce immune toxicity via both a direct and indirect mechanism.

Certain xenobiotics affect the synthesis, release and uptake of neurotransmitters. Carbon disulphide affects NA synthesis via a mechanism involving the metabolite of carbon disulphide, dithiocarbamate (McKenna and Distefano, 1977). Dithiocarbamate chelates copper, a metal that binds to the enzyme dopamine β-hydroxylase to allow for the conversion of dopamine to NA. Noradrenaline uptake into nerve terminals is inhibited by trimethyltin (Doctor *et al.*, 1982). This effect of trimethyltin usually manifests itself centrally and this central effect may disrupt the delivery of critical neurotransmitter signals to the periphery. It is also possible that long-term exposure to trimethyltin may affect the peripheral nervous system. This effect of trimethyltin may allow the concentration of NA in lymphoid organs to remain elevated for longer periods of time following NA release during the course of an immune response and, thus, may modulate a given immune response over a longer period of time. One report shows that rats treated *in vivo* with trimethyltin exhibited atrophy of the thymus, spleen, and lymph nodes within 2 days of treatment (Hioe and Jones, 1984). This report also showed that trimethyltin specifically affected the ability of the rats to produce antibody against a T cell-dependent antigen, without affecting either the delayed hypersensitivity response, natural killer cell activity, or macrophage phagocytic activity. Thus, trimethyltin is one example of an immunotoxicant that may induce immune toxicity indirectly by affecting nervous sytem function directly.

Certain xenobiotics may interfere with the level of expression of neurotransmitter receptors on cells. As discussed in a previous section of this chapter, the expression of the β_2AR on cells is essential for the transduction of signals from NA to the cell to modulate immune cell activity. Although controversial, individuals suffering from 'sick building syndrome' often manifest symptoms of depression, stress, and immune dysfunction. A xenobiotic may induce a state of depression that may subsequently influence some parameter involved with the communication between the sympathetic nervous system and the immune system. For example, individuals with unipolar depression express a lower level of the βAR on cells in comparison with control individuals (Jeanningros *et al.*, 1991). The degree of depression in these individuals appears to correlate with the magnitude of decrease in βAR expression on immune cells. Mononuclear cells isolated from patients suffering from depression demonstrate a decreased ability to respond to mitogenic stimulation (Maes *et al.*, 1989) and a loss of the ability of the adrenergic receptor to induce second messenger production (Halper *et al.*, 1988).

These findings may be relevant to certain chemical exposure situations, since episodes of depression that are associated with chemical exposure are often accompanied by compromized immune responsiveness. Even though the exact mechanisms responsible for the development of depression are complex, it is possible that xenobiotics directly affect the level of expression of the βAR on cells and, by some undefined mechanism, play a role in the onset of depression. Conversely, it is also possible that xenobiotics directly affect central mechanisms to induce depression and that this event indirectly influences immune responsiveness by affecting either the level of neurotransmitter release in the periphery or the level of expression of the β_2AR on cells. Therefore, if a given number of neurotransmitter receptors is important for optimal immune cell functioning in response to NE, then a chemically induced clinical depression may precipitate a change in the level of β_2AR expression on lymphocytes by altering the level of NA in the periphery.

Cessation of cigarette smoking is associated with an increase in the level of expression of the βAR on immune cells (Laustiola *et al.*, 1991). The change in adrenergic receptor expression on immune cells appears to correlate with a decrease in the level of circulating NA in these subjects, a finding that may in itself explain the increase in adrenergic receptor number since NA itself leads to a downmodulation of receptor expression. Although no studies have been designed to determine if a correlation exists between the increase in β_2AR expression and the status of immune responsiveness in ex-smokers, it is interesting to speculate that certain tobacco components may affect nervous system function and, consequently, influence the responsiveness of immune cells by affecting the level of expression of adrenergic receptors on their cell surface that are capable of responding to NA.

Certain chemicals interfere with neurotransmitter receptor function by blocking the binding of the neurotransmitter to the receptor. For example, formamidines, such as chlordimeform, inhibit the *in vivo* and *in vitro* binding of specific ligands to the alpha-2-adrenergic receptor (Costa and Murphy, 1987). This finding may be important for two reasons. First, this xenobiotic may interfere with the feedback mechanism for the termination of NA release which involves stimulation of presynaptic alpha-2-adrenergic receptors. If chlordimeform were to interact with these receptors presynaptically on sympathetic nerve terminals in lymphoid organs, it may possibly prolong cell exposure to NA and, perhaps, lead to an exaggerated modulation of immune responsiveness. Second, pharmacologicol studies indicate the presence of alpha-2-adrenergic receptors on lymphocytes that, when stimulated, appear to suppress immune functions such as antibody production (Sanders and Munson, 1985). Reports have shown that chlordimeform exposure of mice *in vivo* induces an inhibition of T cell-dependent antibody production, without affecting the delayed hypersensitivity response (Shopp *et al.*, 1985; Wiltrout *et al.*, 1978). Interestingly, in order to induce *in vitro* an equivalent toxicity to that induced *in vivo*, directly cytotoxic concentrations of chlordimeform were required to suppress antibody production. Thus, *in vivo* immune toxicity may have been mediated indirectly through another endogenous mechanism that was more sensitive to chlordimeform toxicity and which could exert an effect on immune system function. Thus, if chlordimeform were to interact with neurotransmitter receptors that are expressed on lymphocytes residing in lymphoid organs, it would block the effect of NA at alpha-2-adrenergic binding sites and, perhaps, lead to an exaggerated modulation of immune responsiveness.

Lead-induced immune toxicity may involve lead-induced effects on a parameter associated with the nervous system

One xenobiotic that affects immune cell function is the heavy metal, lead. The effect of lead on immune responsiveness *in vivo* remains controversial, with contradictory results having been reported (Bendich *et al.*, 1981; Blakeley *et al.*, 1980; Gaworski and Sharma, 1978; Kimber *et al.*, 1986; Koller *et al.*, 1979; Lawrence, 1981b; Mudzinski *et al.*, 1986; Muller *et al.*, 1977;Neila *et al.*, 1980; Rabinowitz *et al.*, 1990). In addition, when lead-induced toxicity *in vivo* is reported, no evaluation is made to determine a direct *versus* indirect effect of lead on immune responsiveness. Interestingly, the neurotoxic effects of this metal on neurotransmission are well characterized (Chang, 1977; Cheung and Verity, 1983; Winder and Kitchen, 1984) and include an increase in central and peripheral sympathetic neuronal tone. If a neuroimmune relationship influences the level of immune cell function, then the neurotoxic effects of lead may contribute to some of the changes in immune responsiveness observed after *in vivo* exposure to this metal.

It has been less difficult to demonstrate the immunotoxic effect of lead *in vitro*. Reports show that lead exposure of cells *in vitro* induces an increase in anti-SRBC antibody production (Lawrence, 1981a; McCabe and Lawrence, 1991) and an increase in the level of expression of MHC class II molecules on B cells (McCabe and Lawrence, 1990). More importantly, *in vitro* exposure of $CD4^+$ Th cell clones to lead enhances the ability of Th2 cells to produce IL-4, but decreases the ability of Th1 cells to produce IFN-γ (Heo *et al.*, 1996). The latter study also reported that *in vivo* exposure to lead induces an increase in serum IL-4 levels and a decrease in IFN-γ levels. When the *in vivo*-exposed cells were cultured with anti-CD3 mAb *ex vivo*, they also produced increased levels of IL-4 and decreased levels of IFN-γ. Taken together, these results suggest that lead exerts a direct modulatory effect on Th cell function and that this effect appears to be specific for the Th cell subset involved in the response. It has been suggested that this differential effect of lead on Th1 and Th2 cells may upset the balance between the numbers of these cells in the body, resulting in predominantly Th2 cell-mediated responses that, in combination with the direct effects of lead on B cells, enhance the level of autoantibody production (Heo *et al.*, 1996).

The relationship between these direct effects of lead on B cell/Th cell function and the indirect effects of lead on immune cell function via a direct effect of lead on NA production and/or release is speculative at present. A study reported that exposure of cells to lead and dibutyryl cAMP *in vitro* induced a synergistic enhancement of anti-SRBC antibody production (Lawrence, 1981a). *In vivo*, one might propose that the cAMP signal would be induced within a cell via NA stimulation of the β_2AR expressed on immune cells. In this manner, NA release after immunization, concomitant with exposure to lead, may alter both B cell and Th cell function. Work from our laboratory suggests that only functions associated with Th1 cells and B cells would be affected by NA, since these cells express a detectable level of the β_2AR, while Th2 cells do not (Ramer-Quinn *et al.*, 1997; Sanders *et al.*, 1997). Therefore, one might predict that an increase in Th2 cell-dependent antibody production would result from either 1) a direct effect of lead on Th2 cells to increase IL-4 production to 'help' B cells produce more antibody (Heo *et al.*, 1996), 2) a direct effect of lead on B cells to increase the level of MHC Class II expression to make these B cells more effective antigen-presenting cells to recruit the T cell 'help' they need to produce more antibody (McCabe and Lawrence, 1991), or 3) a combination of the direct effects of lead on both Th2 cells and B cells and the normal participation of NA in the Th2-dependent antibody response which results in an enhancement of the number of B cells made capable to differentiate into antibody-secreting cells (Sanders and Powell-Oliver, 1992).

Conversely, one might also predict that either an increase or decrease in Th1 cell-dependent antibody production is possible and that it would result from either 1) a direct effect of lead on B cells to increase the level of MHC class II expression, 2) the normal participation of NA in a Th cell-dependent antibody response that results in an enhancement of the number of B cells made capable of differentiating into antibody-secreting cells, or 3) a combination of both. Of course, the direct effect of lead to decrease Th1 cell cytokine production together with the indirect effect from β_2AR stimulation by NA differentially to modulate IL-2 and IFN-γ production (Ramer-Quinn *et al.*, 1997; Sanders *et al.*, 1997) may result in differential antibody production, also.

Data addressing the immune status of humans exposed to low levels of lead are sparse. Humans exposed to low levels of lead in either the workplace or their living environment for many years do not show any signs of immune dysfunction or toxicity (Kimber *et al.*, 1986; Rabinowitz *et al.*, 1990). These data may be important in light of the recent finding that humans exposed to lead in the workplace for 6 years displayed increased levels of serum NA, but decreased numbers of the β_2AR on their cells when compared to unexposed controls (Chang *et al.*, 1996). The decrease in the level of β_2AR expression in these lead-exposed workers most likely results from chronic exposure to elevated serum NA levels in these exposed individuals. Although the immune status of these individuals was not reported, it is possible that the effect of lowering β_2ARs may serve as a compensatory mechanism to counteract the synergistic immuno-enhancing effects of lead and NA on cell responses. In this manner, the predicted synergistic enhancement would not occur in lead-exposed workers because the level of β_2AR expression on lymphocytes would be low, and thus would not contribute to enhancing the response via a cAMP-mediated synergy.

Summary

In this chapter, evidence supporting the existence of a functional link between the sympathetic nervous system and the immune system has been reviewed, particularly with regard to the role of NA in modulating CD4$^+$ Th cell function. The possible mechanisms by which a xenobiotic might disturb the neuroimmune relationship and contribute to a disruption of CD4$^+$ Th cell homeostasis have been discussed. Many questions remain to be answered as to whether or not neuroimmune mechanisms play a role in xenobiotic-induced changes in CD4$^+$ Th cell function. For example, what cellular and molecular mechanisms are responsible for neurotransmitter-induced modulation of Th cell function? What are the structural and functional pathways that establish an interdependency between the nervous system and immune system? Which system is targeted by a xenobiotic to disrupt this interdependency? How does the toxicity manifest itself when the xenobiotic induces a change in the functioning of one or both systems? By designing experiments to address these questions, we may begin to understand better the mechanisms by which immune homeostasis is disrupted by a xenobiotic-induced change in nervous system function that alters the microenvironment of immune cells responding to antigen.

References

Ackerman, K.D., Bellinger, D.L., Felten, S.Y. and Felten, D.L. (1991a) Ontogeny and senescence of noradrenergic innervation of the rodent thymus and spleen, in *Psychoneuroimmunology* (eds R. Ader, N. Cohen and D.L. Felten), Academic Press, New York, pp. 71–125.

Ackerman, K.D., Madden, K.S., Livnat, S., Felten, S.Y. and Felten, D.L. (1991b) Neonatal sympathetic denervation alters the development of *in vitro* spleen cell proliferation and differentiation. *Brain Behav. Immun.*, **5**, 235–61.

Alava, M.A., DeBell, K.E., Conti, A., Hoffman, T. and Bonvini, E. (1992) Increased intracellular cyclic AMP inhibits inositol phospholipid hydrolysis induced by perturbation of the T cell receptor/CD3 complex but not by G-protein stimulation. *Biochem. J.*, **284**, 189–99.

Bellinger, D.L., Ackerman, K.D., Felten, S.Y. and Felten, D.L. (1992a) A longitudinal study of age-related loss of noradrenergic nerves and lymphoid cells in the rat spleen. *Exp. Neurol.*, **116**, 295–311.

Bellinger, D.L., Lorton, D., Felten, S.Y. and Felten, D.L. (1992b) Innervation of lymphoid organs and implications in development, aging, and autoimmunity. *Int. J. Immunopharmacol.*, **14**, 329–44.

Bendich, A., Belisle, E.H. and Strausser, H.R. (1981) Immune responses of rats chronically fed subclinical doses of lead. *Clin. Exp. Immunol.* **43**,189–94.

Bishopric, N.H., Cohen, H.J. and Lefkowitz, R.J. (1980) Beta-adrenergic receptors in lymphocyte subpopulations. *J. Allergy Clin. Immunol.*, **65**, 29–33.

Blakeley, B.R., Sisodia, C.S. and Mukkur, T.K. (1980) The effect of methylmercury, tetraethyl lead, and sodium arsenite on the humoral immune response in mice. *Toxicol. Appl. Pharmacol.*, 52, 245–54.

Breneman, S.M., Moynihan, J.A., Grota, L.J., Felten, D.L. and Felten, S.Y. (1993) Splenic norepinephrine is decreased in MRL-lpr/lpr mice. *Brain Behav. Immun.* **7**, 135–43.

Bulloch, K. and Pomerantz, W. (1984) Autonomic nervous system innervation of thymic related lymphoid tissue in wild-type and nude mice. *J, Comp, Neurology*, **228**, 57–68.

Burchiel, S.W. and Melmon, K.L. (1979a) Augmentation of the *in vitro* humoral immune response by pharmacologic agents. I: An explanation of the differential enhancement of humoral immunity via agents that elevate cAMP. *Immunopharmacology*, **1**, 137–50.

Burchiel, S.W. and Melmon, K.L. (1979b) Augmentation of the *in vitro* humoral immune response by pharmacologic agents. II. The comparison of the effects of antiproliferative agents with dibutyryl cAMP. *Immunopharmacology*, **1**, 151–63.

Calvo, W. (1968) The innervation of the bone marrow in laboratory animals. *J. Anat.*, **123**, 315–28.

Carlson, S.L., Beiting D.J., Kiani, C.A., Abell, K.M. and McGillis, J.P. (1996) Catecholamines decrease lymphocyte adhesion to cytokine-activated endothelial cells. *Brain Behav. Immun.*, **10**, 55–67.

Cazaux, C.A, Sterin-Borda, L., Gorelik, G. and Cremaschi, G.A. (1995) Down-regulation of beta-adrenergic receptors induced by mitogen activation of intracellular signaling events in lymphocytes. *FEBS Lett.*, **364**, 120–4.

Chang, H.-R., Chen, S.-S., Chen, T.-J., Ho, C.-K., Chiang, H.-C. and Yu, H.-S. (1996) Lymphocyte beta-2-adrenergic receptors and plasma catecholamine levels in lead-exposed workers. *Toxicol. Appl. Pharmacol.*, **139**, 1–5.

Chang, L.W. (1977) Neurotoxic effects of mercury: a review. *Environ. Res.*, **14**, 329–73.

Chelmicka-Schorr, E., Checinski, M. and Arnason, B.G.W. (1988) Chemical sympathectomy augments the severity of experimental allergic encephalomyelitis. *J. Neuroimmunol.*, **17**, 347–50.

Chelmicka-Schorr, E., Kwasniewski, M.N., Thomas, B.E. and Arnason, B.G.W. (1989) The beta-adrenergic agonist isoproterenol suppresses experimental allergic encephalomyelitis in Lewis rats. *J. Neuroimmunol.*, **25**, 203-207.

Chelmicka-Schorr, E., Kwasniewski, M.N. and Wollmann, R.L. (1992) Sympathectomy augments adoptively transferred experimental allergic encephalomyelitis. *J. Neuroimmunol.*, **37**, 99–103.

Cheung, M.K. and Verity, M.A. (1983) Experimental methyl mercury neurotoxicity: *in vivo* and *in vitro* perturbation of brain cell-free synthesis. *Exp. Mol. Pathol.*, **38**, 230–42.

Conolly, M.E. and Greenacre, J.K. (1977) The beta-adrenoceptor of the human lymphocyte and human lung parenchyma. *Br. J. Pharmacol.*, **59**, 17–23.

Costa, L.G. and Murphy, S.D. (1987) Interaction of the pesticide chlordimeform with adrenergic receptors in mouse brain: an *in vitro* study. *Arch. Toxicol.*, **59**, 323–7.

Cremaschi, G.A., Fisher, P. and Boege, F. (1991) Beta-adrenoceptor distribution in murine lymphoid cell lines. *Immunopharmacology*, **22**,195–206.

Cross, R.J., Jackson, J.C., Brooks, W.H., Sparks, D.L., Markesbery, W.R. and Roszman, T.L. (1986) Neuroimmunomodulation: impairment of humoral immune responsiveness by 6-hydroxydopamine treatment. *Immunology*, **57**, 145–52.

Dahlstrom, A.B. and Zetterstrom, B.E.M. (1965) Noradrenergic stores in nerve terminals of the spleen: changes during hemorhagic shock. *Science*, 1**47**, 1583–5.

DeRubertis, F.R., Zenser, T.V., Adler, W.H. and Hudson, T. (1974) Role of cyclic adenosine $3', 5'$-monophosphate in lymphocyte mitogenesis. *J. Immunol.*, **113**, 151–61.

Diamantstein, T. and Ulmer, A. (1975) The antagonistic action of cyclic GMP and cyclic AMP on proliferation of B and T lymphocytes. *Immunology*, **28**, 113–19.

Doctor, S.V., Costa, L.G., Kendall, D.A. and Murphy, S.D. (1982) Trimethyltin inhibits uptake of neurotransmitters into mouse forebrain synaptosomes. *Toxicology*, **25**, 213–21.

Estes, G., Solomon, S.S. and Norton, W.L. (1971) Inhibition of lymphocyte stimulation by cyclic and non-cyclic nucleotides. *J. Immunol.*, **107**, 1489–92.

Feldman, R.D., Hunninghake, G.W. and McArdle,W. (1987) Beta-adrenergic receptor-mediated suppression of interleukin 2 receptors in human lymphocytes. *J. Immunol.*, **139**(10), 3355–9.

Felten, D.L., Livnat, S., Felten, S.Y., Carlson, S.L., Bellinger, D.L. and Yeh, P. (1984) Sympathetic innervation of lymph nodes in mice. *Brain. Res. Bull.* 1**3**, 693–9.

Felten, D.L., Felten, S.Y., Carlson, S.L., Olschowka, J.A. and Livnat, S. (1985) Noradrenergic and peptidergic innervation of lymphoid tissue. *J. Immunol.*, **135**(2), 755s–65s.

Felten, D.L., Ackerman, K.D., Wiegand, S.J. and Felten, S.Y. (1987a) Noradrenergic sympathetic innervation of the spleen: I. Nerve fibres associate with lymphocytes and macrophages in specific compartments of the splenic white pulp. *J. Neurosci. Res.*, **18**, 28–36.

Felten, D.L., Felten, S.Y., Bellinger, D.L., Carlson, S.L., Ackerman, K.D., Madden, K.S., Olschowki, J.A. and Livnat, S. (1987b) Noradrenergic sympathetic neural interactions with the immune system: Structure and function. *Immunol. Rev.*, **100**, 225–60.

Felten, D.L., Felten, S.Y., Bellinger, D.L. and Lorton, D. (1992) Noradrenergic and peptidergic innervation of secondary lymphoid organs: role in experimental rheumatoid arthritis. *Eur., J., Clin., Invest.*, **22**, 37–41.

Felten, S.Y., Bellinger, D.L., Collier, T.J., Coleman, P.D. and Felten, D.L. (1987) Decreased sympathetic innervation of spleen in aged Fischer 344 rats. *Neurobiol. Aging*, **8**, 159–65.

Felten, S.Y., Felten, D.L., Bellinger, D.L., Carlson, S.L., Ackerman, K.D., Madden, K.S., Olschowka, J.A. and Livnat, S. (1988) Noradrenergic sympathetic innervation of lymphoid organs. *Prog. Allergy*, **43**, 14–36.

Finkelman, F.D., Holmes, J., Katona, I.M., Urban, J.F., Beckmann, J.P., Park, L.S., Schooley, K.A., Coffman, R.L., Mosmann, T.R. and Paul, W.E. (1990) Lymphokine control of *in vivo* immunoglobulin isotype selection. *Ann. Rev. Immunol.*, **8**, 303–33.

Fuchs, B.A., Albright, J.W. and Albright, J.F. (1988a) Beta-adrenergic receptor on murine lymphocytes: Density varies with cell maturity and lymphocyte subtype and is decreased after antigen administration. *Cell. Immunol.*, **114**, 231–45.

Fuchs, B.A., Campbell, K.S. and Munson, A.E. (1988b) Norepinephrine and serotonin content of the murine spleen: Its relationship to lymphocyte beta-adrenergic receptor density and the humoral immune response *in vivo* and *in vitro*. *Cell. Immunol.*, **117**, 339–51.

Gaworski, C.L. and Sharma, R.P. (1978) The effects of heavy metals on $_3$H-thymidine uptake in lymphocytes. *Toxicol. Appl. Pharmacol.*, **46**, 305–13.

Hadden, J.W., Hadden, E.M. and Middleton, E. (1970) Lymphocyte blast transformation. I. Demonstration of adrenergic receptors in human peripheral lymphocytes. *Cell Immunol.*, **1**, 583–95.

Hall, N.R., McClure, J.E., Hu, S., Tare, S., Seals, C.M. and Goldstein, A.L. (1982) Effects of 6-hydroxydopamine upon primary and secondary thymus dependent immune responses. *Immunopharmacology*, **5**, 39–48.

Halper, J.P., Brown, R.P., Sweeney, J.A., Kocsis, J.H., Peters, A. and Mann, J. (1988) Blunted beta-adrenergic responsivity of peripheral blood mononuclear cells in endogenous depression. *Arch. Gen. Psychiatry*, **45**, 241–4.

Heo, Y., Parsons, P.J. and Lawrence, D.A. (1996) Lead differentially modifies cytokine production *in vitro* and *in vivo*. *Toxicol. Appl. Pharmacol.*, **138**, 149–57.

Hioe, K.M. and Jones, J.M. (1984) Effects of trimethyltin on the immune system of rats. *Toxicol. Lett.*, **20**, 317–23.

Irwin, M., Hauger, R.L., Jones, L., Provencio, M. and Britton, K.T. (1990) Sympathetic nervous system mediates central corticotropin-releasing factor induced suppression of natural killer cytotoxicity. *J. Pharmacol. Exp. Ther.*, **255**, 101–7.

Jeanningros, R., Mazzola, P., Azorin, J.M., Samuelian-Massa, C. and Tissot, R. (1991) Beta-adrenoceptor density of intact mononuclear leukocytes in subgroups of depressive disorders. *Biol. Psychiatry*, **29**, 789–98.

Johnson, D.L. and Gordon, M.A. (1981) Effect of chronic beta-adrenergic therapy on the human lymphocyte response to concanavalin A. *Res. Commun. Chem. Pathol. Pharmacol.*, **32**, 377–80.

Johnson, D.L., Ashmore, R.C. and Gordon, M.A. (1981) Effects of beta-adrenergic agents on the murine lymphocyte response to mitogen stimulation. *J. Immunopharmacol.*, **3**, 205–19.

Karaszewski, J.W., Reder, A.T., Maselli, R., Brown, M. and Arnason, B.G.W. (1990) Sympathetic skin responses are decreased and lymphocyte beta-adrenergic receptors are increased in progressive multiple sclerosis. *Ann. Neurol.*, **27**, 366–72.

Kasahara, K., Tanaka, S., Ito, T. and Hamashima, Y. (1977) Suppression of the primary immune response by chemical sympathectomy. *Res. Commun. Chem. Pathol. Pharmacol.*, **16**, 687–94.

Khan, M.M., Sansoni, P., Silverman, E.D., Engleman, E.G. and Melmon, K.L. (1986) Beta-adrenergic receptors on human suppressor, helper, and cytolytic lymphocytes. *Biol. Psychiatry*, **35**, 1137–42.

Kimber, I., Stonard, M.D., Gidlow, D.A. and Niewola, Z. (1986) Influence of chronic low-level exposure to lead on plasma immunoglobulin concentration and cellular immune function in man. *Int. Arch. Occup. Environ. Health*, **57**, 117–25.

Koller, L.D., Roan, J.G. and Kerkvliet, N.I. (1979) Mitogen stimulation of lymphocytes in CBA mice exposed to lead and cadmium. *Environ. Res.*, **19**, 177–88.

Kostrzewa, R.M. and Jacobwitz, D.M. (1974) Pharmacologial actions of 6-hydroxydopamine. *Pharmacol. Rev.*, **26**, 199–288.

Kouassi, E., Li, Y.S., Boukhris, W., Millet, I. and Revillard, J.-P. (1988) Opposite effects of the catecholamines dopamine and norepinephrine on murine polyclonal B-cell activation. *Immunopharmacology*, **16**, 125–37.

Krause, D.S. and Deutsch, C. (1991) Cyclic AMP directly inhibits IL-2 receptor expression in human T cells: expression of both p55 and p75 subunits is affected. *J. Immunol.*, **146**, 2285–94.

Krawietz, W., Werdan, K., Schober, M., Erdmann, E., Rindfleisch, G.E. and Hannig, K. (1982) Different numbers of beta-receptors in human lymphocyte subpopulations. *Biochem. Pharmacol.*, **31**, 133–6.

Kruszewska, B., Felten, S.Y. and Moynihan, J.A. (1995) Alterations in cytokine and antibody production following chemical sympathectomy in two strains of mice. *J. Immunol.*, **155**, 4613–20.

Laustiola, K.E., Kotamaki, M., Lassila, R., Kallioniemi, O.-P. and Manninen, V. (1991) Cigarette smoking alters sympathoadrenal regulation by decreasing the density of beta-2-adrenoceptors. A study of monitored smoking cessation. *J. Cardiovasc. Pharmacol.*, **17**, 923–8.

Lawrence, D.A. (1981a) Heavy metal modulation of lymphocyte activities – II. Lead, an *in vitro* mediator of B cell activation. *Int. J. Immunopharmacol.*, **3**, 153–61.

Lawrence, D.A. (1981b) The *in vivo* and *in vitro* effects of lead on humoral and cell-mediated immunity. *Infect. Immun.*, **31**, 136–43.

Ledbetter, J.A., Parsons, M., Martin, P.J., Hansen, J.A., Rabinovitch, P.S. and June, C.H. (1986) Antibody binding to CD5 (Tp67) and Tp44 T cell surface molecules: effects on cyclic nucleotides, cytoplasmic free calcium, and cAMP-mediated suppression. *J. Immunol.*, **137**, 3299–305.

Liblau, R.S., Singer, S.M. and McDevitt, H.O. (1995) Th1 and Th2 CD4+ T cells in the pathogenesis of organ-specific autoimmune diseases. *Immunol. Today*, **16**, 34–8.

Livnat, S., Felten, S.Y., Carlson, S.L., Bellinger, D.. and Felten, D.L. (1985) Involvement of peripheral and central catecholamine systems in neural-immune interactions. *J. Neuroimmunol.*, **10**, 5–30.

Loveland, B.E., Jarrott, B. and McKenzie, I.F.C. (1981) The detection of beta-adrenoceptors on murine lymphocytes. *Int. J. Immunopharmacol.*, 45–55.

Madden, K.S. and Livnat, S. (1991) Catecholamine action and immunologic reactivity, in *Psychoneuroimmunology* (eds R. Ader, N. Cohen and D.L. Felten), Academic Press, New York, pp. 283–310.

Madden, K.S., Felten, S.Y., Felten, D.L., Sundaresan, P.R. and Livnat, S. (1989) Sympathetic neural modulation of the immune system. I. Depression of T cell immunity *in vivo* and *in vitro* following chemical sympathectomy. *Brain Behav, Immun.*, **3**, 72–89.

Madden, K.S., Bellinger, D.L., Tong, A., Housel, J. and Costello, N. (1992) Plasticity of noradrenergic nerves in spleens from aged F344 rats following chemical sympathectomy. *Soc. Neurosci. Abstr.*, **18**, 1009.

Madden, K.S., Felten, S.Y., Felten, D.L., Hardy, C.A. and Livnat, S. (1994a) Sympathetic nervous system modulation of the immune system. II. Induction of lymphocyte proliferation and migration *in vivo* by chemical sympathectomy. *J. Neuroimmunol.*, **49**, 67–75.

Madden, K.S., Moynihan, J.A., Brenner, G.J., Felten, S.Y., Felten, D.L. and Livnat, S. (1994b) Sympathetic nervous system modulation of the immune system. III. Alterations in T and B cell proliferation and differentiation *in vitro* following chemical sympathectomy. *J. Neuroimmunol.*, **49**, 77–87.

Madden, K.S., Sanders, V.M. and Felten, D.L. (1995) Catecholamine influences and sympathetic neural modulation of immune responsiveness. *Annu. Rev. Pharmacol. Toxicol.*, **35**, 417–48.

Maes, M., Bosmans, E., Suy, E., Minner, B. and Raus, J. (1989) Impaired lymphocyte stimulation by mitogens in severely depressed patients. A complex interface with HPA-axis hyperfunction, noradrenergic activity, and the ageing process. *Br. J. Psych.* **155**, 793–8.

Mary, D., Peyron, J.-F., Auberger, P., Aussel, C. and Fehlmann, M. (1989) Modulation of T cell activation by differential regulation of the phosphorylation of two cytosolic proteins. *J. Biol. Chem.*, **264**, 14498–502.

McCabe, M.J. and Lawrence, D.A. (1990) The heavy metal lead exhibits B cell-stimulatory factor activity by enhancing B cell Ia expression and differentiation. *J. Immunol.*, **145**, 671–7.

McCabe, M.J. and Lawrence, D.A. (1991) Lead, a major environmental pollutant, is immunomodulatory by its differential effects on CD4+ T cell subsets. *Toxicol. Appl Pharmacol.* **111**, 13–23.

McCuskey, R.S. and McCuskey, P.A. (1977) *In vivo* microscopy of the spleen. *Bibliogr. Anat.*, **16**, 121–5.

McKenna, M.J. and Distefano, V. (1977) Carbon disulfide. II. A proposed mechanism for the action of carbon disulfide on dopamine beta-hydroxylase. *J. Pharmacol. Exp. Ther.*, **202**, 253–66.

Meurs, H., Van Den Bogaard, W., Kauffman, H.F. and Bruynzeel, P.L.B. (1982) Characterization of (-)-[^{3}H]dihydroalprenolol binding to intact and broken cell preparations of human peripheral blood lymphocytes. *Eur. J. Pharmacol.*, **85**, 185–94.

Miller, R.A. (1996) The aging immune system: primer and prospectus. *Science*, **273**, 70–4.

Miossec, P., Chomarat, P. and Dechanet, J. (1996) Bypassing the antigen to control rheumatoid arthritis. *Immunol. Today*, **17**, 170–3.

Mosmann, T.R., Schumacher, J.H., Street, N.E., Budd, R., O'Garra, A., Fong, T.A.T., Bond, M.W., Moore, K.W.M., Sher, A. and Fiorentino, D.F. (1991) Diversity of cytokine synthesis and function of mouse $CD4^+$ T cells. *Immunol, Rev.* **123**, 209–29.

Mudzinski, S.P., Rudofsky, U.H., Mitchell, D.G. and Lawrence, D.A. (1986) Analysis of lead effects in *in vivo* antibody-mediated immunity in several mouse strains. *Toxicol. Appl. Pharmacol.*, **83**, 321–30.

Muller, S., Gilbert, K.E., Krause, C., Gross, U., L'Age-Stehr, J. and Diamantstein, T. (1977) Suppression of delayed type hypersensitivity of mice by lead. *Experientia*, **33**, 667–8.

Neila, B.A., Taddeini, L., McJilton, C.E. and Handwerger, B.S. (1980) Decreased T cell function in mice exposed to chronic low levels of lead. *Clin. Exp. Immunol.*, **39**, 746–9.

Park, D.J., Min, H.K. and Rhee, S.G. (1992) Inhibition of CD3-linked phospholipase C by phorbol ester and by cAMP is associated with decreased phosphotyrosine and increased phosphoserine contents of PLC-gamma-1. *J. Biol. Chem.*, **267**, 1496–501.

Pochet, R., Delespesse, G., Gausset, P.W. and Collet, H. (1979) Distribution of beta-adrenergic receptors on human lymphocyte subpopulations. *Clin. Exp. Immunol.* **38**, 578–84.

Rabinowitz, M.B., Allred, E.N., Bellinger, D.C., Leviton, A. and Needleman, H.L. (1990) Lead and childhood propensity to infectious and allergic disorders: Is there an association? *Bull. Environ. Contam. Toxicol.*, **44**, 657–60.

Radojcic, T., Baird, S., Darko, D., Smith, D. and Bulloch, K. (1991) Changes in beta-adrenergic receptor distribution on immunocytes during differentiation: An analysis of T cells and macrophages. *J. Neurosci. Res.* **30**, 328–35.

Ramer-Quinn, D.S., Baker, R.A. and Sanders, V.M. (1997) Activated Th1 and Th2 cells differentially express the beta-2-adrenergic receptor: A mechanism for selective modulation of Th1 cell cytokine production. *J. Immunol.*, **159**, 4857–67.

Reilly, F.D., McCuskey, P.A. and Meineke, H.A. (1976) Studies of the hematopoietic microenvironment. VIII. Adrenergic and cholinergic innervation of the murine spleen. *Anat. Rec.*, **185**, 109–18.

Reilly, F.D., McCuskey, P.A., Miller, M.L., McCuskey, R.S. and Meineke, H.A. (1979) Innervation of the periarteriolar lymphatic sheath of the spleen. *Tissue Cell*, **11**, 121–6.

Riegele, L. (1929) Uber die mikroscopische Innervation der Milz. *Z. Zellforsch. mikrosk. Anat.*, **9**, 511–33.

Roper, R.L., Conrad, D.H., Brown, D.M., Warner, G.L. and Phipps, R.P. (1990) Prostaglandin E2 promotes IL-4-induced IgE and IgG1 synthesis. *J. Immunol.*, **145**, 2644–51.

Sanders, V.M. (1995) The role of adrenoceptor-mediated signals in the modulation of lymphocyte function. *Adv. Neuroimmunol.*, **5**, 283–98.

Sanders, V.M. and Munson, A.E. (1984a) Kinetics of the enhancing effect produced by norepinephrine and terbutaline on the murine primary antibody response *in vitro*. *J. Pharmacol. Exp. Ther.*, **231**(3), 527–31.

Sanders, V.M. and Munson, A.E. (1984b) Beta-adrenoceptor mediation of the enhancing effect of norepinephrine on the murine primary antibody response *in vitro*. *J. Pharmacol. Exp. Ther.*, **230**(1), 183–92.

Sanders, V.M. and Munson, A.E. (1985) Role of alpha adrenoceptor activation in modulating the murine primary antibody response *in vitro*. *J. Pharmacol. Exp. Ther.*, **232**(2), 395–400.

Sanders, V.M. and Powell-Oliver, F.E. (1992) Beta-2-adrenoceptor stimulation increases the number of antigen-specific precursor B lymphocytes that differentiate into IgM-secreting cells without affecting burst size. *J. Immunol.*, **148**, 1822–8.

Sanders, V.M., Baker, R.A., Ramer-Quinn, D.S., Kasprowicz, D.J., Fuchs, B.A. and Street, N.A. (1997) Differential expression of the beta-2-adrenergic receptor by Th1 and Th2 clones: implications for cytokine production and B cell help. *J. Immunol.*, **158**, 4200–10

Seder, R.A. and Paul, W.E. (1994) Acquisition of lymphokine-producing phenotype by CD4+ T cells. *Ann. Rev. Immunol.*, **12**, 635–73.

Selliah, N., Bartik, M.M., Carlson, S.L., Brooks, W.H. and Roszman, T.L. (1995) cAMP accumulation in T-cells inhibits anti-CD3 monoclonal antibody-induced actin polymerization. *J. Neuroimmunol.*, **56**, 107–12.

Shimizu, N., Hori, T. and Nakane, H. (1994) An interleukin-1-beta-induced noradrenaline release in the spleen is mediated by brain corticotropin-releasing factor: an *in vivo* microdialysis study in conscious rats. *Brain Behav. Immun.*, **7**, 14–23.

Shopp, G.M. Jr, McCay, J.A. and Holsapple, M.P. (1985) Suppression of the antibody response by a formamidine pesticide: dependence on the route of exposure. *J. Toxicol. Environ. Hlth.*, **15**, 293–304.

Shreurs, A.J.M., Versteeg, D.H.G. and Nijkamp, F.P. (1982) Involvement of catecholamines in *Haemophilus influenzae* induced decrease of beta-adrenoceptor function. *Naunyn-Schmiedeberg's Arch. Pharmacol.*, **320**, 235–9.

Smith, J.W., Steiner, A.L., Newberry, M. and Parker, C.W. (1971a) Cyclic adenosine 3′, 5′-monophosphate in human lymphocytes. Alterations after phytohemagglutinin stimulation. *J. Clin. Invest.*, **50**, 432–41.

Smith, J.W., Steiner, A.L. and Parker, C.W. (1971b) Human lymphocyte metabolism. Effects of cyclic and noncyclic nucleotides on stimulation by phytohemagglutinin. *J. Clin. Invest.*, **50**, 442–8.

Stein, S.H. and Phipps, R.P. (1991) Antigen-specific IgG2a production in response to prostaglandin E2, immune complexes, and IFN-gamma. *J. Immunol.*, **147**, 2500–6.

Stevens, T.L., Bossie, A., Sanders, V.M., Fernandez-Botran, R., Coffman, R.L., Mosmann, T.R. and Vitetta, E.S. (1988) Regulation of antibody isotype secretion by subsets of antigen-specific helper T cells. *Nature*, **334**, 255–8.

Teh, H.S. and Paetkau, V. (1976) Regulation of immune responses. II. The cellular basis of cyclic AMP effects on humoral immunity. *Cell., Immunol.*, **24**, 220–9.

Thoman, M.L. and Weigle, W.O. (1989) The cellular and subcellular bases of immunosenescence. *Adv. Immunol.*, **46**, 221–61.

Van Tits, L.J.H., Michel, M.C., Grosse-Wilde, H., Happel, M., Eigler, F.-W., Soliman, A. and Brodde, O.-E. (1990) Catecholamines increase lymphocyte beta-2-adrenergic receptors via a beta-2-adrenergic, spleen-dependent process. *Am. J. Physiol.*, **258**, E191–E202.

Van Tits, L.J.H., Michel, M.C., Motulsky, H.J., Maisel, A.S. and Brodde, O.-E. (1991) Cyclic AMP counteracts mitogen-induced inositol phosphate generation and increases in intracellular Ca^{2+} concentrations in human lymphocytes. *Br. J. Pharmacol.*, **103**, 1288–94.

Vischer, T.L. (1976) The differential effect of cAMP on lymphocyte stimulation by T- or B-cell mitogens. *Immunology*, **30**, 735–9.

Wacholtz, M.C., Minakuchi, R. and Lipsky, P.E. (1991) Characterization of the 3′, 5′-cyclic adenosine monophosphate-mediated regulation of IL2 production by T cells and Jurkat cells. *Cell. Immunol.*, **135**, 285–98.

Watson, J., Epstein, R. and Cohn, M. (1973) Cyclic nucleotides as intracellular mediators of the expression of antigen-sensitive cells. *Nature*, **246**, 405–9.

Weinstein, Y., Segal, S. and Melmon, K.L., (1975) Specific mitogenic activity of 8-Br-guanosine 3′,5′-monophosphate (Br-cyclic GMP) on B lymphocytes. *J. Immunol.*, **115**, 112–17.

Westly, H.J. and Kelley, K.W. (1987) Down-regulation of glucocorticoid and beta-adrenergic receptors on lectin-stimulated splenocytes. *Proc. Soc. Exp. Biol. Med.* **185**, 211–18.

Wiegmann, K., Muthyala, S., Kim, D.H., Arnason, B.G.W. and Chelmicka-Schorr, E. (1995) Beta-adrenergic agonists suppress chronic/relapsing experimental allergic encephalomyelitis (CREAE) in Lewis rats. *J. Neuroimmunol.*, **56**, 201–6.

Williams, J.M. and Felten, D.L. (1981) Sympathetic innervation of murine thymus and spleen: a comparative histofluorescence study. *Anat. Rec.*, **199**, 531–42.

Williams, J.M., Peterson, R.G., Shea, P.A., Schmedtje, J.F., Bauer, D.C. and Felten, D.L. (1981) Sympathetic innervation of murine thymus and spleen: Evidence for a functional link between the nervous and immune systems. *Brain Res. Bull.*, **6**(1), 83–94.

Williams, L.T., Snyderman, R. and Lefkowitz, R.J. (1976) Identification of beta adrenergic receptors in human lymphocytes by (-)^{3}H-alprenolol binding. *J. Clin. Invest.*, **57**, 149–55.

Wiltrout, R.W., Ercegovich, C.D. and Ceglowski, W.S. (1978) Humoral immunity in mice following oral administration of selected pesticides. *Bull. Environ. Contam. Toxicol.*, **20**, 423–31.

Winder, C. and Kitchen, I. (1984) Lead neurotoxicity: a review of the biochemical, neurochemical and drug induced behavioural evidence. *Prog. Neurobiol.*, **27**, 7–87.

10 Endocrine Regulation of T Lymphocyte Subpopulations

Graham A.W. Rook, Robert W. Baker, Sahal Al Nakhli and Ali Zumla

Introduction

It is no longer possible to consider immunology independently of endocrinology. There is a continuous dialogue between the two systems that constantly regulates T lymphocyte functions and cytokine profiles. This chapter discusses the impact of glucocorticoid hormones (GC) and of dehydroepiandrosterone sulphate (DHEAS) on T lymphocytes and the ways in which the effects of these steroids are themselves regulated. Other steroid hormones (such as sex steroids) also affect T cells, but GC and DHEAS are chosen for discussion here because they are the most important, present at similar levels in both sexes, and have contrasting effects on the overall function of the immune system. GC are anti-inflammatory, and in spite of more complex short-term effects, tend eventually to bias the lymphocyte response towards a Th2 cytokine profile that is rich in interleukin-4 (IL-4), but inappropriate for the cell-mediated immunity required for resistance to intracellular parasites and cancer. Certain toxic chemicals such as 2,3,7,8-tetrachlorodibenzo-*p*-dioxin (TCCD) bind to a steroid-type receptor leading to changes in gene expression and immunosuppression that in some ways mimic the effects of GC. In contrast, DHEAS, or perhaps an unidentified metabolite of this steroid, is pro-inflammatory, and tends to oppose effects of GC and bias the T lymphocyte cytokine profile towards Th1, with production of interleukin-2 (IL-2) and interferon gamma (IFNγ). The ratio of Th1 to Th2 activity is crucial in murine and probably in human infections. For instance immunization to evoke the wrong cytokine profile before infection with *M. tuberculosis* leads to *increased* susceptibility rather than to protection (Rook and Hernandez-Pando, 1996). Tuberculosis is used to illustrate the arguments throughout.

The effects of these steroids cannot be determined simply by measuring serum levels. Their eventual impact on a given target tissue depends on the

T Lymphocyte Subpopulations in Immunotoxicology. Edited by Ian Kimber and MaryJane K. Selgrade.
 ISBN 0 471 97194 4

diurnal rhythm of the GC, on patterns of metabolism and inactivation within the target organs themselves, and on interaction with other active compounds. Therefore in this chapter we first consider the molecular mechanisms associated with the actions of GC, before describing other physiological factors that influence these actions. We then outline the 'anti-glucocorticoid' functions of DHEAS. The final sections of the chapter deal with the overall impact of the steroids on T cell function and disease.

The physiology of glucocorticoid action

THE MODE OF ACTION OF GLUCOCORTICOIDS

Glucocorticoid hormones (GC) bind to intracellular receptors that exist complexed to the 90 kDa heat shock protein (hsp90) and to other proteins of lower molecular weight. Dissociation of the receptor–GC complex from hsp90 then occurs, and dimers are formed (Wilckens, 1995). These can then interact with 'glucocorticoid response elements' (GRE) that are involved in the regulation of expression of numerous genes including some that encode cytokines and other genes involved in the regulation of inflammation and immunity (Figure 10.1).

However, this relatively simple picture is enormously complicated by several other factors that render the precise consequence of exposing a cell to glucocorticoid difficult to predict. These are outlined in the next few paragraphs, and numerous further examples occur later in the text.

INTERACTION OF GLUCOCORTICOID-RECEPTOR COMPLEXES WITH OTHER TRANSCRIPTION FACTORS

In addition to their ability to interact with glucocorticoid-specific GRE, the receptor–glucocorticoid complexes can modify the functions of transcription factors such as AP-1, NFκB and JAK STAT (Wilckens, 1995; Stöcklin *et al.*, 1996) and the latter are also involved in regulation of transcription of genes relevant to the immune response (Figure 10.1.). Since glucocorticoid signalling pathways interact with other signalling pathways and transcription factors, it is not surprising that the effect of GC can be profoundly modified, or even reversed, by agonists, including compounds of interest to toxicologists, that activate these other pathways. Examples of this will be discussed in greater detail in relation to the control of T cell apoptosis.

MODULATION OF GC ACTION VIA ALTERNATIVE RECEPTORS FOR GLUCOCORTICOIDS

Alternative splicing of the human glucocorticoid receptor (GR) primary transcript produces two isoforms hGRα and hGRβ. The β form can inhibit the activation of gene expression by the α form, so it may have a role as a

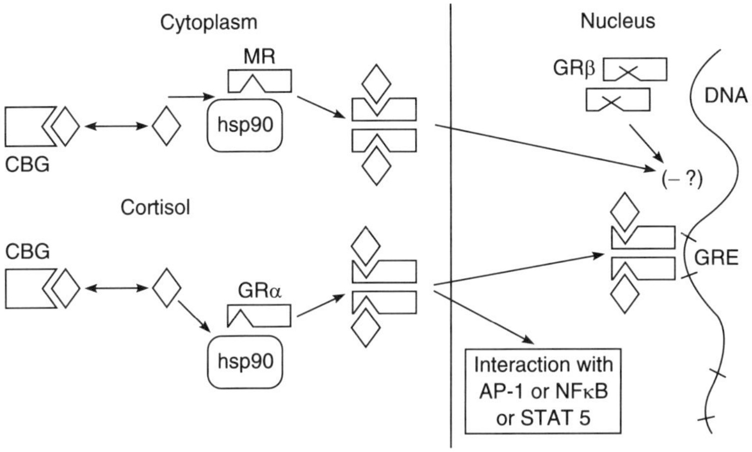

Figure 10.1 Cortisol is mostly bound to corticosteroid binding globulin (CBG). Once free cortisol (represented as diamonds) has entered the cytoplasm it causes dissociation of the glucocorticoid receptor (GR or type 2 receptor) and mineralocorticoid receptor (MR or type 1 receptor) from the 90 kDa heat shock protein (hsp90) and other smaller proteins that are not shown. Receptor dimers are formed that travel to the nucleus and bind to glucocorticoid response elements (GRE) that regulate transcription of the associated genes. MR dimers can act as functional antagonists of the GR, and an alternative transcript of the GR (GRβ) may also act as a physiological inhibitor [indicated by (–?)]. The receptor dimers can also interact with, and modulate the function of several other transcription factors such as AP-1, NFκB and STAT5, or perhaps the cortisol-GR complexes form heterodimers with such transcription factors.

physiological inhibitor (Oakley *et al.*, 1996). Little is yet known about its physiological role (Figure 10.1).

The situation is complicated further by the existence of a second type of receptor for endogenous glucocorticoids. The glucocorticoid receptor (GR) already discussed is also called the type 2 receptor. However GC also bind to the type 1 receptor (Figure 10.1), usually known as the mineralocorticoid receptor (MR). This can have important consequences. For instance, the rat spleen expresses MR as well as GR. Proliferative responses to Con A (a T lymphocyte mitogen) can be *stimulated* by GC acting via MR, but inhibited by GC acting via GR (Wiegers *et al.*, 1993, 1994). Thus the MR–cortisol complex can act as a functional antagonist of the GR–cortisol complex, the extent of which relates to the diurnal rhythm of cortisol release (see below).

Additional factors modulating glucocorticoid function *in vivo*

The eventual extent and nature of glucocorticoid-mediated effects *in vivo* depend not only on the conventionally measured circulating GC levels, but also on diurnal rhythms, and on a series of metabolic events that regulate GC action in individual tissues.

THE DIURNAL RHYTHM OF GLUCOCORTICOID PRODUCTION

In normal individuals, there is a diurnal rhythm of serum cortisol. In man the level peaks in the morning and then declines throughout the day. Since cortisol has a 10-fold higher affinity for the MR than it has for the GR, this rhythm tends to result in occupation of the MR *and* GR in the morning, but only the MR later in the day. T cells in the thymus and peripheral blood express mostly or exclusively GR. These tend to mediate a lymphocyte suppressive effect, so these T cells alternate between steroid-suppressed and non-steroid-suppressed states in each 24-h cycle.

In other organs such as spleen, the situation is more complex, as explained in the previous section. In an organ that expresses MR as well as GR, the trough in the diurnal cycle of GC levels can result not just in diminished GC effects, but even in their replacement with contrary effects mediated via the MR. This has been demonstrated in rodents (Wiegers *et al.*, 1993, 1994) which have an equally pronounced diurnal rhythm that is approximately the reverse of that seen in man The balance of these effects must vary naturally during the 24-h cycle.

THE DIURNAL RHYTHM OF CORTISOL, AND LYMPHOCYTE FUNCTION

There is suggestive evidence that this physiological cycle is sufficient to cause a corresponding cyclical variation in lymphocyte function in humans (Fukuda *et al.*, 1994; Hiemke *et al.*, 1995; Petrovsky and Harrison, 1995; Petrovsky *et al.*, 1994). For instance, release of IFNγ in response to tetanus toxoid or PPD (Petrovsky *et al.*, 1994), and the proliferative response of peripheral blood T cells to tetanus toxoid (Hiemke *et al.*, 1995), both show an inverse correlation with plasma cortisol levels. Diurnal rhythms can also be shown to be relevant *in vivo*, and there is cyclical variation in the response to challenge with oxazolone in sensitized rats (Pownall *et al.*, 1979).

Clearly, these correlations could be coincidental. Some further support for a role for the cortisol rhythm comes from the observation that diurnal changes in circulating T cell subsets are comparable to those induced by prednisolone (Fukuda *et al.*, 1994). However, there are several other immunologically relevant diurnal rhythms that may not be secondary to the cortisol rhythms. There is a diurnal rhythm of interleukin-6 (IL-6), with the highest levels throughout the night (peaking at 1.00 a.m.) with a nadir at 10.00 a.m. (Sothern *et al.*, 1995). The complexity of the diurnal rhythms found with tumour necrosis factor alpha (TNFα), interleukin-10 (IL-10) and granulocyte–macrophage colony-stimulating factor (GMCSF) was considerable, with unique biphasic patterns for each one, while interleukin-2 (IL-2) showed a single peak at noon (Young *et al.*, 1995). Soluble IL-2 receptors show a peak at 12.29 and a nadir at 4.14, so the peak at least correlates with IL-2 itself (Lemmer *et al.*, 1992).

DISTURBANCES OF THE CORTISOL DIURNAL RHYTHM IN DISEASE

If the diurnal rhythm of cortisol is an important physiological clock that affects T cell and macrophage function, then changes in that rhythm must have important effects on the immune system. Unfortunately the endocrinological 'fashion' has been to measure the early morning cortisol only, whereas from an immunological point of view, that is the least interesting time point. A sample at 8.00 p.m. is much more revealing. However, abnormalities of the rhythm have been documented in several situations where there are profound changes in the immune system; stress and chronic infection.

The diurnal rhythm of cortisol is lost during acute stress caused by strenuous physical exertion accompanied by sleep and energy deprivation (Opstad, 1994). There is also very striking loss of the diurnal rhythm in plasma cortisol in tuberculosis patients (Sarma *et al.*, 1990). This may be a feature of chronic infection, and the diurnal rhythms of ACTH and cortisol are blunted in some human immunodeficiency virus (HIV) seropositive subjects (Lortholary *et al.*, 1996). This may relate to the observation that the diurnal rhythm in $CD4^+$ cell numbers (rise between 0800 and 2200 h) is also blunted in HIV (Malone *et al.*, 1990) or according to one report, lost altogether (Martini *et al.*, 1988).

The regulatory role of cortisol metabolism in the periphery

The local concentration of cortisol in any cell or tissue is not dependent only on the concentration reaching that tissue. Most tissues need to regulate local concentrations to suit their own particular needs.

GLUCOCORTICOID CATABOLISM BY TARGET CELLS

Clearly, differences in receptor numbers, or affinity, or in the signalling pathways can alter the GC-sensitivity of individual cells. The ability to catabolize the cortisol may be another factor. It was noted many years ago that strains of fibroblasts that were resistant to cortisol-mediated growth inhibition were able to catabolize the cortisol via a variety of oxidations and reductions (Berliner and Dougherty, 1961; Dougherty *et al.*, 1960). There may be a similar variation in the sensitivity of individual lymphocytes to GC. Catabolism of cortisol by lymphocytes from patients with systemic lupus erythematosus (SLE) was found to be significantly increased relative to normal. Lymphocytes from RA patients were not different from normal (Klein *et al.*, 1990).

The cortisol–cortisone shuttle

In the kidney, the enzyme 11 β-hydroxysteroid type 2 (11βHSD-2) converts cortisol into inactive cortisone (Figure 10.2), and so stops the MR from binding cortisol (Walker, 1994). This allows mineralocorticoid functions to be mediated almost exclusively by aldosterone, although this hormone is present at much lower levels. In normal individuals an oxidoreductase in the liver known as 11βHSD-1 then rapidly converts the resulting cortisone back to cortisol, so that the circulating cortisol is maintained 5–10 times higher than the circulating cortisone (Figure 10.2).

In reality, the situation is more complex. Other organs such as lymph nodes and the lung also possess 11βHSD (Berliner and Dougherty, 1961; Daynes *et al.*, 1995; Hubbard *et al.*, 1994), but they may not be either the conventional type 1 or type 2 enzymes, and their regulation is interesting. The enzyme in

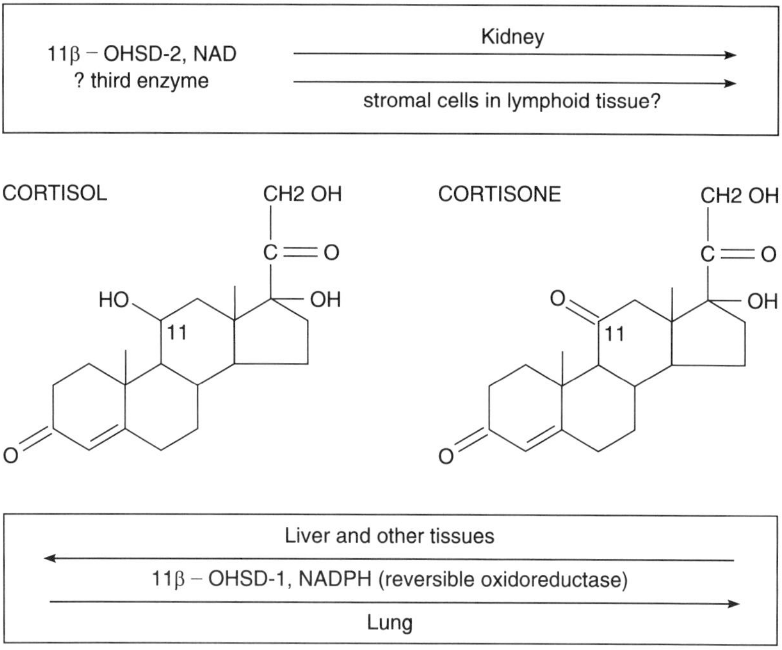

Figure 10.2 The cortisol–cortisone shuttle. Cortisol is rapidly converted to cortisone in the kidney. It is converted back to cortisol in the liver. However, most tissues contain enzymes of this family, and therefore can regulate the local cortisol concentration independently of the circulating levels. Such enzymes in large organs such as the lung must also affect the systemic ratio of cortisol to cortisone. The enzyme 11βHSD-1 is a reversible oxidoreductase that can be regulated, though this regulation is not understood. We postulate that its direction of action in the lung may change in disease states such as tuberculosis. Lymphoid tissue contains an enzyme that may be neither 11β-HSD-1 nor 11β-HSD-2.

lymph nodes may resemble a novel isoform recently observed in a choriocarcinoma line (Deckx and De Moor, 1996; Gomez-Sanchez *et al.*, 1996; Hennebold *et al.*, 1996), while the enzyme in the lung resembles the one in the liver (Figure 10.2), though in the lungs of normal donors it is running in reverse, converting cortisol into inactive cortisone (Hubbard *et al.*, 1994). Therefore the type 1 enzyme is clearly a reversible and regulated oxidoreductase. It has been observed that in granulosa cell preparations, removal of contaminating T cells alters the activity of the enzyme, suggesting that it may be regulated by cytokines (Evangelatou *et al.*, 1996).

THE CORTISOL–CORTISONE SHUTTLE IN DISEASE

In tuberculosis, analysis of GC metabolites in urine reveals a massive switch in the cortisol/cortisone balance towards the active compound (Baker *et al.*, 1996; Rook *et al.*, 1996). The anatomical site of the abnormal metabolism is not known but in view of the points made in the previous paragraph, a cytokine-mediated effect on the type 1 enzyme in the lung is an obvious possibility (Hubbard *et al.*, 1994) .

It is possible that this change in the systemic cortisol:cortisone balance also plays a role in the changes in diurnal rhythm discussed earlier (Sarma *et al.*, 1990).

Antiglucocorticoid functions of dehydroepiandrosterone

The most abundant steroid produced by the human adrenal after adrenarche is dehydroepiandrosterone sulphate (DHEAS). Either DHEA itself, or unidentified metabolites clearly oppose some functions of GC (Figure 10.3) and are therefore essential to the theme of this review (Blauer *et al.*, 1991; Daynes *et al.*, 1995; Wright *et al.*, 1992). However, DHEA also acts as a precursor for oestrogens and testosterone, and has several other categories of function, most of which are poorly understood.

The human adrenal secretes 10–15 mg DHEAS/day in young adults, but serum levels then fall steadily with increasing age. It is strongly bound to albumin and undergoes renal tubular reabsorption. In young adults it is present at concentrations close to 4 μg/ml. Eventually most of it is converted to DHEA. Thus in humans the major source of DHEA is DHEAS. Most DHEA circulates free, but some is bound weakly to albumin.

DHEA IN RODENTS; DIFFERENCES BETWEEN RODENTS AND HUMANS

One problem that must be faced when assessing the literature on DHEA is the fact that its physiology in rodents is clearly quite different from its physiology in humans. Thus whereas there are about 4 μg/ml of DHEAS in adult human

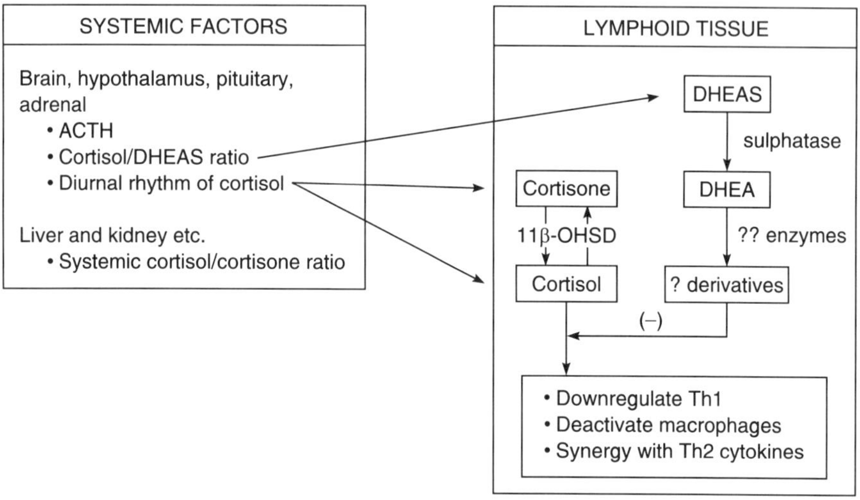

Figure 10.3 Factors that determine the effective cortisol function in lymphoid tissue. Systemic factors (in the box on the left) include the output of the adrenal, the diurnal rhythm of this output, and the ratio of cortisol to DHEAS. The cortisol/cortisone shuttle then modulates the actual delivery of cortisol to the tissues. Within lymphoid tissues, both DHEAS and cortisol are subject to regulated local metabolism. The pathway of conversion of DHEA to the putative active antiglucocorticoid is unknown, because we do not know which are the active metabolites.

plasma, rat plasma contains < 1 ng/ml of this hormone (Robel and Baulieu, 1995). Unexpectedly, in rodents the brain is the site where concentrations of DHEAS (and, at a lower level, DHEA) are highest. Thus rat brain contained an average of 3.5 ng/g of tissue over a 24 h period (Robel and Baulieu, 1995). DHEAS in brain has clear neurological functions that are outside the scope of this review. The DHEAS in brain is thought to be locally synthesized, and it is possible that it is not synthesized at all in rodent adrenals, since this organ lacks 17α-hydroxylase. Nevertheless, DHEAS and fatty acid esters of DHEA were found in the adrenals of rats at 5.1 ± 2.6 and 16.1 ± 6.4 ng/g of tissue, respectively (Robel and Baulieu, 1995). If the rodent adrenal does make DHEA, it must use an unusual pathway. Alternatively, the immunological functions of DHEA may be performed by some other derivative that can be reached via a different route.

EFFECTS OF DHEA ON IMMUNE RESPONSES IN RODENTS

In view of the facts outlined in the previous paragraph, it is clear that much of the literature on the effects of DHEA or DHEAS in rodents should be interpreted with caution, or discounted altogether. Some authors have used doses of 1 g/kg (Morfin and Courchay, 1994), in a species with physiological levels < 1 ng/ml of plasma or gram of tissue! Doses of this magnitude result in rapid formation of androgens and oestrogens, with permanent priapism in male mice. Sex hormones are known to have immunological effects of their own (reviewed in Wilder, 1995) so such protocols clearly cause confusion. Other authors using almost equally excessive doses (10 mg DHEA/mouse, i.e. about 0.5 g/kg) showed that DHEA enhances influenza immunization in aged mice (Danenberg *et al.*, 1995). In rats immunosuppressed with dexamethasone, 120 mg/kg per day of DHEA was able to partially restore spleen cell mitogenic responses to endotoxin and concanavalin A, and *in vitro* IgG production. DHEA also reduced the colonization of the gut by *Cryptosporidium parvum* in these animals (Rasmussen *et al.*, 1993). Similarly, three daily injections of 1.2 mg DHEA (i.e. 50–60 mg/kg per day) were found to stop dexamethasone from rendering peripheral lymphocytes of mice unresponsive to mitogens, or causing involution of the thymus (Blauer *et al.*, 1991).

In fact, these vast doses are neither required nor optimal, and doses in the μg range are effective. For instance, optimal inhibition of GC-induced thymic involution in mice requires only 10–20 μg/mouse of DHEA or of its derivative 3,17-androstenediol (i.e. 0.5 mg/kg; Al-Nakhli *et al.*, manuscript in preparation). This is 100th of the dose used by Blauer and colleagues (Blauer *et al.*, 1991). Similarly, immunization with pneumococcal polysaccharide yielded higher antibody titres and increased numbers of plaque-forming cells, in old (but not young) mice if they were given a single s.c. injection of 100 μg DHEAS in propylene glycol, or 10 μg DHEA dissolved in ethanol, 3 h before the immunization (Garg and Bondada, 1993). This is a T cell-independent response, so the mode of action is unclear. In general DHEA, using acceptable doses in the low μg range, appears to oppose GC, and to promote a Th1 cytokine pattern. It has been shown to restore immune functions in aged mice, and to correct the dysregulated spontaneous cytokine release seen in old animals (Daynes *et al.*, 1993; Garg and Bondada, 1993). It has been tested for similar properties in aged humans (Morales *et al.*, 1994). It also enhances production of Th1 cytokines such as IL-2 and interferon gamma (IFNγ) (Daynes and Araneo, 1989; Daynes *et al.*, 1990, 1991, 1995). This is the reverse of the effect of glucocorticoids, which often enhance Th2 activity and synergize with Th2 cytokines as discussed in greater detail later (Fischer and Konig, 1991; Guida *et al.*, 1994; Padgett *et al.*, 1995; Wu *et al.*, 1991).

DHEA also enhances IL-2 secretion from human peripheral blood T cells (Suzuki *et al.*, 1991), and we have found that DHEA or 3β,17β-androstenediol (AED) will enhance mitogen-stimulated production of IFNγ from murine or human T cells in the range 10^{-7}–10^{-8}M (Al-Nakhli *et al.*, manuscript in preparation).

MECHANISM OF THE 'ANTI-GLUCOCORTICOID' ACTIONS OF DEHYDROEPIANDROSTERONE

We do not know how DHEA exerts its anti-glucocorticoid effects. It is not a competitive antagonist of the GC receptor. There is a report of a specific DHEA-binding protein (? receptor) in T lymphocytes (Meikle *et al.*, 1992) but this remains unconfirmed.

In its role as a 'neurosteroid', DHEAS clearly binds to several cell membrane-associated receptors in the brain, and it is an antagonist of $GABA_A$ receptors (Robel and Baulieu, 1995). The only hint at the possibility of cell membrane-associated receptors for DHEAS in the periphery has emerged from the observation that DHEAS very rapidly inhibits archidonic acid-induced platelet aggregation (Jesse *et al.*, 1995). The effect was seen at concentrations that are within the physiological range in man but not in rodents.

When large doses of DHEAS are given to rodents, it acts as a peroxisome proliferator. To do this, DHEAS requires the presence of peroxisome proliferator-activated receptor alpha (PPARα) (Peters *et al.*, 1996). There is no direct evidence that it is a ligand for this receptor, but it is interesting that there is a protein that binds DHEAS in liver, and PPARα is expressed at particularly high levels in liver and in the immune system (reviewed in Spencer *et al.*, 1995; Devchand *et al.*, 1996). The PPARs are a family of transcription factors analogous to the steroid receptors. They form heterodimers with RXR, the receptor for 9-*cis*-retinoic acid, and bind to DNA motifs known as PPAR-response elements. This results in upregulation of expression of enzymes responsible for lipid homoeostasis, fatty acid degradation, and destruction of leukotriene B4 (LTB4). It has been suggested that this effect could explain the restoration of immunological competence by DHEAS in old animals, via changes in cell membrane fluidity, phospholipid-dependent cell signalling pathways, and arachidonate-dependent mediators (Spencer *et al.*, 1995). This is an interesting possibility, though PPARα upregulates the enzymes that degrade LTB4 and and this in turn limits inflammation (Devchand *et al.*, 1996), whereas DHEA actually increases non-specific inflammation. Therefore some other PPAR-dependent pathway would need to be found.

A further obvious explanation is that the active 'anti-glucocorticoid' is an undiscovered metabolite of DHEA, acting via, perhaps, one of the many 'orphan' steroid receptors. It would be 'biologically logical' for there to be anti-glucocorticoid metabolites that are beyond the point in the metabolic pathways that can give rise to sex steroids, so that regulation of inflammation and immunity could occur independently of sexual functions. It has been suggested that 7-hydroxylated derivatives perform this function, but the supporting *in vivo* data are unacceptable, and the *in vitro* data are not convincing (Morfin and Courchay, 1994; Padgett and Loria, 1994).

METABOLISM OF DHEA IN DISEASE

A fall in DHEA levels is a good marker of progression to AIDS in HIV-infected individuals (Wisniewski *et al.*, 1993) and the deficit in DHEA relative to cortisol can be striking in tuberculosis (Rook *et al.*, 1996). In both diseases, this correlates with a defect in Th1-mediated immunity. The pattern of metabolism of DHEA also changes in tuberculosis, so that less than normal is reduced to androsterone or aetiocholanolone, while more of it is converted to 16α-hydroxylated derivatives (even in the absence of antituberculosis therapy; this is not an effect induced by rifampicin, though rifampicin can activate this enzyme) (Rook *et al.*, 1996). Since we do not know precisely which metabolites exert the anti-glucocorticoid effects, interpretation of this change is difficult, but it may be significant that these 16α-hydroxylated derivatives are not active *in vivo* as antiglucocorticoids (Al Nakhli *et al.*, manuscript in preparation). Moreover the same 16α-hydroxylated derivatives are formed during pregnancy and by premature neonates (Kitada *et al.*, 1987). Our hypothesis is that this represents one of several mechanisms that tend to drive the immune response towards Th2 in pregnancy (Wegmann *et al.*, 1993), since Th1 responses to placental antigens are associated with abortion (Hill *et al.*, 1995). Perhaps tuberculosis is 'attempting' to do the same thing.

The effect of stress on T cell function

Psychological and physical stress can activate the hypothalamo–pituitary–adrenal axis, and so lead to an increase in the production of cortisol. However, the immunological effects of stress are complicated by the activation of other pathways, such as the adrenal medulla, so not everything observed is due to cortisol. Moreover, most studies of the immunological effects of stress have tended not to include the types of immunological investigation that cast any light on the Th1/Th2 balance. Nevertheless, it is clear that stress increases Th2 and decreases Th1 activity. In a study of military trainees, excessive exercise and deprivation of food and sleep resulted in raised cortisol, but no fall in DHEA (though testosterone fell to castrate levels). The DHEA/GC ratio therefore fell and correlated with a fall in DTH responsiveness (Bernton *et al.*, 1995). The same study documented a stress-induced rise in serum IgE levels.

In chronic illness or after burn injury, DHEA levels fall dramatically (Araneo *et al.*, 1993), so DHEA/GC ratios can be more severely affected than in these stressed but otherwise healthy army personnel (Bernton *et al.*, 1995). In another study of such personnel, Opstad found a rise in DHEAS (? due to high ACTH and low insulin) accompanied by a fall in DHEA, showing that these levels can be dissociated (Opstad, 1992).

A classical example of the Th1$\rightarrow$Th2 switching effect of stress is the increase in antibody to Epstein Barr virus in students reacting in a stressed manner to their exams. This virus is usually controlled by a type 1 response and cytotoxic

T cells, and loss of control results in virus replication and increased antibody (Zwilling, 1992). Similarly, peripheral blood leukocytes from medical students during exam periods showed lower mRNA for IFNγ and for the glucocorticoid receptor (Glaser *et al.*, 1993).

Similar points can be demonstrated in a more controlled manner in animals. Stress due to crowding or restraint can increase mycobacterial growth in tuberculosis in mice (Tobach and Bloch, 1956; Brown *et al.*, 1993). This is a model which is acutely sensitive to the presence of even a small Th2 component (Rook and Hernandez-Pando, 1996).

The effects of glucocorticoids on T cells in the thymus

The fundamental importance of GC as regulators of T lymphocyte function is highlighted by the discovery that radioresistant thymic epithelial cells contain steroidogenic enzymes (Vacchio *et al.*, 1994). These cells actively secrete steroids that may play a crucial role in thymic T lymphocyte repertoire selection (Vacchio *et al.*, 1994). Exposure to GC, or engagement of the T cell receptor both provide signals for apoptosis in immature thymocytes. However when both signals are present together they can antagonize each other, and if the balance of the two signals is within a critical range, apoptosis does not occur. These observations have led Ashwell's group to propose the following hypothesis. T cells of which the receptors fail to bind MHC/peptide with significant avidity (i.e. useless cells) die as a result of apoptosis triggered by GC alone. T cells that engage MHC/peptide with moderate avidity are preserved by the antagonism of the two signals, while T cells that engage MHC/peptide with very high avidity (i.e. potentially auto-reactive T cells) are eliminated because the GC signal is unable wholly to antagonize the signal for apoptosis mediated via the antigen receptor (Vacchio *et al.*, 1994; King *et al.*, 1995).

These delicate regulatory pathways may be particularly susceptible to dysregulation by toxic substances. 2,3,7,8-Tetrachlorodibenzo-*p*-dioxin (TCCD) also kills immature thymocytes via induction of apoptosis. Several of the biochemical features associated with TCCD-induced thymocyte apoptosis are indistinguishable from those associated with GC-induced thymocyte apoptosis. This raises the obvious concern that a toxic chemical could upset this pivotal event in the regulation of the T cell repertoire. In fact, it has been suggested that TCCD-induced thymic atrophy may be due to synergism between GC and TCCD (McConkey *et al.*, 1994).

When apoptosis of mouse thymocytes is induced *in vitro* with GC, without engagement of the T cell receptor, programmed cell death can be inhibited by cytokines, particularly by interleukin-4 (IL-4), which in this system was more effective than IL-2 or IL-1 (Migliorati *et al.*, 1992).

These GC-mediated effects must also be modulated by the thymic 11-β-HSD discussed above (Dougherty *et al.*, 1960). Moreover, it is important to remember that GC are not the only steroid hormones to influence thymic function.

Gonadectomy increases thymic bulk, while oestrogen and testosterone decrease it. This may be because testosterone increases expression of TGFβ in the thymus (Olsen *et al.*, 1993). Nevertheless, in summary, it is clear that GC can either cause or inhibit apoptosis of T cells, depending on the presence of other signals, and on the nature of the T cells.

The effects of GC on T cell function *in vitro*

The overall physiological effect of GC, like the effect of stress (Bernton *et al.*, 1995; Brown *et al.*, 1993; Zwilling, 1992), seems to be to drive the immune response towards Th2 (Fischer and Konig, 1991; Guida *et al.*, 1994; Padgett *et al.*, 1995; Wu *et al.*, 1991). A recent example is the demonstration that GC plus IL-4 will synergistically increase IgE production from human peripheral blood mononuclear cells (Bohle *et al.*, 1995). However, the data from *in vitro* work using co-culture of mixed T cell populations with GC are conflicting and difficult to interpret. This is often due to the use of vastly supraphysiological levels of GC. However, the major problem can be deduced from the previous section on the role of GC in the thymus. The precise role of GC clearly depends on the nature of the T cells being considered, and on the other stimuli present.

MODULATION OF THE EFFECT OF GC ON T CELL FUNCTION AND APOPTOSIS IN THE PRESENCE OF OTHER AGONISTS

In vivo, the T cells are exposed to GC in the presence of multiple agonists, including xenobiotic compounds, and a few examples will serve to illustrate the complexity of the resulting interactions. For instance, low concentrations of dexamethasone or prostaglandin E2 that did not inhibit the responsiveness of human T cells to stimulation by anti-CD3, acted synergistically to inhibit both IL-2 secretion and subsequent proliferation (Elliott *et al.*, 1996). Moreover, anti-CD28 overcame much of this suppression (Elliott *et al.*, 1992).

The interactions of GC with other mediators are particularly apparent when GC are considered as inducers of T cell apoptosis. Thus GC-induced apoptosis of T cells can be inhibited by ligands for CD44 (Ayroldi *et al.*, 1995). It is sometimes stated that peripheral T cells are relatively insensitive to GC-induced apoptosis. However, peripheral blood T cells showed sensitivity to GC-induced apoptosis soon after the proliferative response to mitogen stimulation. They could then be protected from apoptosis by several cytokines (IL-2 > IL-4 > IL-10), while puromycin and cycloheximide increased it (Brunetti *et al.*, 1995).

T CELL SUBPOPULATIONS AND SUSCEPTIBILITY TO THE EFFECTS OF GC *IN VITRO*

In a particularly illuminating study, T cells were sorted by fluorescence activated cell sorter into CD4+CD45RO- (naive) and CD4+CD45RO+ (memory) subsets, and then primed with anti-CD3 (solid phase) in the presence of IL-2 for 9 days, resulting in considerable clonal expansion. They were then washed, put into fresh medium, and restimulated for 72 h with anti-CD3 and IL-2. GC was added to these cultures, either in the priming phase, or during the restimulation. Then supernatants from the restimulation phase were assayed for cytokines (Brinkmann and Kristofic, 1995). It should be noted that transcription of the IL-2 gene is downregulated by the GC receptor (Northrop *et al*., 1992) so addition of IL-2 to this system may have been essential, but does somewhat complicate interpretation. Nevertheless, this study showed that the results were quite different with naive and primed cells, and also depended critically on whether the GC were present during priming or restimulation (Table 10.1). If cultured without GC, both T cell types showed a Th0-like pattern, secreting IL-5, IL-10, IFNγ, and in the case of the CD45RO+ memory cells, IL-4 as well. However culturing naive cells with GC tended to drive them towards Th2. They eventually secreted IL-10 if the GC was present during the priming phase, and both IL-4 and IL-10 if GC was present during the restimulation phase.

Table 10.1 Differing effects of GC on the cytokine profiles of naive and memory T cells (adapted from Brinkmann and Kristofic, 1995)

	Proliferation	IL-4[a]	IL-5	IL-10	IFNγ
CD4+CD45RO$^-$ T cells (naive)					
No GC	++	–	++	++	++
+ GC during priming phase (anti-CD3+IL-2, 9 days)	–	–	–	+++	–
+ GC during restimulation (anti-CD3+IL-2, 72 h)	++	+++	–	++	–
CD4+ CD45RO$^+$ T cells (memory)					
No GC	++	++	++	++	++
+ GC during priming phase (anti-CD3+IL-2, 9 days)	+	–	–	+++	–
+ GC during restimulation (anti-CD3+IL-2, 72 h)	++	–	–	–	+[b]

[a]Cytokines were measured in the supernatants of the restimulation cultures. GC was present either during the 9 days of priming or during the 72 hrs of restimulation, but not both. The stimulus was solid phase anti-CD3 plus IL-2.
[b]A reduction related to the concentration of GC, but always >50% of the levels seen in the absence of GC.

The memory cells also switched to exclusive IL-10 production if the GC was present during the priming phase (Table 10.1), but if present during *restimulation*, the only cytokine produced was IFNγ (Brinkmann and Kristofic, 1995).

This careful study resolves many of the previous controversies that arose when mixed cell populations were studied. It explains why misleading conclusions can be drawn from short-term experiments with whole peripheral blood mononuclear cell populations stimulated with PHA. In such experiments, IL-5 production is suppressed more easily than IL-2 or IFNγ (Van-Wauwe *et al.*, 1995). Similarly, it explains why when T cell clones were grown from bronchoalveolar lavage (BAL) samples using PHA and IL-2 as the stimulus, dexamethasone (at supraphysiologic doses) inhibited anti-CD3-induced production of IL-4 and IL-5 more than it inhibited production of IFNγ (Krouwels *et al.*, 1996). BAL T cells are known to be activated memory cells, so this is compatible with the results of Brinkmann *et al.*, and does not alter the fact that if an immune response in naive cells is allowed to develop in the presence of GC, a Th2 line will develop. This has been rather clearly shown with spleen cells from 'clean' laboratory rodents (Ramirez *et al.*, 1996), which have few memory cells under normal circumstances.

Conclusions

Overall, the 'bottom line' may be that GC favour the *development* of a Th2 cytokine profile from naive cells, although cytokine secretion by established Th2 cells is readily inhibited. The ability of some Th1 cells to continue to secrete Th1 cytokines after exposure to GC does not alter the fact that Th1 *function* is blocked by GC. This is partly due to other GC effects such as blockage of macrophage function (Brown *et al.*, 1993), and enhanced expression of TGFβ (Batuman *et al.*, 1991), and partly due to the fact that in a dynamic *in vivo* situation what is important is the diversion of subsequently recruited T cells towards Th2. Thus conventional treatments for Th2-mediated diseases such as eczema, asthma and hayfever may work via anti-inflammatory effects, and by reducing cytokine production by Th2 cells (Corrigan *et al.*, 1995), and by increasing apoptosis of eosinophils (Meagher *et al.*, 1996) and yet at the same time encourage perpetuation of the underlying problem by driving newly recruited T cells towards Th2.

Perhaps in the future it will be possible to exploit the anti-inflammatory effects of GC without simultaneously causing immunosuppression, or subjecting the immune system to an inappropriate Th1-to-Th2 switch. In this review we have attempted to point out the extreme complexity of the regulation of GC action on the T lymphocyte system, and the naive nature of attempts to assess this only through the 8.00 a.m. plasma cortisol level. Clearly there are numerous targets for immunotoxic chemicals as well as for the pharmaceutical interventions of the future.

Acknowledgements

We are grateful to the Wellcome Trust for supporting R.B.'s studies of the endocrinology of tuberculosis.

References

Araneo, B.A., Shelby, J., Li, G.Z., Ku, W. and Daynes, R.A. (1993) Administration of dehydroepiandrosterone to burned mice preserves normal immunologic competence. *Arch. Surg.*, **128**, 318–25.

Ayroldi, E., Cannarile, L., Migliorati, G., Bartoli, A., Nicoletti, I. and Riccardi, C. (1995) CD44 (Pgp-1) inhibits CD3 and dexamethasone-induced apoptosis. *Blood*, **86**, 2672–8.

Baker, R., Zumla, A. and Rook, G.A.W. (1996) Tuberculosis, steroid metabolism and immunity. *Q. J. Med.*, **89**, 387–94.

Batuman, O.A., Ferrero, A.P., Diaz, A. and Jimenez, S.A. (1991) Regulation of transforming growth factor-beta 1 gene expression by glucocorticoids in normal human T lymphocytes. *J. Clin. Invest.*, **88**, 1574–80.

Berliner, D.L. and Dougherty, T.F. (1961) Hepatic and extrahepatic regulation of corticosteroids. *Pharmacol. Rev.*, **13**, 329–59.

Bernton, E., Hoover, D., Galloway, R. and Popp, K. (1995) Adaptation to chronic stress in military trainees. Adrenal androgens, testosterone, glucocorticoids, IGF-1 and immune fucntion. *Ann. N.Y. Acad. Sci.*, **774**, 217–31.

Blauer, K.L., Poth, M., Rogers, W.M. and Bernton, E.W. (1991) Dehydroepiandrosterone antagonises the suppressive effects of dexamethasone on lymphocyte proliferation. *Endocrinology*, **129**, 3174–9.

Bohle, B., Willheim, M., Baier, K., Stadler, B., Spitzauer, S., Scheiner, O., Kraft, D. and Ebner, C. (1995) Hydrocortisone enhances total IgE levels – but not the synthesis of allergen-specific IgE – in a monocyte-dependent manner. *Clin. Exp. Immunol.*, **101**, 474–9.

Brinkmann, V. and Kristofic, C. (1995) Regulation by corticosteroids of Th1 and Th2 cytokine production in human CD4+ effector T cells generated from CD45RO- and CD45RO+ subsets. *J. Immunol.*, **155**, 3322–8.

Brown, D.H., Sheridan, J., Pearl, D. and Zwilling, B.S. (1993) Regulation of mycobacterial growth by the hypothalamus–pituitary–adrenal axis: differential responses of *Mycobacterium bovis* BCG-resistant and -susceptible mice. *Infect. Immun.*, **61**, 4793–800.

Brunetti, M., Martelli, N., Colasante, A., Piantelli, M., Musiani, P. and Aiello, F.B. (1995) Spontaneous and glucocorticoid-induced apoptosis in human mature T lymphocytes. *Blood*, **86**, 4199–205.

Corrigan, C.J., Hamid, Q., North, J., Barkans, J., Moqbel, R., Durham, S., Gemou, E.V. and Kay, A.B. (1995) Peripheral blood CD4 but not CD8 T-lymphocytes in patients with exacerbation of asthma transcribe and translate messenger RNA encoding cytokines which prolong eosinophil survival in the context of a Th2-type pattern: effect of glucocorticoid therapy. *Am. J. Respir. Cell. Mol. Biol.*, **12**, 567–78.

Danenberg, H.D., Ben-Yehuda, A., Zakay-Rones, Z. and Friedman, G. (1995) Dehydroepiandrosterone enhances influenza vaccination in aged mice. *Ann. N.Y. Acad. Sci.*, **774**, 297–9.

Daynes, R.A. and Araneo, B.A. (1989) Contrasting effects of glucocorticoids on the capacity of T cells to produce the growth factors IL-2 and IL-4. *Eur. J. Immunol.*, **19**, 2319–25.

Daynes, R.A., Araneo, B.A., Dowell, T.A., Huang, K. and Dudley, D. (1990) Regulation of murine lymphokine production *in vivo*. III. The lymphoid tissue microenvironment exerts regulatory influences over T helper cell function. *J. Exp. Med.*, **171**, 979–96.

Daynes, R.A., Meikle, A.W. and Araneo, B.A. (1991) Locally active steroid hormones may facilitate compartmentalization of immunity by regulating the types of lymphokines produced by helper T cells. *Res. Immunol.*, **142**, 40–5.

Daynes, R.A., Araneo, B.A., Ershler, W.B., Maloney, C., Li, G.-Z. and Ryu, S.-Y. (1993) Altered regulation of IL-6 production with normal aging; possible linkage to the age-associated decline in dehydroepiandrosterone and its sulphated derivative. *J. Immunol.*, **150**, 5219–30.

Daynes, R.A., Araneo, B.A., Hennebold, J., Enioutina, J. and Mu, H.H. (1995) Steroids as regulators of the mammalian immune response. *J. Invest. Dermatol.*, **105**, 14S–19S.

Deckx, R. and De Moor, P. (1996) Study of 11-beta hydroxysteroid dehydrogenase *in vitro*. 1. Biochemical characterisation in spleen homogenate. *Pflugers Arch.*, **289**, 59.

Devchand, P.R., Keller, H., Peters, J.M., Vasquez, M., Gonzalez, F.J. and Wahli, W. (1996) The PPARα-leukotriene B4 pathway to inflammation control. *Nature*, **384**, 39–43.

Dougherty, T.F., Berliner, M.L. and Berliner, D.L. (1960) 11β-Hydroxy dehydrogenase system activity in thymi of mice following prolonged cortisol treatment. *Endocrinology*, **66**, 550–8.

Elliott, L., Brooks, W. and Roszman, T. (1992) Inhibition of anti-CD3 monoclonal antibody-induced T-cell proliferation by dexamethasone, isoproterenol, or prostaglandin E2 either alone or in combination. *Cell. Mol. Neurobiol.*, **12**, 411–27.

Elliott, L.H., Levay, A.K., Sparks, B., Miller, M. and Roszman, T.L. (1996) Dexamethasone and prostaglandin E2 modulate T-cell receptor signaling through a cAMP-independent mechanism. *Cell. Immunol.*, **169**, 117–24.

Evangelatou, M., Antoniw, J. and Cooke, B.A. (1996) The effect of leukocytes on 11β-HSD activity in human granulosa cell cultures. *J. Endocrinol.*, **148** (Supplement), Abstract P55.

Fischer, A. and Konig, W. (1991) Influence of cytokines and cellular interactions on the glucocorticoid-induced Ig (E, G, A, M) synthesis of peripheral blood mononuclear cells. *Immunology*, **74**, 228–33.

Fukuda, R., Ichikawa, Y., Takaya, M., Ogawa, Y. and Masumoto, A. (1994) Circadian variations and prednisolone-induced alterations of circulating lymphocyte subsets in man. *Int. Med.*, **33**, 733–8.

Garg, M. and Bondada, S. (1993) Reversal of age-associated decline in immune response to Pnu-imune vaccine by supplementation with the steroid hormone dehydroepiandrosterone. *Infect. Immun.*, **61**, 2238–41.

Glaser, R., Lafuse, W.P., Bonneau, R.H., Atkinson, C. and Kiecolt, G.J. (1993) Stress-associated modulation of proto-oncogene expression in human peripheral blood leukocytes. *Behav. Neurosci.*, **107**, 525–9.

Gomez-Sanchez, E.P., Cox, D., Foecking, M., Ganjam, V. and Gomez-Sanchez, C.E. (1996) 11-Beta hydroxysteroid dehydrogenases of the choriocarcinoma line JEG-3 and their inhibition by glycyrrhitinic acid and other natural substances. *Steroids*, **61**, 110.

Guida, L., O'Hehir, R.E. and Hawrylowicz, C.M. (1994) Synergy between dexamethasone and interleukin-5 for the induction of major histocompatibility complex class II expression by human peripheral blood eosinophils. *Blood*, **84**, 2733–40.

Hennebold, J.D., Ryu, S.Y., Mu, H.H., Galbraith, A. and Daynes, R.A. (1996) 11-Beta hydroxysteroid dehydrogenase modulation of glucocorticoid activities in lymphoid organs. *Am. J. Physiol.*, **270**, R1296.

Hiemke, C., Brunner, R., Hammes, E., Muller, H., Meyer, Z.B.K. and Lohse, A.W. (1995) Circadian variations in antigen-specific proliferation of human T lymphocytes and correlation to cortisol production. *Psychoneuroendocrinology*, **20**, 335–42.

Hill, J.A., Polgar, K. and Anderson, D.J. (1995) T-helper 1-type immunity to trophoblast in women with recurrent spontaneous abortion. *JAMA*, **273**, 1933–6.

Hubbard, W.C., Bickel, C. and Schleimer, R.P. (1994) Simultaneous quantitation of endogenous levels of cortisone and cortisol in human nasal and bronchoalveolar lavage fluids and plasma via gas chromatography-negative ion chemical ionization mass spectrometry. *Anal. Biochem.*, **221**, 109–17.

Jesse, R.L., Loesser, K., Eich, D.M., Qian, Y.Z., Hess, M.L. and Nestler, J.E. (1995) Dehydroepiandrosterone inhibits human platelet aggregation *in vitro* and *in vivo*. *Ann. N.Y. Acad. Sci.*, **774**, 281–90.

King, L.B., Vacchio, M.S., Dixon, K., Hunziker, R., Margulies, D.H. and Ashwell, J.D. (1995) A targeted glucocorticoid receptor antisense transgene increases thymocyte apoptosis and alters thymocyte development. *Immunity*, **3**, 647–56.

Kitada, M., Kamataki, T., Itahashi, K., Rikihisa, T. and Kanakubo, Y. (1987) P-450 HFLa, a form of cytochrome P-450 purified from human foetal livers, is the 16α-hydroxylase of dehydroepiandrosterone 3-sulphate. *J. Biol. Chem.*, **262**, 13534–7.

Klein, A., Buskila, D., Gladman, D., Bruser, B. and Malkin, A. (1990) Cortisol catabolism by lymphocytes of patients with systemic lupus erythematosus and rheumatoid arthritis. *J. Rheumatol.*, **17**, 30–3.

Krouwels, F.H., van der Heijden, J.F., Lutter, R., van Neerwen, R.J.J., Jansen, H.M. and Out, T.A. (1996) Glucocorticoids affect functions of airway- and blood-derived human T cell clones, favoring the Th1 profile through two mechanisms. *Am. J. Respir. Cell. Mol. Biol.*, **14**, 388–97.

Lemmer, B., Schwulera, U., Thrun, A. and Lissner, R. (1992) Circadian rhythm of soluble interleukin-2 receptor in healthy individuals. *Eur. Cytokine Netw.*, **3**, 335–6.

Lortholary, O., Christeff, N., Casassus, P., Thobie, N., Veyssier, P., Trogoff, B., Torri, O., Brauner, M., Nunez, E.A. and Guillevin, L. (1996) Hypothalamo–pituitary–adrenal function in human immunodeficiency virus-infected men. *J. Clin. Endocrinol. Metab.*, **81**, 791–6.

Malone, J.L., Simms, T.E., Gray, G.C., Wagner, K.F., Burge, J.R. and Burke, D.S. (1990) Sources of variability in repeated T-helper lymphocyte counts from human immunodeficiency virus type 1-infected patients: total lymphocyte count fluctuations and diurnal cycle are important. *J. Acquir. Immune Defic. Syndr.*, **3**, 144–51.

Martini, E., Muller, J.Y., Doinel, C., Gastal, C., Roquin, H., Douay, L. and Salmon, C. (1988) Disappearance of CD4-lymphocyte circadian cycles in HIV-infected patients: early event during asymptomatic infection. *Aids*, **2**, 133–4.

McConkey, D.J., Jondal, M.B. and Orrenium, S.G. (1994) Chemical-induced apoptosis in the immune system, in *Immunotoxicology and Immunopharmacology*, 2nd edition (eds J.H. Dean, M.I. Luster, A.E. Munson and I. Kimber), Raven Press, New York, pp. 473–85.

Meagher, L.C., Cousin, J.M., Seckl, J.R. and Haslett, C. (1996) Opposing effects of glucocorticoids on the rate of apoptosis in neutrophilic and eosinophilic granulocytes. *J. Immunol.*, **156**, 4422–8.

Meikle, A.W., Dorchuck, R.W., Araneo, B.A., Stringham, J.D., Evans, T.G., Spruance, S.L. and Daynes, R.A. (1992) The presence of a dehydroepiandrosterone-specific receptor-binding complex in murine T cells. *J. Steroid Biochem. Molec. Biol.*, **42**, 293–304.

Migliorati, G., Pagliacci, C., Moraca, R., Crocicchio, F., Nicoletti, I. and Riccardi, C. (1992) Interleukins modulate glucocorticoid-induced thymocyte apoptosis. *Int. J. Clin. Lab. Res.*, **21**, 300–3.

Morales, A.J., Nolan, J., Nelson, J. and Yen, A.S. (1994) Effects of replacement dose of DHEA in men and women of advancing age. *J. Clin. Endocrinol. Metab.*, **78**, 1360–7.
Morfin, R. and Courchay, G. (1994) Pregnenolone and dehydroepiandrosterone as precursors of native 7-hydroxylated metabolites which increase the immune response in mice. *J. Steroid Biochem. Molec. Biol.*, **50**, 91–100.
Northrop, J.P., Crabtree, G.R. and Mattila, P.S. (1992) Negative regulation of interleukin-2 transcription by the glucocorticoid receptor. *J. Exp. Med.*, **175**, 1235–41.
Oakley, R.H., Sar, M. and Cidlowski, J.A. (1996) The human glucocorticoid receptor beta isoform. Expression, biochemical properties, and putative function. *J. Biol. Chem.*, **271**, 9550–9.
Olsen, N.J., Zhou, P. and Ong, H. (1993) Testosterone induces expression of transforming growth factor-β in the murine thumus. *Steroid. Biochem. Mol. Biol.*, **45**, 327–32.
Opstad, P.K. (1992) Androgenic hormones during prolonged physical stress, sleep and energy deficiency. *J. Clin. Endocrinol. Metab.*, **74**, 1176–83.
Opstad, P.K. (1994) Circadian rhythm of hormones is extinguished during prolonged physical stress, sleep and energy deficiency in young men. *Eur. J. Endocrinol.*, **131**, 56–66.
Padgett, D.A. and Loria, R.M. (1994) *In vitro* potentiation of lymphocyte activation by dehydroepiandrosterone, androstenediol and androstenetriol. *J. Immunol.*, **153**, 1544–52.
Padgett, D.A., Sheridan, J.F. and Loria, R. (1995) Steroid hormone regulation of a polyclonal Th2 immune response. *Ann. N.Y. Acad. Sci.*, **774**, 323–5.
Peters, J.M., Yuan-Chun, Z., Ram, P.A., Lee, S.S.T., Gonzalez, F.J. and Waxman, D.J. (1996) Peroxisome proliferator-activated receptor α required for gene induction by dehydroepiandrosterone-3β-sulphate. *Mol. Pharmacol.*, **50**, 67–74.
Petrovsky, N. and Harrison, L.C. (1995) Th1 and Th2: swinging to a hormonal rhythm. *Immunol. Today*, **16**, 605.
Petrovsky, N., McNair, P. and Harrison, L.C. (1994) Circadian rhythmicity of interferon-gamma production in antigen-stimulated whole blood. *Chronobiologia*, **21**, 293–300.
Pownall, R., Kabler, P.A. and Knapp, M.S. (1979) The time of day of antigen encounter influences the magnitude of the immune response. *Clin. Exp. Immunol.*, **36**, 347–54.
Ramirez, F., Fowell, D.J., Puklavec, M., Simmonds, S. and Mason, D. (1996) Glucocorticoids promote a Th2 cytokine response by CD4+ T cells *in vitro*. *J. Immunol.*, **156**, 2406–12.
Rasmussen, K.R., Martin, E.G. and Healey, M.C. (1993) Effects of dehydroepiandrosterone in immunosuppressed rats infected with *Cryptosporidium parvum*. *J. Parasitol.*, **79**, 364–70.
Robel, P. and Baulieu, E.E. (1995) Dehydroepiandrosterone (DHEA) is a neuroactive neurosteroid. *Ann. N.Y. Acad. Sci.*, **774**, 82–110.
Rook, G.A.W. and Hernandez-Pando, R. (1996) The pathogenesis of tuberculosis. *Annu. Rev. Microbiol.*, **50**, 259–84.
Rook, G.A.W., Honour, J., Kon, O.M., Wilkinson, R.J., Davidson, R. and Shaw, R.J. (1996) Urinary steroid metabolites in tuberculosis; a new clue to pathogenesis. *Q. J. Med.*, **89**, 333–41.
Sarma, G.R., Chandra, I., Ramachandran, G., Krishnamurthy, P.V., Kumaraswami, V. and Prabhakar, R. (1990) Adrenocortical function in patients with pulmonary tuberculosis. *Tubercle*, **71**, 277–82.
Sothern, R.B., Roitman, J.B., Kanabrocki, E.L., Yager, J.G., Roodell, M.M., Weatherbee, J.A., Young, M.R., Nenchausky, B.M. and Scheving, L.E. (1995) Circadian characteristics of circulating interleukin-6 in men. *J. Allergy Clin. Immunol.*, **95**, 1029–35.

Spencer, N.F.L., Poynter, M.E., Henebold, J.D., Mu, H.H. and Daynes, R.A. (1995) Does DHEAS restore immune competence in aged animals through its capacity to function as a natural modulator of peroxisome activities? *Ann. N.Y. Acad. Sci.*, **774**, 201–16.

Stöcklin, E., Vissler, M., Gouilleux, F. and Groner, B. (1996) Functional interactions between Stat5 and the glucocorticoid receptor. *Nature*, **383**, 726–8.

Suzuki, T., Suzuki, N., Daynes, R.A. and Engleman, E.G. (1991) Dehydroepiandrosterone enhances IL2 production and cytotoxic effector function of human T cells. *Clin. Immunol. Immunopathol.*, **61**, 202–11.

Tobach, E. and Bloch, H. (1956) Effect of stress by crowding prior to and following tuberculous infection. *Am. J. Physiol.*, **187**, 399–402.

Vacchio, M.S., Papdopoulos, V. and Ashwell, J.D. (1994) Steroid production in the thymus: implications for thymocyte selection. *J. Exp. Med.*, **179**, 1835–46.

Van-Wauwe, J., Aerts, F., Walter, H. and de-Boer, M. (1995) Cytokine production by phytohemagglutinin-stimulated human blood cells: effects of corticosteroids, T cell immunosuppressants and phosphodiesterase IV inhibitors. *Inflamm. Res.*, **44**, 400–5.

Walker, B.R. (1994) Organ-specific actions of 11 beta-hydroxysteroid dehydrogenase in humans: implications for the pathophysiology of hypertension. *Steroids*, **59**, 84–9.

Wegmann, T.G., Lin, H., Guilbert, L. and Mosmann, T.R. (1993) Bidirectional cytokine interactions in the maternal–fetal relationship: is successful pregnancy a Th2 phenomenon? *Immunol. Today*, **14**, 353–6.

Wiegers, G.J., Croiset, G., Reul, J.M., Holsboer, F. and de Kloet, E.R. (1993) Differential effects of corticosteroids on rat peripheral blood T-lymphocyte mitogenesis *in vivo* and *in vitro*. *Am. J. Physiol.*, **265**, E825–E830.

Wiegers, G.J., Reul, J.M., Holsboer, F. and de Kloet, E.R. (1994) Enhancement of rat splenic lymphocyte mitogenesis after short-term preexposure to corticosteroids *in vitro*. *Endocrinology*, **135**, 2351–7.

Wilckens, T. (1995) Glucocorticoids and immune function: physiological relevance and pathogenic potential of hormonal dysfunction. *Trends Pharmacol. Sci.*, **16**, 193–7.

Wilder, R.L. (1995) Neuroendocrine-immune system interactions and autoimmunity. *Annu. Rev. Immunol.*, **13**, 307–38.

Wisniewski, T.L., Hilton, C.W., Morse, E.V. and Svec, F. (1993) The relationship of serum DHEA-S and cortisol levels to measures of immune function in human immunodeficiency virus-related illness. *Am. J. Med. Sci.*, **305**, 79–83.

Wright, B.E., Porter, J.R., Browne, E.S. and Svec, F. (1992) Antiglucocorticoid action of dehydroepiandrosterone in young obese Zucker rats. *Int. J. Obes.*, **16**, 579–83.

Wu, C.Y., Sarfati, M., Heusser, C., Fournier, S., Rubio-Trujillo, M., Peleman, R. and Delespesse, G. (1991) Glucocorticoids increase the synthesis of immunoglobulin E by interleukin 4-stimulated human lymphocytes. *J. Clin. Invest.*, **87**, 870–7.

Young, M.R., Matthews, J.P., Kanabrocki, E.L., Sothern, R.B., Roitman, J.B. and Scheving, L.E. (1995) Circadian rhythmometry of serum interleukin-2, interleukin-10, tumor necrosis factor-alpha, and granulocyte-macrophage colony-stimulating factor in men. *Chronobiol. Int.*, **12**, 19–27.

Zwilling, B.S. (1992) Stress affects disease outcomes. *Am. Soc. Microbiol. News*, **58**, 23–5.

11 T Lymphocyte Subpopulations and Immune Responses to Chemical Allergens

Ian Kimber and Rebecca J. Dearman

Introduction

Allergic disease resulting from exposure of susceptible individuals to chemicals or proteins is an important occupational health problem. For the purposes of this chapter, allergy is defined as the adverse health effects that may result from the stimulation of specific immune responses. Attention will focus here upon occupational allergy induced by chemicals. In this context, there are two forms of allergic disease that command most attention; allergic contact dermatitis or skin sensitization and sensitization of the respiratory tract leading to rhinitis and asthma. In both instances, allergic sensitization in susceptible individuals is provoked following exposure to sufficient quantities of the inducing chemical via an appropriate route. The sensitized individual is then primed to mount an accelerated and more aggressive secondary immune response following subsequent encounter with the same allergen. An important feature of this process is that allergic hypersensitivity reactions may be induced in the relevant target organs, skin or respiratory tract, by exposure of the sensitized individual to only small amounts of the inducing allergen; amounts smaller than those required to induce sensitization initially.

Chemical allergy poses a number of intriguing questions, not least among them being the nature of heritable and environmental factors that render individuals susceptible to particular allergens. Of central importance also is consideration of the variables that dictate the form or forms that allergic reactions will take. It is apparent in humans that some chemical allergens are associated primarily or exclusively with the induction of skin sensitization and allergic contact dermatitis, whereas others preferentially cause sensitization of the respiratory tract (Kimber, 1995). Among the former are chemicals such as

T Lymphocyte Subpopulations in Immunotoxicology. Edited by Ian Kimber and MaryJane K. Selgrade.
 ISBN 0 471 97194 4

phenylenediamine, cinnamic aldehyde, mercaptobenzthiazole, formaldehyde, thiurams and 2,4-dinitrochlorobenzene and the metals nickel, chromium and cobalt. Chemicals known in humans to cause occupational respiratory allergy include anhydrides, such as trimellitic anhydride and phthalic anhydride, isocyanates, such as toluene diisocyanate and diphenylmethane diisocyanate, certain reactive dyes and some platinum salts.

The purpose of this chapter is to consider the roles of T lymphocytes and T lymphocyte subpopulations in the induction and elicitation of chemical-induced allergic disease and to explore whether the nature of T lymphocyte responses influences the form allergic reactions will take.

Allergic responses to chemicals

The development and expression of skin sensitization and allergic contact dermatitis are considered normally to be a form of delayed-type hypersensitivity in which T lymphocytes play a predominant role (Kimber and Dearman, 1996). While this is undoubtedly the case, it must be emphasized that other immunological effector mechanisms may contribute toward the elicitation of allergic reactions in the skin. There is less certainty about the immunobiological mechanisms that effect respiratory sensitization to chemicals and cause occupational asthma (Kimber and Dearman, 1997). There is every reason to believe that allergic reactions to proteins are mediated via IgE antibody-dependent mechanisms. Protein allergens will provoke in susceptible individuals specific IgE antibody production. This antibody distributes systemically and associates with mast cells, including mast cells in the respiratory tract. Following subsequent inhalation exposure to the inducing protein allergen, mast cells will be stimulated to degranulate and release a variety of inflammatory mediators that in turn cause vasodilation, bronchoconstriction and the symptoms of immediate-type respiratory hypersensitivity (Garssen *et al.*, 1996). Whether IgE antibody plays such a central role in the development of respiratory sensitization to chemicals remains contentious. This uncertainty is partly a reflection of the fact that in many instances, particularly with respect to occupational asthma caused by the diisocyanates, it has proven difficult to detect specific IgE antibody in the serum of symptomatic individuals (Chan-Yeung, 1995). Also, allergic reactions may be of immediate or delayed onset. At a superficial level at least the manifestation of symptoms of allergic respiratory hypersensitivity to chemicals in the absence of demonstrable IgE antibody may be considered consistent with the view from some recent experimental studies that antibody is not always required. It was found, for instance, that sensitization and challenge of B lymphocyte-deficient (and thereby immunoglobulin-deficient) mice with ovalbumin resulted in an eosinophil-rich allergic reaction in the respiratory tract (Korsgren *et al.*, 1997). Nevertheless, associations between IgE antibody and chemical respiratory sensitization have been found (Drexler *et al.*, 1993; Karol *et al.*, 1994; Moller *et al.*, 1985; Nielsen *et al.*,

1994; Salvaggio *et al.*, 1988). The question is whether there exists a universal obligatory role for IgE antibody in chemical respiratory hypersensitivity, or whether there are other immunological mechanisms that in some instances result in pulmonary reactions and occupational asthma. In considering this question, it must be borne in mind that, in some cases at least, the failure to detect specific IgE antibody in the serum of patients may be secondary to the use of inappropriate methods, perhaps employing protein-hapten conjugates that are not recognized by IgE antibody. Alternatively, it is possible also that mast cell-binding (homocytotropic) antibody of an isotype other than IgE (IgG4 in humans) may, in some instances, be the main mediator of respiratory sensitization. There is available some evidence to support a role for IgG4 antibody in the pathogenesis of chemical respiratory allergy (Park and Hong, 1991; Topping *et al.*, 1989).

Although the exact role of IgE antibody in the development of chemical sensitization has yet to be resolved, there is no doubt that this class of antibody is a potent effector of hypersensitivity reactions and that the ability of a chemical to induce IgE will influence markedly the nature of the allergic reaction that will be provoked. Not all immunogenic stimuli result in IgE antibody production and this applies also to chemical allergens. The potent contact allergen 2,4-dinitrochlorobenzene (DNCB) appears not to cause respiratory sensitization in humans and fails in guinea pigs to induce allergic respiratory hypersensitivity reactions (Blaikie *et al.*, 1995; Botham *et al.*, 1989). In the absence of adjuvant, and under conditions where specific IgG anti-hapten antibody is induced, DNCB is unable to stimulate IgE responses in mice (Dearman and Kimber, 1991; Dearman *et al.*, 1991). A different pattern of reactivity is seen with trimellitic anhydride (TMA), a chemical that in humans is associated with respiratory sensitization and occupational asthma (Bernstein *et al.*, 1982; Topping *et al.*, 1986; Zeiss *et al.*, 1977). This chemical has been shown to induce respiratory hypersensitivity reactions in guinea pigs (Blaikie *et al.*, 1995; Botham *et al.*, 1989) and in mice stimulates both specific IgE antibody production and an increase in the serum concentration of IgE (Dearman and Kimber, 1991; Dearman *et al.*, 1991). Before considering the contribution of discrete T lymphocyte subpopulations to the characteristics of immune responses induced by chemical allergens, it is necessary to review the regulation by cytokines of IgE antibody production.

Cytokines and IgE antibody responses

It is clear that the stimulation and maintenance of IgE antibody responses are dependent upon T lymphocytes and regulated actively by a number of factors including cytokines. Of critical importance to the development of IgE responses is interleukin 4 (IL-4), a cytokine that causes a switch to, and the increased expression of, $C\epsilon$ germline transcripts (Azuma *et al.*, 1987; Finkelman *et al.*, 1986, 1988b, 1990). Mice that lack a functional gene for

IL-4 fail to display IgE responses (Kuhn *et al.*, 1991), whereas mice that have a transgene for this cytokine exhibit increased serum levels of IgE and mount more vigorous IgE antibody responses (Burstein *et al.*, 1991; Tepper *et al.*, 1990). Despite these data, it has been found recently that class switching to IgE can be effected in IL-4-deficient mice by certain stimuli, although the levels of productive ∈ transcripts and of serum IgE were considerably lower than those observed in wild type controls (Morawetz *et al.*, 1996). Although IL-4 may not be an absolute prerequisite for some form of IgE response in all circumstances, there is no doubt that this cytokine is normally the major promotional factor (Coffman *et al.*, 1993). A controlling influence is provided by interferon γ (IFN-γ), a cytokine that antagonizes the stimulation of IgE responses (Finkelman *et al.*, 1988a). These same cytokines also serve to regulate IgE antibody production in humans (Chretien *et al.*, 1990; Del Prete *et al.*, 1988; Pene *et al.*, 1988; Romagnani *et al.*, 1989). It is worth remarking also that in humans, IgG4 is similarly subject to the reciprocal regulatory effects of IL-4, IFN-γ and other cytokines (Armitage *et al.*, 1993; Aversa *et al.*, 1993; Ishizaka *et al.*, 1990; Punnonen *et al.*, 1993). Furthermore, there is evidence for the coregulation of IgE and IgG4 antibody responses in certain disease states (Hammerstrom and Smith, 1987; Hussain and Ottesen, 1986). Another cytokine, interleukin 13 (IL-13), a molecule that shares some homology with IL-4, may also be important for the regulation of IgE and IgG4 expression in humans (Cocks *et al.*, 1993; Minty *et al.*, 1993; Punnonen *et al.*, 1993). Although it is IL-4 (and IL-13 in humans) and IFN-γ that are considered to have the greatest influence on IgE antibody production, there are other cytokines that influence, either directly or indirectly, the integrity of IgE responses. Among those cytokines that have been found to promote the synthesis of human IgE *in vitro* are interleukins 3, 5, 6, and 9 (IL-3, IL-5, IL-6 and IL-9) and tumour necrosis factor α (TNF-α) (Delespesse *et al.*, 1989; Dugas *et al.*, 1991; Herrod, 1989; Maggi *et al.*, 1989; Rousset *et al.*, 1991). It has also been demonstrated recently that macrophage inflammatory protein 1α (MIP-1α) and RANTES (*R*egulated on *A*ctivation, *N*ormal *T* cell *E*xpressed and *S*ecreted) selectively promote the production by activated human B lymphocytes of IgE and IgG4, but not of other immunoglobulin isotypes (Kimata *et al.*, 1996). In contrast, other cytokines including interleukins 2, 7 and 8 (IL-2, IL-7 and IL-8) and transforming growth factor β (TGF-β) all appear to have the ability to down-regulate the production by human B cells of IgE (Aebischer and Stadler, 1996; Borger *et al.*, 1996; Coffman *et al.*, 1993; Kimata *et al.*, 1992; Knoller *et al.*, 1989).

Although other processes, such as the complex interactions between the immune and neuroendocrine systems, can influence IgE production and IgE responses (Aebischer *et al.*, 1994; Rook *et al.*, 1994), it is cytokines that appear to have the greatest impact. To an important extent, the relative availability of the cytokines identified above that serve to regulate the induction and vigour of

IgE responses is a function of their secretion by discrete subpopulations of differentiated $CD4^+$ and $CD8^+$ T lymphocytes.

T lymphocyte subpopulations

As described in Chapter 2 of this volume, a functional heterogeneity among $CD4^+$ murine T helper (Th) cells was first described in 1986 (Mosmann *et al.*, 1986). Two types of Th cell clones were recognized, designated Th1 and Th2, that differed with respect to cytokine secretion patterns. A clear dichotomy of Th cell responses is now known to be a considerable over-simplification of what in fact is a very complex picture of functional heterogeneity, the nature of which is subject to regulation in both time and space. Nevertheless, recognition of diverse Th effector cell function provided a basis for consideration of qualitative regulation of immune responses. Maintenance of host resistance requires that the immune system responds in an appropriate way to many and diverse pathogenic microorganisms. The ability to tailor more or less polarized responses to bring to bear particular effector mechanisms that best provide protective immunity against specific microbial challenges clearly confers selective advantage. The issue considered here is whether this same functional heterogeneity of T lymphocytes also determines the characteristics of immune responses provoked by chemical allergens.

$CD4^+$ T LYMPHOCYTES

Functional subpopulations of $CD4^+$ T lymphocytes are defined according to their secretion of cytokines. Both Th1 and Th2 cells have been shown to secrete IL-3 and granulocyte/macrophage colony-stimulating factor (GM-CSF). However, Th1 cells produce IL-2, IFN-γ and tumour necrosis factor β (TNF-β), whereas Th2 cells elaborate IL-4, IL-5, IL-6 and IL-10 (Mosmann and Coffman, 1989; Mosmann *et al.*, 1986, 1991; Chapter 2 this volume). As such, these populations reflect at the cellular level what is known of the regulation of IgE responses by cytokines. Under conditions where the development of Th1 cell responses is favoured, the cytokines produced will fail to promote the production of IgE antibody, the reverse being the case when Th2-type responses predominate. The differential stimulation of Th cell populations will influence the characteristics of allergic responses in ways other than via the regulation of IgE antibody production. Not only do Th2 cells favour isotype switching to IgE and IgG4; their cytokine products serve also as growth and differentiation factors for, and activators of, those cells (mast cells and eosinophils) that are known to play central roles in the elicitation of immediate-type allergic reactions and respiratory allergic disease (Kimber and Dearman, 1997). Conversely, it is Th1-type cells that are associated usually with experimental contact hypersensitivity; the development of such reactions being promoted by IFN-γ (Cher and Mosmann, 1987; Diamanstein *et al.*, 1988; Fong and Mosmann, 1989).

It is instructive to consider whether there is evidence that experimental sensitization with chemical allergens results in the stimulation of selective T lymphocyte responses. As described above, it has been found in mice that TMA will provoke hapten-specific IgE antibody responses and an elevation in the total serum concentration of IgE. Under the same conditions of exposure, and where there was evidence for comparable overall immunogenicity, DNCB failed to stimulate IgE responses (Dearman and Kimber, 1991; Dearman *et al.*, 1991). It has been demonstrated subsequently, with a number of chemical sensitizers, that there is clear variation in the ability to induce in mice increases in the serum concentration of IgE. Chemicals that are known in humans to cause sensitization of the respiratory tract and occupational asthma were found, following topical administration, to stimulate in BALB/c strain mice a substantial increase in serum IgE levels. In contrast, chemical allergens that are known or suspected not to cause respiratory hypersensitivity, but which nevertheless are able to effect contact sensitization, failed to induce elevated serum IgE, or did so to a lesser degree (Dearman *et al.* 1992a; Hilton *et al.*, 1995, 1996b; Kimber *et al.*, 1996). However, consistent with the selective stimulation of Th1 responses, DNCB and oxazolone (another potent contact allergen), induced in mice strong IgG2a antibody responses; antibody of this isotype being favoured by IFN-γ (Dearman and Kimber, 1991, 1992). Using TMA and DNCB as examples, respectively, of respiratory and contact chemical allergens, the nature of cutaneous hypersensitivity reactions induced in mice was examined. In animals sensitized topically to TMA, subsequent dermal challenge resulted in the elicitation of both immediate (1 h) and delayed (24 h) hypersensitivity reactions. Similar sensitization and challenge with DNCB was associated only with a delayed-type hypersensitivity reaction (Dearman *et al.*, 1992b). In parallel experiments, it was shown that serum from mice sensitized with TMA, but not from untreated or DNCB-sensitized mice, was able to transfer immediate hypersensitivity to naive syngeneic recipients. The kinetics of passive sensitization with TMA-immune serum, together with the fact that immediate hypersensitivity to DNCB could be induced with a monoclonal IgE anti-dinitrophenol (DNP) antibody, provided strong evidence that the immediate cutaneous reactions that characterize sensitization and challenge with TMA are effected by hapten-specific IgE antibody (Dearman *et al.*, 1992b). That the immediate reaction seen with TMA was a reflection of specific IgE antibody production and the sensitization of cutaneous mast cells was supported further by the observation that exposure to this chemical was associated also with the sensitization of mast cells *in situ*, as measured by the release of serotonin *in vitro* induced by a TMA-protein conjugate (Holliday *et al.*, 1993).

Further investigation of the association of Th cell subpopulations with responses stimulated in mice by topical administration of chemical allergens focused upon the kinetics and characteristics of induced cytokine secretion patterns. The results of these experiments, described below, must be considered in the context of what is known of the development with time of discrete Th cell subsets. Th1 and Th2 cells are believed to be the most differentiated populations of $CD4^+$ cells, with their phenotypes of polarized cytokine secretion developing during the evolution of immune responses. They arise from common precursors, the immediate progenitors being cells designated Th0 that are characterized by the ability to produce both Th1- and Th2-type cytokines. Such Th0 cells may themselves develop from a more primitive precursor, the Thp cell, that is believed to secrete only IL-2 (Bendelac and Schwartz, 1991; Mosmann *et al.*, 1991; Chapter 2, this volume). Direct evidence for the development of Th1 and Th2 cells from a common precursor came from studies of lineage ablation. It was found that elimination of cells that expressed the gene for IL-4 resulted in the failure of both IL-4- and IFN-γ-producing cells to develop (Kamagowa *et al.*, 1993). Superimposed upon the need for populations displaying polarized cytokine secretion profiles to differentiate during immune responses is the fact that within such populations individual T lymphocytes may exhibit considerable variation in the amounts of particular cytokines secreted. Only a proportion, and perhaps only a small proportion, of individual cells displays coordinate expression of cytokines that characterize differentiated Th populations (Bucy *et al.*, 1995; Kelso, 1995; Mosmann and Sad, 1996). Notwithstanding the fact that within populations there may be considerable variation with respect to the coexpression of cytokines by lymphocytes at the single cell level, it is relevant to examine the nature of total cytokine responses induced in mice by chemical allergens.

Consistent with the view that differentiated phenotypes develop with time during immune responses, it was found that following acute dermal exposure of mice to sensitizing chemicals, draining lymph node cells (LNC) displayed comparable cytokine secretion profiles in response to both contact and respiratory allergens. Following repeated treatment of mice with chemicals, however, there was evidence for the emergence of selective cytokine secretion patterns. Exposure to TMA over a 2-week period was associated with the production by LNC of comparatively high levels of IL-10 and mitogen-inducible IL-4, but only low levels of IFN-γ. The opposite profile was observed after exposure of mice to oxazolone using an identical treatment regime. In this case vigorous IFN-γ secretion was stimulated, but with only relatively low levels of IL-4 or IL-10. A similar divergence of cytokine production was found following exposure of mice to other contact and respiratory allergens (Dearman *et al.*, 1994, 1995, 1996a, 1997a,b; Hilton *et al.*, 1996a) (Figure 11.1).

There is evidence that in rats also chemical allergens of different types provoke qualitatively divergent immune responses. Rats of Brown-Norway strain were exposed topically to concentrations of oxazolone or TMA that stimulated similar levels of proliferative activity by draining LNC. After treatment for approximately 2 weeks, it was found that although both chemicals induced

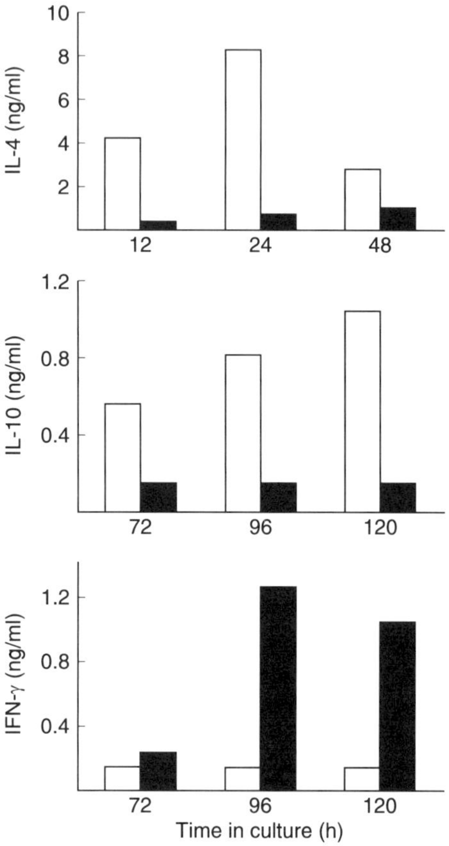

Figure 11.1 Cytokine secretion profiles of draining lymph node cells (LNC) cultured for various periods following repeated topical exposure of BALB/c strain mice to either 10% trimellitic anhydride (TMA; □) or 1% 2,4-dinitrochlorobenzene (DNCB; ■); concentrations of TMA and DNCB known to provoke comparable levels of overall immunogenicity. Concentrations of interleukin 4 (IL-4), interleukin 10 (IL-10) and interferon γ (IFN-γ) were measured (using enzyme-linked immunosorbent assays) in supernatants of LNC cultured for 72, 96 or 120 h in the absence (IL-10 and IFN-γ), or for 12, 24 and 48 h in the presence (IL-4), of the T lymphocyte mitogen concanavalin A.

specific IgG antibody, exposure only to TMA resulted in the appearance of IgE antibody. Moreover, TMA, but not oxazolone, caused an increase in the concentration of serum IgE and the active sensitization of mast cells. Cytokine analyses were performed also. Draining LNC from rats sensitized to oxazolone exhibited elevated expression of mRNA for IFN-γ, while in contrast, LNC from rats exposed to TMA had increased IL-5 mRNA. Finally, mitogen-activated LNC from TMA-, but not oxazolone-sensitized rats secreted significantly increased concentrations of IL-4 (Vento *et al.* 1996). In an independent study, using the same strain of rat, Arts *et al.* (1997) examined the ability of several chemicals, including TMA and DNCB, to elicit increases in the serum concentration of IgE. Of the chemicals tested only TMA caused a significant elevation of IgE levels; DNCB was without effect.

The experimental studies summarized above indicate that in both mice and rats chemical allergens of different types stimulate qualitatively variable immune responses. Chronic exposure to respiratory allergens is characterized by the preferential expression of type 2 cytokines in draining LNC. Under the same conditions, contact allergens induce the expression by LNC of high levels of IFN-γ, but only comparatively low levels of cytokines such as IL-4 and IL-10. The assumption is that these differences represent the basis for the variable ability of chemical allergens to stimulate IgE responses.

CD8$^+$ T LYMPHOCYTES

It should not be assumed, however, that the qualitatively divergent immune responses induced in mice by different classes of chemical allergens are necessarily attributable solely to the function of CD4$^+$ T lymphocytes. More recently it has become apparent that a functional heterogeneity, similar to that which characterizes CD4$^+$ Th cells, is found among CD8$^+$ T lymphocyte populations (Kemeny *et al.*, 1994; Kimber and Dearman, 1997; Chapter 3 this volume; Le Gros and Erard, 1994; Seder and Le Gros, 1995). Two populations, designated Tc1 and Tc2, have been described that exhibit cytokine profiles comparable, respectively, with Th1 and Th2 cells (Croft *et al.*, 1994). As will be discussed later, these subpopulations of CD8$^+$ lymphocytes are considered to have similar requirements to CD4$^+$ subsets for their development from more primitive precursors (Croft *et al.*, 1994; Noble and Kemeny, 1995; Noble *et al.*, 1995; Sad *et al.*, 1995).

There is growing interest in the role of CD8$^+$ cells in the expression of contact hypersensitivity; the relevant effector population being presumably Tc1-type cells that elaborate IFN-γ (Bour *et al.*, 1995; Fehr *et al.*, 1994; Gocinski and Tigelaar, 1990; Xu *et al.*, 1996, 1997). It has been proposed also that CD8$^+$ T lymphocytes which produce type 2 cytokines may in some instances cause unresponsiveness and immunological tolerance to contact allergens in mice (Steinbrink *et al.*, 1996; Xu *et al.*, 1996). Although CD8$^+$ cells may contribute both positively and negatively to contact hypersensitivity it is important to acknowledge that there is an undoubted role for CD4$^+$ cells. It is known that in mice lacking the CD4 gene contact sensitization is suboptimal (Kondo *et al.*, 1996). It should be recognized also that under certain circumstances both Tc1 and Tc2 cells may effect inflammatory delayed-type hypersensitivity reactions. Investigations by Li *et al.* (1997) have revealed that allospecific Tc1 and Tc2 cells induce comparable levels of swelling, and with similar kinetics, when injected into the footpads of naive allogeneic recipients.

There is interest also in the roles played by $CD8^+$ cells in allergic respiratory hypersensitivity reactions. It has been found that allergen-induced airways hyperresponsiveness in rats and mice may be inhibited by Tc1-type cells (Renz *et al.*, 1994; Underwood *et al.*, 1995). Conversely, other populations of $CD8^+$ cells appear in some systems to be necessary for airway reactions. In mice sensitized with ovalbumin and depleted of $CD8^+$ cells, it was shown that neither airway hyperresponsiveness, nor the accumulation of eosinophils within the lung parenchyma could be stimulated. These responses could be restored by the transfer of $CD8^+$ cells from naive mice during sensitization. The conclusion drawn was that, in the model used, type 2 $CD8^+$ T lymphocytes were required as a source of IL-5 for attraction of eosinophils (Hamelmann *et al.*, 1996). The situation is complicated, however, as Haczku and colleagues (1997) have reported recently that the adoptive transfer of purified $CD4^+$ T lymphocytes from sensitized Brown-Norway rats would effectively induce airway inflammation and hyperresponsiveness in syngeneic recipients. Under the same conditions sensitized $CD8^+$ cells and naive $CD4^+$ cells failed to induce respiratory responses. In this instance, it appeared that $CD4^+$, rather than $CD8^+$, cells were necessary for eosinophilic airway inflammation (Haczku *et al.*, 1997). Currently, there is no information about the role of $CD8^+$ T lymphocytes, or of $CD8^+$ T cell subpopulations, in the elicitation of respiratory hypersensitivity reactions to chemical allergens. There is, however, evidence that $CD8^+$ cells are able to regulate, usually in negative fashion, IgE antibody responses (Kemeny and Diaz-Sanchez, 1993; Kemeny *et al.*, 1992; McMenamin and Holt, 1993; Renz *et al.*, 1992; Underwood *et al.*, 1995). It has been found that the ability of $CD8^+$ cells to down-regulate IgE responses to inhaled antigen is largely or wholly a function of a small population of $CD8^+$ lymphocytes that produces high levels of IFN-γ and that display $\gamma\delta$ T cell receptors (McMenamin *et al.*, 1994, 1995). Transfer to syngeneic recipients of as few as 1000 of such $\gamma\delta^+$, $CD8^+$ T cells from Brown-Norway rats exposed repeatedly to antigen by inhalation caused antigen-specific, and IgE isotype-specific, tolerance (McMenamin *et al.*, 1995). The role of such cells in regulating antibody responses to chemical allergens encountered via the respiratory tract is not known. Nor is it clear whether $\gamma\delta^+$, $CD8^+$ T cells influence the integrity of IgE antibody production following exposure to antigens by other routes.

Taken together, the information available to date indicates that $CD8^+$ T lymphocyte subpopulations may have the potential to play influential roles in chemical allergy. Both $CD4^+$ and $CD8^+$ cells are induced during contact sensitization, with the latter possibly contributing to the regulation of responses and the elicitation of cutaneous hypersensitivity reactions. There is now some evidence that in mice chemical respiratory allergens are also able to induce both $CD4^+$ and $CD8^+$ cell responses. Comparative experiments have been performed to examine the contribution of T lymphocyte subpopulations to the cytokine secretion patterns displayed by draining LNC following repeated exposure of mice to DNCB or TMA. With both allergens, the production by

LNC of IL-10 and of mitogen-inducible IL-4 was attributable solely to $CD4^+$ cells (Th2 type cells). In contrast, the small amount of IFN-γ secreted by draining LNC prepared from mice treated chronically with the respiratory allergen TMA derived exclusively from $CD8^+$ cells. The much higher levels of IFN-γ produced by LNC from DNCB-sensitized mice were found to derive from both $CD4^+$ and $CD8^+$ cells (Dearman *et al.*, 1996b; Selgrade *et al.*, 1997). Both $CD4^+$ and $CD8^+$ cells presumably contribute to the cytokine environment that develops following exposure of mice to the chemical allergens. The characteristics of allergic responses will therefore be a function of the balance achieved between subsets of these populations (Figure 11.2).

Development of T lymphocyte subpopulations

It is instructive to consider the mechanisms through which exposure to chemical allergens may result in the selective development of functional subpopulations of T lymphocytes. Although most of the information available, and summarized below, derives from investigations of Th cell differentiation, it may be that the same or similar influences affect the development of Tc1 and Tc2 cells; in many instances, however, this remains unproven.

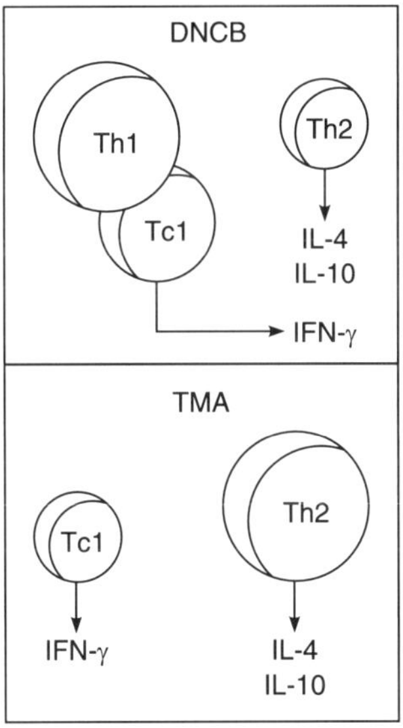

Figure 11.2 The relative contributions of $CD4^+$ (Th1 and Th2) and $CD8^+$ (Tc1 and Tc2) lymphocytes to patterns of cytokine (interleukin 4, IL-4; interleukin 10, IL-10; interferon γ, IFN-γ) production by draining lymph node cells isolated following repeated topical exposure of BALB/c strain mice to 2,4-dinitrochlorobenzene (DNCB) or trimellitic anhydride (TMA). A diagrammatic representation.

It is probable that the relative availability of cytokines in the reactive microenvironment plays a critical, and arguably the most critical, role in determining the development of differentiated $CD4^+$ and $CD8^+$ T cell function (Abehsira-Amar *et al.*, 1992; Coffman *et al.*, 1991; Croft *et al.*, 1994; Hsieh *et al.*, 1992; Mosmann and Sad, 1996; Noble and Kemeny, 1995; Noble *et al.*, 1995; Sad *et al.*, 1995; Swain *et al.*, 1991). In general terms it is type 1 cytokines that promote the development of Th1 and Tc1 cells, whereas IL-4, IL-10 and other type 2 cytokines drive the differentiation of Th2 (and possibly Tc2) cells. In this context, it has been found that Th2 cell responses fail to develop in IL-4 gene knockout mice (Kopf *et al.*, 1993). These same cytokines serve to regulate reciprocally Th cell development such that IL-10 inhibits cytokine synthesis by Th1 cells and the proliferation of Th2 cells is inhibited by IFN-γ. The cytokines that are probably of greatest significance in the initial orientation and tuning of adaptive immune responses are IL-10 and interleukin 12 (IL-12). Interleukin 10 inhibits the ability of dendritic cells to stimulate the production by T lymphocytes of IFN-γ (Macatonia *et al.*, 1993). Furthermore, IL-10 inhibits the elicitation phase of contact hypersensitivity in mice (Ferguson *et al.*, 1994; Kondo *et al.*, 1994), possibly secondary to suppression of IFN-γ production by T lymphocytes at the challenge site (Kondo *et al.*, 1994; Li *et al.*, 1994).

The major promoter of Th1 cell responses is IL-12, a heterodimeric cytokine produced by phagocytic cells, dendritic cells, keratinocytes and some B lymphocytes (Aragne *et al.*, 1994; Kang *et al.*, 1996; Trinchieri, 1993; Trinchieri and Scott, 1994). This cytokine has been shown to both stimulate the development of Th1 cells and inhibit Th2 cell responses (Heufler *et al.*, 1996; Hsieh *et al.*, 1993; McKnight *et al.*, 1994; Ohshima and Delespesse, 1997; Reiner and Seder, 1995). The negative regulation of Th2 cell function by IL-12 may be largely attributable to its ability to stimulate the secretion by natural killer (NK) cells of IFN-γ. However, it is likely that promotion by IL-12 of Th1-type immune responses operates via both IFN-γ-dependent and IFN-γ-independent mechanisms (Morris *et al.*, 1994; Reiner and Seder, 1995; Schmitt *et al.*, 1994; Seder *et al.*, 1993; Wenner *et al.*, 1996).

To a large extent, it is the expression of receptors for IL-12 that regulates the differential development of Th cell responses. The IL-12 receptor complex comprises two β-type cytokine receptor subunits (Presky *et al.*, 1996). In 1995, Szabo *et al.* proposed that developmental commitment to one or other Th lineage might be regulated via the maintenance or extinction of IL-12 signalling capability. It was found that murine Th1 cells retain their potential to signal in response to IL-12 (and IL-4), but that the IL-12 signalling pathway is rapidly extinguished in Th2 cells (Szabo *et al.*, 1995). More recently, it has been found in mice that whereas both Th1 and Th2 cells express the IL-12 receptor (IL-12R) $\beta1$ subunit, only Th1 cells have the $\beta2$ subunit. As a consequence, only Th1 cells respond productively to IL-12 with the phosphorylation of Jak2 and the STAT (*S*ignal *T*ransducers and *A*ctivators of

*T*ranscription) proteins Stat1, Stat3 and Stat4 (Szabo *et al.*, 1997). In humans, selective expression of the IL-12R $\beta 2$ chain may also provide a pivotal determinant of Th cell development. It has been found that human allergen-specific Th2 cells lack IL-12 signalling function (similarly characterized by the absence of Stat4 phosphorylation and, as a consequence, the failure of IL-12 to induce in these clones DNA-binding activity of Stat4 complexes). As with studies in mice, Th2 clones expressed normal levels of the $\beta 1$ subunit of the IL-12R (Hilkens *et al.*, 1996). Consistent with the above, Rogge *et al.* (1997) have also found that the $\beta 2$ chain of the IL-12R is expressed by human Th1, but not Th2, clones. It has been postulated that in naive, unstimulated $CD4^+$ T lymphocytes there is low level expression of functional IL-12R and that in cells which develop along a Th1 pathway (presumably in the presence of IL-12 itself, IFN-γ and in the absence of high levels of IL-4) this expression is maintained. In contrast, in other microenvironmental conditions (with high levels of IL-4), cells destined to develop a polarized Th2 phenotype sustain a selective loss of $\beta 2$ chain expression with the attendant failure of IL-12 to signal through Stat4 (Gallagher, 1997). The significance of the latter being that Stat4 null mice display an impaired production of IFN-γ upon antigenic stimulation, consistent with a defect of Th1 cell differentiation (Kaplan *et al.*, 1996; Thierfelder *et al.*, 1996). Conversely, Stat6 deficiency is associated with a failure of Th2 cell differentiation (Shimoda *et al.*, 1996). The loss of signalling through the IL-12R may represent a stable commitment to the Th2 lineage. This in turn suggests that IL-12 may have only a limited potential to influence established Th2 responses in allergic disease.

In conclusion, there is strong evidence that the cytokine microenvironment that pertains at the time of antigenic stimulation during primary immune responses will influence markedly the subsequent development of differentiated subpopulations of Th (and possibly Tc) cells. In the case of chemical allergy, the supposition is that the initial response to some chemicals favours the development of Th2 cells and that this is sustained through loss of the IL-12R $\beta 2$ subunit. In such circumstances, IgE antibody responses will be induced and maintained. Alternatively, where IL-4 is limiting, and under conditions where IFN-γ predominates, then Th1 responses will develop which are non-permissive for IgE production. The early events that influence initial commitment to one or the other lineage and the sources of cytokines that affect lineage development are considered below.

TRIGGERS FOR DIVERGENT T LYMPHOCYTE SUBPOPULATION DEVELOPMENT

The clear influence of cytokines on the selective development of discrete subpopulations of T lymphocytes raises intriguing questions about the possible cellular sources of IL-4, IL-12, IFN-γ and other cytokines before a full primary adaptive immune response has been initiated. One suggestion is that cellular components of the innate or natural immune system may provide one source of cytokines early following first exposure to an antigen (Romagnani, 1992). The general argument is that the differentiation of Th1 cells would be favoured by the action of IL-12 (derived presumably from dendritic cells, macrophages or other cell types) on NK cells stimulating their production of IFN-γ. In the case of Th2-type responses then the proposal is that mast cells represent an important local source of IL-4. While it is possible, theoretically, that NK cells and mast cells could both contribute to the cytokine environment, it is difficult to reconcile exclusive roles for these cells with their anatomical locations. Other candidate cells have now been identified. An atypical population of T lymphocytes has been described in mice that appears to be programmed for the rapid synthesis of IL-4 following primary encounter with antigen. These cells are characterized by expression of the NK cell antigen NK1.1. They are largely $CD4^+$, have a limited $\alpha\beta$ T cell receptor repertoire, are restricted to the non-major histocompatibility complex (MHC)-linked CD1 antigen and are deficient in β2-microglobulin (β2m)-null mice (Vicari and Zlotnik, 1996; Yoshimoto *et al.*, 1994, 1995a,b). However, more recent investigations have suggested that this population of cells may not represent an absolute requirement for the development of Th2-type responses and that IgE antibody production may proceed in their absence. It has been found that β2m-deficient mice (which lack the relevant NK1.1, $\alpha\beta^+$ T cell population) produce normal levels of IgE antibody, mount normal eosinophil responses and exhibit type 2 cytokine expression that is comparable with wild-type animals (Brown *et al.*, 1996; von der Weid *et al.*, 1996; Zhang *et al.*, 1996). Similar conclusions were drawn from investigations of mice lacking IL-4-secreting CD1-dependent cells. Such animals were found to display typical Th2-type immune responses following appropriate stimulation, with the normal production of IgE (Smiley *et al.*, 1997). Another candidate population, distinct from $NK1.1^+$, $CD4^+$ T cells, has now been identified in BALB/c strain mice. These also are $CD4^+$ T cells, but with a memory-type phenotype of $CD62L^-$, $CD45RB^-$. These cells were found in BALB/c mice to accelerate the development of IL-4-producing memory-type $CD4^+$ T cells from naive populations (Nishimura *et al.*, 1997). Yet another possibility is that B lymphocytes themselves provide one important signal for the drive toward Th2 cell differentiation. The conclusion drawn from recent studies in mice was that B lymphocytes facilitate the induction of IL-4 gene expression in the T cells that provide helper activity for antibody production (Macaulay *et al.*, 1997; Stockinger *et al.*, 1996). It may prove that one or other population of the $CD4^+$ T lymphocytes described above ($CD4^+$, $NK1.1^+$ cells and/or $CD4^+$, $CD62L^-$, $CD45RB^-$ cells) in B cell follicles stimulates the IL-4 driven expansion of antigen-stimulated B lymphocytes and that the latter then prime classical $CD4^+$ T cells for further IL-4 synthesis (Mason, 1996). At a rather simplistic level the argument is that B cells represent the central antigen-presenting population for the stimulation of IL-4 production by $CD4^+$ T lymphocytes and for the development of Th2 cells and thereby, the provision of help for antibody production. Such would appear to be consistent with the demonstration in IL-4 gene deletion transgenic mice that the expression by $CD4^+$ T lymphocytes of IL-4 is independent of the provision of this cytokine by $CD4^-$ cells. Reconstitution of these mice with $CD4^+$ cells from wild-type congenic animals, such that only $CD4^+$ cells were able to express IL-4, resulted in an antigen-specific IgE response (Schmitz *et al.*, 1994). The ability of chemical allergens to stimulate the necessary primary response by B lymphocytes and/or the production of IL-4 by non-classical $CD4^+$ T lymphocytes is not known. However, since certain chemical respiratory sensitizers such as TMA are known to provoke in mice the development of selective Th2-type immune responses, the implication is that they are able to stimulate the rapid release of IL-4 necessary to initiate and maintain the differentiation of Th2 cells.

Important for the development of Th1 cells are immunostimulatory, antigen-presenting dendritic cells (DC) and the precursors from which they derive. The activity of these cells is demonstrated most readily following topical exposure of mice to chemical allergens where Langerhans cells (LC) and other cutaneous DC are known to play important roles (Kimber, 1996; Kimber and Cumberbatch, 1992; Kimber and Dearman, 1996; Lappin *et al.*, 1996). Following topical application of chemical allergens LC, many of which bear high levels of antigen, are induced to leave the skin and travel via afferent lymphatics to lymph nodes draining the site of exposure. The cells that arrive in the nodes have acquired the characteristics of immunostimulatory DC and effectively present antigen to responsive T lymphocytes (Cumberbatch *et al.*, 1991, 1992; Jones *et al.*, 1989; Kinnaird *et al.*, 1989; Knight *et al.*, 1985; Macatonia *et al.*, 1986, 1987). There is reason to believe that in many instances the presentation of antigen to T lymphocytes by DC will favour the development of Th1 cells. Certainly both LC and DC produce IL-12 (Heufler *et al.*, 1996; Kang *et al.*, 1996). The suggestion is that under conditions where the interaction of antigen-bearing DC with responsive T lymphocytes results in production by the former of high levels of IL-12 then the selective development of Th1 responses will be promoted (Ohshima and Delespesse, 1997). It is of interest that in some circumstances DC may, theoretically, exert the opposite effect. Rincon *et al.* (1997) have shown recently that IL-6, which derives presumably from antigen-presenting cells, is able to polarize T helper cell responses to those of Th2-type by inducing the initial production of IL-4 by $CD4^+$ T lymphocytes. As there is evidence that, in addition to IL-12, LC and DC synthesize and secrete IL-6 (Cumberbatch *et al.*, 1996; Hope *et al.*, 1995), it is possible that in some situations the relative production by DC of IL-6 and IL-12 may influence the balance between Th1 and Th2 cell differentiation. In this context, the level of expression by DC of various costimulatory molecules may be relevant. LC express heat-stable antigen (HSA), a costimulatory molecule for $CD4^+$ T lymphocytes that is believed to enhance the antigen-driven development of Th1 cells (Enk and Katz, 1994). Of particular importance in the activation of T lymphocytes are membrane determinants of the B7 family, B7-1 (CD80) and B7-2 (CD86), that serve as ligands for the CTLA-4/CD28 T lymphocyte signalling pathway. The expression of both B7-1 and B7-2 by LC is upregulated during culture, the assumption being that a similar increase in expression will be associated with the functional maturation of LC as they migrate from the epidermis to draining lymph nodes (Caux *et al.*, 1994; Girolomoni *et al.*, 1994; Hart *et al.*, 1993; Inaba *et al.*, 1994; Larssen *et al.*, 1992; Razi-Wolf *et al.*, 1994). It has been proposed that costimulation by B7-1 during antigen presentation favours Th1 cell development, whereas B7-2 promotes Th2 responses (Kuchroo *et al.*, 1995). It is of interest that costimulation through B7-2 is necessary for lung mucosal Th2 responses and altered airway responsiveness (Tsuyuki *et al.*, 1997). However, the relationship between B7-1- and B7-2-mediated interactions with CD28/CTLA-4 and the selective stimulation of discrete populations of Th cells may be more complex than conceived originally (Gause *et al.*, 1997; Greenwald *et al.*, 1997; Palmer and van Seventer, 1997).

Dendritic cells in the respiratory tract may also play a key role in determining the nature of immune responses to inhaled chemical allergens. Within the airway epithelium is a network of DC and these presumably fulfil a function similar to LC found in the epidermis (Holt *et al.*, 1989, 1990a,b; Lipscomb *et al.*, 1995; Pollard and Lipscomb,1990; Schon-Hegrad *et al.* 1991; Sertl *et al.*, 1986; Tunon-de-Lara *et al.*, 1996). The initiation of pulmonary immune responses and acute inflammatory reactions in the respiratory tract is associated with an accumulation of DC within the airway epithelium and the subsequent migration of these cells to draining lymph nodes (McWilliam *et al.*, 1994; Xia *et al.*, 1995). The ability of airway DC to stimulate IgE responses appears to be constrained by alveolar macrophages; the experimental elimination of the latter having been shown to potentiate selectively IgE antibody production to inhaled allergen (Holt *et al.*, 1988; Thepen *et al.*, 1989, 1991, 1992).

The conclusions drawn are that the characteristics of immune responses stimulated by dendritic cells, and by other populations of antigen-presenting cells, will be dependent upon the nature of the cells themselves, their expression of costimulatory molecules, their pattern of cytokine production and by interactions with other cell types that may serve to enhance or modify their functional activity. However, superimposed upon these considerations is the question of the antigen itself and the way in which it is displayed to responsive T lymphocytes by antigen-presenting cells.

ANTIGEN AND ANTIGEN PRESENTATION

Chemical allergens are haptens (or prohaptens). In their native state they are unable, on their own, to provoke immune responses. For this, they must first associate with protein, in the case of prohaptens following metabolic conversion to a protein-reactive species. The best evidence available indicates that immune responses induced by exposure to chemical allergens result from the recognition by T lymphocytes of the inducing hapten in association with peptides anchored to MHC determinants (Cavani *et al.*, 1995; Kohler *et al.*, 1995; Martin and Weltzien, 1994; Martin *et al.*, 1993; Weltzien *et al.*, 1996). What is not clear, however, is the stage at which chemical allergen associates with protein or peptide. It is possible that a protein-reactive hapten might interact directly with peptide that is already displayed with an MHC molecule. The alternative is that chemical associates with extracellular protein at the portal of entry and that this modified protein is then internalized, processed and presented in such a way that the haptenic determinant is recognized on the surface of an anchored peptide by responsive T lymphocytes. Whichever route or routes of hapten acquisition by antigen-presenting cells occurs in practice, it is likely that the orientation and availability of the relevant epitope will exert a strong influence on the vigour and characteristics of T lymphocyte responses. Certainly, it is known that the strength of association between the immunogenic peptide and the MHC class II molecule and the affinity of interaction between the MHC class II-ligand complex and the T cell receptor (TCR) for antigen both appear to affect the selectivity of Th cell responses. In recent investigations, it was shown that the frequency of $CD4^+$ T lymphocytes expressing either type 1 or type 2 cytokines was dependent upon the binding affinity of peptide ligands for the MHC class II molecule (Kumar *et al.*, 1995). Employing a panel of ligands each with identical specificity, but differing in terms of MHC binding constants, it was observed that those peptides which exhibited highest affinity binding to MHC class II molecules increased the frequency of cells producing IFN-γ, but not of cells producing IL-4 or IL-5. Using a slightly different approach, it was found that MHC class II-peptide ligand complexes that associate strongly with the TCR favour Th1 cells (Pfeiffer *et al.*, 1995). Further complexity is provided by the fact that antigen dose may be a critical variable; the possibility being that the extent of TCR ligation influences the selective differentiation of $CD4^+$ T lymphocytes. Using $CD4^+$ cells expressing a transgene-encoded TCR, it has been found that exposure *in vitro* to high doses of antigen stimulated IFN-γ-producing Th1 cells, whereas low doses of the same antigenic peptide induced instead IL-4 producing Th2 cells (Constant *et al.*, 1995; Secrist *et al.*, 1995; Xu *et al.*, 1996). In recent studies of house dust mite allergen-specific human T lymphocytes, it was found that amino acid sequence variants of an immunodominant epitope had variable effects on the ability of Th0 cells to produce IFN-γ (Tsitoura *et al.*, 1996). Finally, it has been shown also that peptide affinity for the MHC may influence not only the characteristics of Th cells responses, but also the extent of apoptosis among antigen activated cells (Pearson *et al.*, 1997).

The available data, although in some instances apparently contradictory, point to the nature of the antigenic complex formed between the peptide (and in the case of chemical allergy, the haptenated peptide) and MHC determinants and the association of this complex with the TCR having important influences on Th cell selection. In addition, the characteristics of the antigen-presenting cells themselves, their location and activation status and the availability of chemical will all impact on the nature of induced immune responses.

At present, it is far from clear why chemical allergens such as DNCB and TMA provoke in mice qualitatively divergent immune responses. Both are immunogenic, stimulating T lymphocyte responses and antibody production, yet the characteristics of the responses induced differ substantially. It will be instructive to determine whether these haptens are handled in different ways by the immune system and whether they exhibit a variable potential to induce the rapid secretion of relevant cytokines by cells local to the site of first encounter. Clarification of the important molecular determinants of selective Th cell responses to chemical allergens may in turn be of value in the diagnosis and treatment of contact and respiratory sensitization in humans. It is instructive, therefore, to consider the extent to which human allergic disease is characterized by Th cell selectivity.

Human allergic disease

In humans also a functional heterogeneity among $CD4^+$ T lymphocytes has been characterized which is similar, but probably not identical, to that described in mice (Romagnani, 1991; Romagnani *et al.*, 1992). Attempts have been made to consider human allergic disease in the context of $CD4^+$ cell phenotype (see also Chapter 12 this volume). The picture that emerges from analyses of allergic contact dermatitis is of a preferential, although not absolute, involvement of Th1-type cells. Most T lymphocyte clones generated from the peripheral blood of nickel dermatitis patients were found to produce high levels of IFN-γ, but only comparatively small amounts of IL-4 and IL-5 (Kapsenberg *et al.*, 1991, 1992). At present, it is not possible to conclude that Th2 cells, or at least cells that produce IL-4, have no role to play in the development of allergic contact dermatitis. It has been reported that mRNA for IL-4 is expressed strongly in contact dermatitis skin lesions (Ohmen *et al.*, 1995). In addition, there are experimental investigations that suggest IL-4 plays a role in the elicitation phase of contact hypersensitivity (Salerno *et al.*, 1995; Weigmann *et al.*, 1997). The interpretation of these experimental data is that IL-4 may modify the activity and/or facilitate the localization of the Th1-type effector cells during the challenge phase of allergic contact reactions in mice. If contact sensitivity is associated usually with cells of a Th1 phenotype, then a different pattern is seen in respiratory hypersensitivity and other forms of allergic reaction in which IgE responses are implicated.

The symptoms of allergic hypersensitivity of the respiratory tract induced by chemicals are various and may include immediate and late phase responses, as well as what are known as dual reactions. Chronic inflammation plays a critical role in asthma and is associated with the destruction and sloughing of airway epithelial cells, the production of mucus, subepithelial fibrosis secondary to collagen deposition and the accumulation in the bronchial mucosa of leukocytes. Central to the development of chronic bronchial inflammation and injury are eosinophils acting in concert with T lymphocytes (Barnes and Lee, 1992; Corrigan, 1992; Corrigan and Kay, 1992; Gleich, 1990). Consistent with an important role for eosinophils in the pathogenesis of allergic respiratory hypersensitivity are investigations that reveal a predominance of Th2 cells in affected tissue. Biopsies of nasal turbinates from pollen sensitized patients were examined following challenge with the relevant allergen. Using *in situ* hybridization, it was shown that provocation with allergen was associated with an increased number of cells expressing mRNA for IL-3, IL-4 and IL-5 and that this increase in the frequency of Th2-type cells correlated positively with tissue eosinophilia (Durham *et al.*, 1992). A predominance of Th2-type cells associated with eosinophil recruitment has been shown also in bronchoalveolar lavage fluid prepared from patients with atopic asthma (Robinson *et al.*, 1992, 1993). It is also the case that atopic dermatitis lesional skin is often associated with a high frequency of IL-4-producing Th2-type cells (van der Heijden *et al.*, 1991). There is also evidence that patients with atopic allergy have an increase in Th2-like cells within the circulation. Parronchi *et al.* (1991) demonstrated that clones of human T lymphocytes specific for house dust mite or pollen aeroallergens produced type 2 cytokines, but not IFN-γ. Other studies also have shown a predominance of type 2 cytokine-producing $CD4^+$ (and $CD8^+$) cells within the peripheral blood of patients with atopic dermatitis or atopic rhinitis (Nakazawa *et al.*, 1997; Till *et al.*, 1997a,b). It is unfortunate that similar studies have not been conducted with patients who have chemical respiratory allergy. There are, however, few investigations which are of relevance. In one, it was found that isocyanate-induced occupational asthma, and other forms of extrinsic asthma, were characterized by very similar patterns of inflammatory cell infiltrates with the presence of activated T lymphocytes and the recruitment of eosinophils (Bentley *et al.*, 1992).

On balance, the data available reveal that whereas allergic contact dermatitis has many features consistent with a Th1-type pathogenesis, atopic allergy (including asthma, rhinitis and atopic eczema) is associated primarily with a predominance of Th2-like cells. This is probably an oversimplification, and Borish and Rosenwasser (1997) have warned recently that the true nature of cytokine responses in human allergic disease may be of a complexity that cannot be accommodated fully within a clear dichotomy between type 1 and type 2 cells. Notwithstanding this, the evidence points to Th2 cells having a unique role in the development and expression of human atopic disease. Clearly, this is consistent with the important contribution made by eosinophils to allergic inflammation and the requirement for IgE antibody responses. In the case of chemical respiratory allergy, where (as discussed previously) an essential role for IgE remains unproven and controversial, the nature of T lymphocyte responses and of T cell infiltrates within inflammatory sites remains largely untested. It might be argued, however, that in chemical respiratory allergy also a selective Th2-type response would be consistent with the development of eosinophilic inflammation and bronchial asthma.

Conclusions

One of the most fascinating questions in chemical allergy is why sensitizing agents differ with respect to the type (and severity) of allergic reactions they will provoke in susceptible individuals. There is now experimental evidence, summarized above, that in mice and rats chemical allergens of different types stimulate qualitatively distinct immune responses consistent with the selective development of functional subpopulations of Th and Tc cells. A similar variability in responses to chemical allergens in humans has still to be confirmed, particularly with regard to the characteristics of T lymphocytes involved in sensitization of the respiratory tract.

References

Abehsira-Amar, O., Gilbert, M., Joliy, M., Theze, J. and Jankovic, D.L. (1992) IL-4 plays a dominant role in the differential development of Th0 into Th1 and Th2 cells. *J. Immunol.*, **148**, 3820–9.

Aebischer, I. and Stadler, B.M. (1996) TH1-TH2 cells in allergic responses: at the limits of a concept. *Adv. Immunol.*, **61**, 341–403.

Aebischer, I., Stampfli, M.R., Zurcher, A., Miescher, S., Urwyler, A., Frey, B., Luger, T., White, R.R. and Stadler, B.M. (1994) Neuropeptides are potent modulators of human *in vitro* immunoglobulin E synthesis. *Eur. J. Immunol.*, **24**, 1908–13.

Aragne, Y., Reimann, H., Bhardwaj, R.S., Schwarz, A., Sawada, Y., Yamada, H., Luger, T.A., Kubin, M., Trinchieri, G. and Schwartz, T. (1994) IL-12 is expressed and released by human keratinocytes and epidermoid carcinoma cell lines. *J. Immunol.*, **153**, 5366–72.

Armitage, R.J., Macduff, B.M., Spriggs, M.K. and Fanslow, W.C. (1993) Human B cell proliferation and Ig secretion induced by recombinant CD40 ligand are modulated by soluble cytokines. *J. Immunol.*, **150**, 3671–9.

Arts, J.H.E., Droge, S.C.M., Spanhaak, S., Bloksma, N., Penninks, A.H. and Kuper, F.C. (1997) Local lymph node activation and IgE responses in Brown Norway and Wistar rats after dermal application of sensitizing and non-sensitizing chemicals. *Toxicology*, **117**, 229–37.

Aversa, G., Punnonen, J., Cocks, B.G., de Waal Malefyt, R., Vega, F. Jr, Zurawski, S.M., Zurawski, G. and de Vries, J.E. (1993) An interleukin 4 (IL-4) mutant protein inhibits both IL-4 or IL-13 induced human immunoglobulin G4 (IgG4) and IgE synthesis and B cell proliferation: support for a common component shared by IL-4 and IL-13 receptors. *J. Exp. Med.*, **178**, 2213–18.

Azuma, M., Hirano, T., Miyajima, H., Watanabe, N., Yagita, H., Enomoto, S., Furusawa, S., Ovary, Z., Kinashi, T., Honja, T. and Okumura, K. (1987) Regulation of IgE production in SJA/9 and nude mice. Potentiation of IgE production by recombinant interleukin 4. *J. Immunol.*, **139**, 2538–44.

Barnes, P.J. and Lee, T.H. (1992) Recent advances in asthma. *Postgrad. Med. J.*, **68**, 942–53.

Bendelac, A. and Schwartz, R.H. (1991) Th0 cells in the thymus. The question of T-helper lineages. *Immunol. Rev.*, **123**, 169–88.

Bentley, A.M., Maestrelli, P., Saetta, M., Fabbri, L.M., Robinson, D.S., Bradley, B.L., Jeffrey, P.K., Durham, S.R. and Kay, A.B. (1992) Activated T-lymphocytes and eosinophils in the bronchial mucosa in isocyanate-induced asthma. *J. Allergy Clin. Immunol.*, **89**, 821–9.

Bernstein, D.I., Patterson, R. and Zeiss, C.R. (1982) Clinical and immunological evaluation of trimellitic anhydride- and phthalic anhydride-exposed workers using a questionnaire and comparative analysis of enzyme-linked immunosorbent and radioimmunoassay studies. *J. Allergy Clin. Immunol.*, **69**, 311–18.

Blaikie, L., Morrow, T., Wilson, A.P., Hext, P., Hartop, P.J., Rattray, N.J., Woodcock, D. and Botham, P.A. (1995) A two-centre study for the evaluation and validation of an animal model for the assessment of the potential of small molecular weight chemicals to cause respiratory allergy. *Toxicology*, **96**, 37–50.

Borger, P., Kauffman, H.F., Postma, D.S and Vellenga, E. (1996) IL-7 differentially modulates the expression of IFN-γ and IL-4 in activated human T lymphocytes by transcriptional and post-transcriptional mechanisms. *J. Immunol.*, **156**, 1333–8.

Borish, L. and Rosenwasser, L. (1997) TH1/TH2 lymphocytes: doubt some more. *J. Allergy Clin. Immunol.*, **99**, 161–4.

Botham, P.A., Rattray, N.J., Woodcock, D.R., Walsh, S.T. and Hext, P.M. (1989) The induction of respiratory allergy in guinea pigs following intradermal injection of trimellitic anhydride: a comparison with the response to 2,4-dinitrochlorobenzene. *Toxicol. Lett.*, **47**, 25–9.

Bour, H., Peyron, E., Gaucherand, M., Garrigue, J.-L., Desvignes, C., Kaiserlian, D., Revillard, J.-P. and Nicolas, J.-F. (1995) Major histocompatibility complex class I-restricted $CD8^+$ T cells and class II-restricted $CD4^+$ T cells, respectively mediate and regulate contact sensitivity to dinitrofluorobenzene. *Eur. J. Immunol.*, **25**, 3006–10.

Brown, D.R., Fowell, D.J., Corry, D.B., Wynn, T.A., Moskowitz, N.H., Cheever, A.W., Locksley, R.M. and Reiner, S.L. (1996) β2-Microglobulin-dependent $NK1.1^+$ T cells are not essential for T helper cell 2 immune responses. *J. Exp. Med.*, **184**, 1295–304.

Bucy, R.P., Karr, L., Huang, G.-Q., Li, J., Carter, D., Honjo, K., Lemons, J.A., Murphy, K.M. and Weaver, C.T. (1995) Single cell analysis of cytokine gene co-expression during $CD4^+$ T-cell phenotype development. *Proc. Natl Acad. Sci. USA*, **92**, 7565–9.

Burstein, H.J., Tepper, R.I., Leder, P. and Abbas, A.K. (1991) Humoral immune functions in IL-4 transgenic mice. *J. Immunol.*, **147**, 2950–6.

Caux, C., Vandervliet, B., Massacrier, C., Azuma, M., Okumura, K., Lanier, L.L. and Banchereau, J. (1994) B70/B7-2 is identical to CD86 and is the major functional ligand for CD28 expressed on human dendritic cells. *J. Exp. Med.*, **180**, 1841–7.

Cavani, A., Hackett, C.J., Wilson, K.J., Rothbard, J.B. and Katz, S.I. (1995) Characterization of epitopes recognized by hapten-specific $CD4^+$ T cells. *J. Immunol.*, **154**, 1232–8.

Chan-Yeung, M. (1995) Occupational asthma. *Environ. Health Perspect.*, **103**, 249–52.

Cher, D.J. and Mosmann, T.R. (1987) Two types of murine helper T cell clones. II. Delayed type hypersensitivity is mediated by Th1 clones. *J. Immunol.*, **138**, 3688–94.

Chretien, I., Pene, J., Briere, F., de Waal Malefyt, R., Rousset, F. and de Vries, J. (1990) Regulation of human IgE synthesis. 1. Human IgE synthesis *in vitro* is determined by the reciprocal antagonistic effects of interleukin-4 and interferon-γ. *Eur. J. Immunol.*, **20**, 243–51.

Cocks, B.G., de Waal Malefyt, R., Galizzi, J.-P., de Vries, J.E. and Aversa, G. (1993) IL-13 induces proliferation and differentiation of human B cells activated by the CD40 ligand. *Int. Immunol.*, **5**, 657–61.

Coffman, R.L., Varkila, K., Scott, P. and Chatelain, R. (1991) Role of cytokines in the differentiation of $CD4^+$ T-cell subsets *in vivo*. *Immunol. Rev.*, **123**, 189–207.

Coffman, R.L., Lehman, D.A. and Rothman, P.A. (1993) Mechanism and regulation of immunoglobulin class switching. *Adv. Immunol.*, **54**, 229–70.

Constant, S., Pfeiffer, C., Woodard, A., Pasqualini, T. and Bottomly, K. (1995) Extent of T cell receptor ligation can determine the functional differentiation of naïve $CD4^+$ T cells. *J. Exp. Med.*, **182**, 1591–6.

Corrigan, C.J. (1992) Allergy of the respiratory tract. *Curr. Opin. Immunol.*, **4**, 798–804.

Corrigan, C.J. and Kay, A.B. (1992) T cells and eosinophils in the pathogenesis of asthma. *Immunol. Today*, **13**, 501–7.

Croft, M., Carter, L., Swain, S.L. and Dutton, R.W. (1994) Generation of polarized antigen-specific CD8 effector populations: reciprocal action of interleukin (IL)-4 and IL-12 in promoting type 2 versus type 1 cytokine profiles. *J. Exp. Med.*, **180**, 1715–28.

Cumberbatch, M., Gould, S.J., Peters, S.W. and Kimber, I. (1991) MHC class II expression by Langerhans cells and lymph node dendritic cells. Possible evidence for maturation of Langerhans cells following contact sensitization. *Immunology*, **74**, 414–19.

Cumberbatch, M., Peters, S.W., Gould, S.J. and Kimber, I. (1992) Intercellular adhesion molecule-1 (ICAM-1) expression by lymph node dendritic cells: comparison with epidermal Langerhans cells. *Immunol. Lett.*, **32**, 105–10.

Cumberbatch, M., Dearman, R.J. and Kimber, I. (1996) Constitutive and inducible expression of interleukin 6 by Langerhans cells and lymph node dendritic cells. *Immunology*, **87**, 513–18.

Dearman, R.J. and Kimber, I. (1991) Differential stimulation of immune function by respiratory and contact chemical allergens. *Immunology*, **72**, 563–70.

Dearman, RJ. and Kimber, I. (1992) Divergent immune responses to respiratory and contact chemical allergens: antibody elicited by phthalic anhydride and oxazolone. *Clin. Exp. Allergy.*, **22**, 241–50.

Dearman, R.J., Hegarty, J.M. and Kimber, I. (1991) Inhalation exposure of mice to trimellitic anhydride induces both IgG and IgE anti-hapten antibody. *Int. Arch. Allergy Appl. Immunol.*, **95** 70–6.

Dearman, R.J., Basketter, D.A. and Kimber, I. (1992a) Variable effects of chemical allergens on serum IgE concentration in mice. Preliminary evaluation of a novel approach to the identification of respiratory sensitizers. *J. Appl. Toxicol.* **12**, 317–23.

Dearman, R.J., Mitchell, J.A., Basketter, D.A. and Kimber, I. (1992b) Differential ability of occupational chemical contact and respiratory allergens to cause immediate and delayed dermal hypersensitivity reactions in mice. *Int. Arch. Allergy Immunol.* **97**, 315–21.

Dearman, R.J., Ramdin, L.S.P., Basketter, D.A. and Kimber, I. (1994) Inducible IL-4-secreting cells provoked in mice during chemical sensitization. *Immunology*, **81**, 551–7.

Dearman, R.J., Basketter, D.A. and Kimber, I. (1995) Differential cytokine production following chronic exposure of mice to chemical respiratory and contact allergens. *Immunology*, **86**, 545–50.

Dearman, R.J., Basketter, D.A. and Kimber, I. (1996a) Characterization of chemical allergens as a function of divergent cytokine secretion profiles induced in mice. *Toxicol. Appl. Pharmacol.*, **138**, 308–16.

Dearman, R.J., Moussavi, A., Kemeny, D.M. and Kimber, I. (1996b) Contribution of $CD4^+$ and $CD8^+$ T lymphocyte subsets to the cytokine secretion patterns induced in mice during sensitization to contact and respiratory chemical allergens. *Immunology*, **89**, 502–10.

Dearman, R.J., Cumberbatch, M., Hilton, J., Fielding, I., Basketter, D.A. and Kimber, I. (1997a) A re-appraisal of the skin-sensitizing activity of 2,4-dinitrothiocyanobenzene. *Food Chem. Toxicol.*, **35**, 261–9.

Dearman, R.J., Smith, S., Basketter, D.A. and Kimber, I. (1997b) Classification of chemical allergens according to cytokine secretion profiles of murine lymph node cells. *J. Appl. Toxicol.*, **71**, 53–62.

Delespesse, G., Sarfati, M. and Heusser, C. (1989) IgE synthesis. *Curr. Opin. Immunol.*, **2**, 506–12.

Del Prete, G.F., Maggi, E., Parronchi, P., Chretien, I., Tiri, A., Macchia, D., Ricci, M., Banchereau, J., de Vries, J. and Romagnani, S. (1988) IL-4 is an essential factor for the IgE synthesis induced *in vitro* by human T cell clones and their supernatants. *J. Immunol.*, **140**, 4193–8.

Diamanstein, T., Eckert, R., Volk, H.-D. and Kupier-Weglinski, J.-W. (1988) Reversal by interferon-γ of inhibition of delayed-type hypersensitivity induction by anti-CD4 or anti-interleukin 2 receptor (CD25) monoclonal antibodies. Evidence for the physiological role of the $CD4^+$ $Th1^+$ subset in mice. *Eur. J. Immunol.*, **181**, 2101–3.

Drexler, H., Schaller, K.-H., Weber, A., Letzel, S. and Lehnert, G. (1993) Skin prick tests with solutions of acid anhydrides in acetone. *Int. Arch. Allergy Immunol.*, **100**, 251–5.

Dugas, B., Renauld, J.C., Pene, J., Bonnefoy, J.Y., Petit-Frere, C., Braquet, P., Bousquet, J., Van Snick, J. and Mencia-Huerta, J.M. (1993) Interleukin 9 potentiates the interleukin-4 induced immunoglobulin (IgG, IgM and IgE) production by normal human B lymphocytes. *Eur. J. Immunol.* **23**, 1687–92.

Durham, S.R., Ying, S., Varney, V.A., Jacobson, M.R., Sudderick, R.M., Mackay, I.S., Kay, A.B. and Hamid, Q. (1992) Cytokine messenger RNA expression for IL-3, IL-4, IL-5 and granulocyte/ macrophage colony-stimulating factor in the nasal mucosa after local allergen provocation: relationship to eosinophilia. *J. Immunol.*, **148**, 2390–4.

Enk, A.H. and Katz, S.I. (1994) Heat-stable antigen is an important costimulatory molecule on epidermal Langerhans cells. *J. Immunol.*, **152**, 3264–70.

Fehr, B.S., Takashima, A., Matsue, H., Gerometta, J.S., Bergstresser, P.R. and Cruz, P.D. Jr (1994) Contact sensitization induces proliferation of heterogeneous populations of hapten-specific T lymphocytes. *Exp. Dermatol.*, **3**, 189–97.

Ferguson, T.A., Dube, P. and Griffith, T.S. (1994) Regulation of contact hypersensitivity by interleukin 10. *J. Exp. Med.*, **179**, 1597–604.

Finkelman, F.D., Katona, I.M., Urban, J.F. Jr, Snapper, C.N., Ohara, J. and Paul, W.E. (1986) Suppression of *in vivo* polyclonal IgE production by monoclonal antibody to the lymphokine B-cell stimulatory factor 1. *Proc. Natl. Acad. Sci. USA*, **83**, 9675–8.

Finkelman, F.D., Katona, I.M., Mosmann,T.R. and Coffman, R.L. (1988a) IFN-γ regulates the isotypes of Ig secreted during *in vivo* humoral immune responses. *J. Immunol.*, **140**, 1022–7.

Finkelman, F.D., Katona, I.M., Urban, J.F. Jr, Holmes, J., Ohara, T., Tung, A.S., Sample, J.G. and Paul, W.E. (1988b) IL-4 is required to generate and sustain *in vivo* IgE responses. *J. Immunol.*, **141**, 2335–41.

Finkelman, F.D., Holmes, J., Katona, I.M., Urban, J.F., Beckman, M.P., Park, L.S., Schooley, K.A., Coffman, R.L., Mosmann, T.R. and Paul, W.E. (1990) Lymphokine control of *in vivo* immunoglobulin isotype selection. *Annu. Rev. Immunol.*, **8**, 303–33.

Fong, T.A.T. and Mosmann, T.R. (1989) The role of IFN-γ in delayed-type hypersensitivity mediated by Th1 clones. *J. Immunol.*, **143**, 2887–93.

Gallagher, R. (1997) Tagging T cells: TH1 or TH2. *Science*, **275**, 1615.

Garssen, J., Vandebriel, R.J., Kimber, I. and Van Loveren, H. (1996) Hypersensitivity reactions: definitions, basic mechanisms and localizations, in *Allergic Hypersensitivities Induced by Chemicals, Recommendations for Prevention* (eds J.C. Vos, M. Younes and E. Smith), CRC Press, Boca Raton, pp. 19–58.

Gause, W.C., Halvorson, M.J., Lu, P., Greenwald, R., Linsley, P., Urban, J.F. and Finkelman, F.D. (1997) The function of costimulatory molecules and the development of IL-4-producing T cells. *Immunol. Today*, **18**, 115–20.

Girolomoni, G., Zambruno, G., Manfredini, R., Zacchi,V., Ferrai, S., Cossarizza, A. and Gianetti, A. (1994) Expression of B7 costimulatory molecule in cultured human epidermal Langerhans cells is regulated at the mRNA level. *J. Invest. Dermatol.*, **103**, 54–9.

Gleich, G.J. (1990) The eosinophil and bronchial asthma: current understanding. *J. Allergy Clin. Immunol.*, **85**, 422–36.

Gocinski, B.L. and Tigelaar, R.E. (1990) Roles of $CD4^+$ and $CD8^+$ T cells in murine contact sensitivity revealed by *in vivo* monoclonal antibody depletion. *J. Immunol.*, **144**, 4121–8.

Greenwald, R.J., Lu, P., Halvorson, M.J., Zhou, X.-D., Chen, S.-J., Madden, K.B., Perrin, P.J., Morris, S.C., Finkelman, F.D., Peach, R., Linsley, P.S., Urban, J.F. Jr and Gause, W.C. (1997) Effects of blocking B7-1 and B7-2 interactions during a type 2 *in vivo* immune response. *J. Immunol.*, **158**, 4088–96.

Haczku, A., Macary, P., Huang, T.-J., Tsukagoshi, H., Barnes, P.J., Kay, A.B., Kemeny, D.M., Chung, K.F. and Moqbel, R. (1997) Adoptive transfer of allergen-specific $CD4^+$ T cells induces airway inflammation and hyperresponsiveness in Brown-Norway rats. *Immunology*, **91**, 176–85.

Hamelmann, E., Oshiba, A., Paluh, J., Bradley, K., Loader, J., Potter, T.A., Larsen, G.L. and Gelfand, E.W. (1996) Requirement for $CD8^+$ T cells in the development of airway hyperresponsiveness in a murine model of airway sensitization. *J. Exp. Med.* **183**, 1719–29.

Hammarstrom, L. and Smith, C.I.E. (1987) Immunoglobulin subclass distribution of specific antibodies in allergic patients. *Allergy*, **42**, 529–36.

Hart, D.N.J., Starling, G.C., Calder, V.L. and Fernando, N.S. (1993) B7/ BB-1 is a leukocyte differentiation antigen on human dendritic cells induced by activation. *Immunology*, **79**, 616–20.

Herrod, H.G. (1989) Interleukins in immunologic and allergic disease. *Ann. Allergy*, **63**, 269–72.

Heufler, C., Koch, F., Stanzl, U., Topar, G., Wysocka, M., Trinchieri, G., Enk, A., Steinman, R.M., Romani, N. and Schuler, G. (1996) Interleukin-12 is produced by dendritic cells and mediates T helper 1 development as well as interferon-γ production by T helper 1 cells. *Eur. J. Immunol.*, **26**, 659–68.

Hilkens, C.M.U., Messer, G., Tesselaar, K., van Rietschoten, A.G.I., Kapsenberg, M.I. and Wierenga, E.A. (1996) Lack of IL-12 signalling in human allergen-specific Th2 cells. *J. Immunol.*, **157**, 4316–21.

Hilton, J., Dearman, R.J., Basketter, D.A. and Kimber, I. (1995) Identification of chemical respiratory allergens: dose-response relationships in the mouse IgE test. *Toxicol. Meth.*, **5**, 51–60.

Hilton, J., Dearman, R.J., Basketter, D.A., Scholes, E.W. and Kimber, I. (1996a) Experimental assessment of the sensitizing properties of formaldehyde. *Food Chem. Toxicol.*, **34**, 571–8.

Hilton, J., Dearman, R.J., Boylett, M.S., Fielding, I., Basketter, D.A. and Kimber, I. (1996b) The mouse IgE test for the identification of potential chemical respiratory allergens: considerations of stability and controls. *J. Appl. Toxicol.*, **16**, 165–70.

Holliday, M.R., Coleman, J.W., Dearman, R.J. and Kimber, I. (1993) Induction of mast cell sensitization by chemical allergens: a comparative study. *J. Appl. Toxicol.*, **13**, 137–42.

Holt, P.G., Schon-Hegrad, M.A. and Oliver, J. (1988) MHC class II antigen-bearing dendritic cells in pulmonary tissues of the rat: regulation of antigen presentation activity by endogenous macrophage populations. *J. Exp. Med.*, **167**, 262–74.

Holt, P.G., Schon-Hegrad, M.A., Phillips, M.J. and McMenamin, P.G. (1989) Ia-positive dendritic cells form a tightly meshed network within the human airways epithelium. *Clin. Exp. Allergy*, **19**, 597–601.

Holt, P.G., Schon-Hegrad, M.A. and McMenamin, P.G. (1990a) Dendritic cells in the respiratory tract. *Int. Rev. Immunol.*, **6**, 139–49.

Holt, P.G., Schon-Hegrad, M.A., Oliver, J., Holt, B.J. and McMenamin, P.G. (1990b) A contiguous network of dendritic antigen-presenting cells within the respiratory epithelium. *Int. Arch. Allergy Appl. Immunol.*, **91**, 155–9.

Hope, J., Cumberbatch, M., Fielding, I., Dearman, R.J., Kimber, I. and Hopkins, S.J. (1995) Identification of dendritic cells as a major source of interleukin-6 in draining lymph nodes following skin sensitization of mice. *Immunology*, **86**, 441–7.

Hsieh, C.-S., Heimberger, A.B., Gold, J.S., O'Garra, A. and Murphy, K.M. (1992) Differential regulation of T helper phenotype development by interleukins 4 and 10 in an $\alpha\beta$ T-cell-receptor transgenic system. *Proc. Natl Acad. Sci. USA*, **89** 6065–9.

Hsieh, C.-S., Macatonia, S.E., Tripp, C.S., Wolf, S.F., O'Garra, A. and Murphy, K.M. (1993) Development of Th1 $CD4^+$ T cells through IL-12 produced by Listeria-induced macrophages. *Science*, **260**, 547–9.

Hussain, R. and Ottesen, E.A. (1986) IgE responses in human filariasis. IV. Parallel antigen recognition by IgE and IgG4 subclass antibodies. *J. Immunol.*, **136**, 1859–63.

Inaba, K., Witmer-Pack, M., Inaba, M., Hathcock, K.S., Sakuta,H., Azuma, M., Yagita, H., Okumura, K., Linsley, P.S., Ikehara, S., Muramatsu, S., Hodes, R.J. and Steinman, R.M. (1994) The tissue distribution of the B7-2 costimulator in mice: abundant expression on dendritic cells *in situ* and during maturation *in vitro*. *J. Exp. Med.*, **180**, 1849–60.

Ishizaka, A., Sakiyama, Y., Nakanishi, M., Tomizawa, K., Oshika, E., Kojima, K., Taguchi, Y., Kandil, E. and Matsumoto, S. (1990) The inductive effect of interleukin 4 on IgG4 and IgE synthesis in human peripheral blood lymphocytes. *Clin. Exp. Immunol.*, **79**, 392–6.

Jones, D.A., Morris, A.G. and Kimber, I. (1989) Assessment of the functional activity of antigen-bearing dendritic cells isolated from the lymph nodes of contact-sensitized mice. *Int. Arch. Allergy Appl. Immunol.*, **90**, 230–6.

Kamagowa, Y., Minasi, L.E., Carding, S.R., Bottomly, K. and Flavell, R. (1993) The relationship of IL-4- and IFN-γ-producing T cells by lineage ablation of IL-4-producing cells. *Cell*, **75**, 985–95

Kang, K., Kubin, M., Cooper, K.D., Lessin, S.R., Trinchieri, G. and Rook, A.H .(1996) IL-12 synthesis by human Langerhans cells. *J. Immunol.*, **156**, 1402–7.

Kaplan, M.H., Sun, Y.-L., Hoey, T. and Grusby, M.J. (1996) Impaired IL-12 responses and enhanced development of Th2 cells in Stat4-deficient mice. *Nature*, **382**, 174–7.

Kapsenberg, M.L., Wierenga, E.A., Bos, J.D. and Jansen, H.M. (1991) Functional subsets of allergen-reactive $CD4^+$ cells. *Immunol. Today*, **12**, 392–5.

Kapsenberg, M.L., Wierenga, E.A., Stiekma, F.E.M., Tiggelman, A.M.B.C. and Bos, J.D. (1992) TH1 lymphokine production profiles of nickel-specific $CD4^+$ T lymphocyte clones from nickel allergic and non-allergic individuals. *J. Invest. Dermatol.*, **98**, 59–63.

Karol, M.H., Tollerud, D.J., Campbell, T.P., Fabbri, L., Maestrelli, P., Saetta, M. and Mapp, C.E. (1994) Predictive value of airways hyperresponsiveness and circulating IgE for identifying types of responses to toluene diisocyanate inhalation challenge. *Am. J. Respir. Crit. Care Med.*, **149**, 611–15.

Kelso, A. (1995) Th1 and Th2 subsets: paradigms lost? *Immunol. Today*, **16**, 374–9.

Kemeny, D.M. and Diaz-Sanchez, D. (1993) The role of $CD8^+$ T cells in the regulation of IgE. *Clin. Exp. Allergy*, **23**, 466–70.

Kemeny, D.M., Noble, A., Diaz-Sanchez, D., Staynov, D. and Lee, T.H. (1992) Ricin-sensitive early activated $CD8^+$ T cells suppress IgE responses and regulate the production of IFN-γ and IL-4 by splenic $CD4^+$, $CD8^+$ T cells. *Int. Arch. Allergy Immunol.*, **99**, 362–5.

Kemeny, D.M., Noble, A., Holmes, B.J. and Diaz-Sanchez, D. (1994) Immunoregulation: a new role for the $CD8^+$ T cell. *Immunol. Today*, **15**, 107–10.

Kimata, H., Yoshida, A., Ishioka, C., Lindley, I. and Mikawa, H. (1992) Interleukin 8 (IL-8) actively inhibits immunoglobulin E production by IL-4 in human B cells. *J. Exp. Med.*, **176**, 1227–31.

Kimata, H., Yoshida, A., Ishioka, C., Fujimoto, M., Lindley, I. and Furusho, K. (1996) RANTES and macrophage inflammatory protein 1α selectively enhance immunoglobulin (IgE) and IgG4 production by human B cells. *J. Exp. Med.*, **183**, 2397–402.

Kimber, I. (1995) Contact and respiratory sensitization by chemical allergens: uneasy relationships. *Am. J. Contact Derm.*, **6**, 34–9.

Kimber, I. (1996) The skin immune system, in *Dermatology*, 5th edn (eds F.N. Marzulli and H.I. Maibach), Taylor and Francis, Washington DC, pp. 131–41.

Kimber, I. and Cumberbatch, M. (1992) Dendritic cells and cutaneous immune responses to chemical allergens. *Toxicol. Appl. Pharmacol.*, **117**, 137–46.

Kimber, I. and Dearman, R.J. (1996) Contact hypersensitivity: immunological mechanisms, in *Toxicology of Contact Hypersensitivity* (eds I. Kimber and T. Maurer), Taylor and Francis, London, pp. 4–25.

Kimber, I. and Dearman, R.J. (1997) Immunobiology of chemical respiratory sensitization, in *Toxicology of Chemical Respiratory Hypersensitivity* (eds I. Kimber and R.J. Dearman), Taylor and Francis, London, pp. 73–106.

Kimber, I., Hilton, J., Basketter, D.A. and Dearman, R.J. (1996) Predictive testing for respiratory sensitization in the mouse. *Toxicol. Lett.*, **86**, 193–8.

Kinnaird, A., Peters, S.W., Foster, J.R. and Kimber, I. (1989) Dendritic cell accumulation in draining lymph nodes during the induction phase of contact allergy in mice. *Int. Arch. Allergy Appl. Immunol.*, **89**, 202–10.

Knight, S.C., Krejci, J., Malkovsky, M., Colizzi, V., Gautam, A. and Asherson, G.L. (1985) The role of dendritic cells in the initiation of immune responses to contact sensitizers. *In vivo* exposure to antigen. *Cell Immunol.*, **94**, 427–34.

Knoller, I., Bujanowski-Weber, J., Brings, B. and Konig, W. (1989) Influence of IL-2 and IL-4 on the IgE synthesis and the IgE-binding factor (sCD23) production by human lymphocytes *in vitro*. *Immunology*, **66**, 368–75.

Kohler, J., Martin, S., Pflugfelder, U., Ruh, H., Vollmer, J. and Weltzien, H.U. (1995) Cross-reactive trinitrophenylated peptides as antigens for class II major histocompatibility complex-restricted T cells and inducers of contact sensitivity in mice. *Eur. J. Immunol.*, **25**, 92–101.

Kondo, S., McKenzie, R.C. and Sauder, D.N. (1994) Interleukin-10 inhibits the elicitation phase of allergic contact hypersensitivity. *J. Invest. Dermatol.*, **103**, 811–14.

Kondo, S., Beissert, S., Wang, B., Fujisawa, H., Kooshesh, F., Stratigos, A., Granstein, R.D., Mak, T.W. and Sauder, D.N. (1996) Hyporesponsiveness in contact hypersensitivity and irritant contact dermatitis in CD4 gene targeted mouse. *J. Invest. Dermatol.*, **106**, 993–1000.

Kopf, M., Le Gros, G., Bachmann, M., Lamers, M.C., Bleuthmann, H. and Kohler, G. (1993) Disruption of the murine IL-4 gene blocks Th2 cytokine responses. *Nature*, **362**, 245–8.

Korsgren, M., Erjefalt, J.S., Korsgren, O., Sundler, F. and Persson, C.G.A. (1997) Allergic eosinophil-rich inflammation develops in lungs and airways of B cell-deficient mice. *J. Exp. Med.*, **185**, 885–92.

Kuchroo, V.K., Das, M.P., Brown, J.A., Ranger, A.M., Zamvil, S.S., Sobel, R.A., Weiner, H.L., Nabavi, N. and Glimcher, L.H. (1995) B7-1 and B7-2 costimulatory molecules activate differentially the Th1/Th2 development pathways: application to autoimmune disease therapy. *Cell*, **80**, 707–18.

Kuhn, R., Rajewsky, K. and Muller,W. (1991) Generation and analysis of interleukin-4 deficient mice. *Science*, **254**, 707–10.

Kumar, V., Bhardwaj, V., Soares, L., Alexander, J., Sette, A. and Sercarz, E. (1995) Major histocompatibility complex binding affinity of an antigenic determinant is crucial for the differential secretion of interleukin 4/5 or interferon γ by T cells. *Proc. Natl Acad. Sci. USA*, **92**, 9510–14.

Lappin, M.B., Kimber, I. and Norval, M. (1996) The role of dendritic cells in cutaneous immunity. *Arch. Dermatol. Res.*, **288**, 109–21.

Larsen, C.P., Ritchie, S.C., Pearson, T.C., Linsley, P.S. and Lowry, R.P. (1992) Functional expression of the costimulatory molecule, B7/BB1, on murine dendritic cell populations. *J. Exp. Med.*, **176**, 1215–20.

Le Gros, G. and Erard, F. (1994) Non cytotoxic, IL-4, IL-5, IL-10 producing $CD8^+$ T cells: their activation and effector functions. *Curr. Opin. Immunol.*, **6**, 453–7.

Li, L., Elliott, J.F. and Mosmann, T.R. (1994) IL-10 inhibits cytokine production, vascular leakage and swelling during T helper 1 cell-induced delayed-type hypersensitivity. *J. Immunol.*, **153**, 3967–78.

Li, L., Sad, S., Kagi, D. and Mosmann, T.R. (1997) CD8 Tc1 and Tc2 cells secrete distinct cytokine patterns *in vitro* and *in vivo* but induce similar inflammatory reactions. *J. Immunol.*, **158**, 4152–61.

Lipscomb, M.F., Bice, D.E., Lyons, C.R., Schuyler, M.R. and Wilkes, D. (1995) The regulation of pulmonary immunity. *Adv. Immunol.*, **59**, 369–455.

Macatonia, S.E, Edwards, A.J. and Knight, S.C. (1986) Dendritic cells and the initiation of contact sensitivity to fluorescein isothiocyanate. *Immunology*, **59**, 509–14.

Macatonia, S.E., Knight, S.C., Edwards, A.J., Griffiths, S. and Fryer, P. (1987) Localization of antigen on lymph node dendritic cells after exposure to the contact sensitizer fluorescein isothiocyanate. Functional and morphological studies. *J. Exp. Med.*, **166**, 1654–67.

Macatonia, S.E., Doherty, T.M., Knight, S.C. and O'Garra, A. (1993) Differential effect of IL-10 on dendritic cell-induced T cell proliferation and IFN-γ production. *J. Immunol.*, **150**, 3755–65.

Macaulay, A.E., DeKruyff, R.H., Goodnow, C.C. and Umetsu, D.T. (1997) Antigen-specific B cells preferentially induce $CD4^+$ T cells to produce IL-4. *J. Immunol.*, **158**, 4171–9.

Maggi, E., Del Prete, G.F., Parronchi, P., Tiri, A., Macchia, D., Biswas, P., Simonelli, C., Ricci, M. and Romagnani, S. (1989) Role for T cells, IL-2 and IL-6 in the IL-4-dependent *in vitro* human IgE synthesis. *Immunology*, **68**, 300–6.

Martin, S. and Weltzien, H.U. (1994) T cell recognition of haptens: a molecular view. *Int. Arch. Allergy Immunol.*, **104**, 10–16.

Martin, S., von Bonin, A., Fessler, C., Pflugfelder, U. and Weltzien, H.U. (1993) Structure complexity of antigenic determinants for class I MHC-restricted, hapten-specific T cells: two qualitatively differing types of H-2K^b-restricted TNP epitopes. *J. Immunol.*, **151**, 678–87.

Mason, D. (1996) The role of B cells in the programming of T cells for IL-4 synthesis. *J. Exp. Med.*, **183**, 717–19.

McKnight, A.J., Zimmer, G.J., Fogelman, I., Wolf, S.F. and Abbas, A.K. (1994) Effects of IL-12 on helper T cell-dependent immune responses *in vivo*. *J. Immunol.* **152**, 2172–9.

McMenamin, C. and Holt, P.G. (1993) The natural immune response to inhaled soluble protein allergens involves major histocompatibility complex (MHC) class I-restricted CD8$^+$ T cell-mediated but MHC class II-restricted CD4$^+$ T cell dependent immune deviation resulting in selective suppression of immunoglobulin E production. *J. Exp. Med.*, **178**, 889–99.

McMenamin, C., Pimm, C., McKersey, M. and Holt, P.G. (1994) Regulation of IgE responses to inhaled antigen in mice by antigen-specific $\gamma\delta$ T cells. *Science*, **265**, 1869–71.

McMenamin, C., McKersey, M., Kuhnlein, P., Hunig, T. and Holt, P.G. (1995) $\gamma\delta$ T cells down-regulate primary IgE responses in rats to inhaled soluble protein antigens. *J. Immunol.*, **154**, 4390–4.

McWilliam, A.S., Nelson, D., Thomas, J.A. and Holt, P.G. (1994) Rapid dendritic cell recruitment is a hallmark of the acute inflammatory response at mucosal surfaces. *J. Exp. Med.*, **179**, 1331–6.

Minty, A., Chalon, P., Derocq, J.-M., Dumont, X., Guillemot, J.-C., Kaghad, M., Labit, C., Leplatois, P., Liauzun, P., Miloux, B., Minty, C., Casellas, P., Loison, G., Lupker, J., Shire, D., Ferrara, P. and Caput, D. (1993) Interleukin 13 is a new human lymphokine regulating immune and inflammatory responses. *Nature*, **362**, 248–50.

Moller, D.R., Gallagher, J.S., Bernstein, D.I., Wilcox, T.G., Burroughs, H.E. and Bernstein, I.L. (1985) Detection of IgE-mediated respiratory hypersensitivity in workers exposed to hexahydrophthalic anhydride. *J. Allergy Clin. Immunol.*, **76**, 663–72.

Morawetz, R.A., Gabriele, L., Rizzo, L.V., Noben-Trauth, N., Kuhn, R., Rajewsky, K., Muller, W., Doherty, T.M., Finkelman, F., Coffman, R.L. and Morse, R.C. III (1996) Interleukin (IL)-4-independent immunoglobulin class switch to immunoglobulin (Ig) E in the mouse. *J. Exp. Med.*, **184**, 1651–61.

Morris, S.C., Madden, K.B., Adamovicz, J.L., Gause, W.C., Hubbard, B.R., Gately, M.K. and Finkelman, F.D. (1994) Effects of IL-12 on *in vivo* cytokine gene expression and Ig isotype selection. *J. Immunol.*, **152**, 1047–56.

Mosmann, T.R. and Coffman, R.L. (1989) Heterogeneity of cytokine secretion patterns and functions of helper T cells. *Adv. Immunol.* **46**, 111–45.

Mosmann, T.R. and Sad, S. (1996) The expanding universe of T cell subsets: Th1, Th2 and more. *Immunol. Today*, **17**, 138–46.

Mosmann, T.R., Cherwinski, H., Bond, M.W., Giedlin, M.A. and Coffman, R.L. (1986) Two types of murine helper T cell clone. 1. Definition according to profiles of lymphokine activities and secreted proteins. *J. Immunol.*, **136**, 2348–57.

Mosmann, T.R., Schumacher, J.H., Street, N.F., Budd, R., O'Garra, A., Fong, T.A.T., Bond, M.W., Moore, K.W.M., Sher, A. and Fiorentino, D.F. (1991) Diversity of cytokine synthesis and function of mouse CD4$^+$ T cells. *Immunol. Rev.*, **123**, 209–29.

Nakazawa, M., Sugi, N., Kawaguchi, H., Ishii, N., Nakajima, H. and Minami, M. (1997) Predominance of type 2 cytokine-producing CD4$^+$ and CD8$^+$ cells in patients with atopic dermatitis. *J. Allergy Clin. Immunol.*, **99**, 673–82.

Nielsen, J., Welinder, H., Ottoson, H., Bensryd, I., Venge, P. and Skerfving, S. (1994) Nasal challenge shows pathogenetic relevance of specific IgE serum antibodies for nasal symptoms caused by hexahydrophthalic auhydride. *Clin. Exp. Allergy*, **24**, 440–9.

Nishimura, T., Santa, K., Yahata, T., Sato, N., Ohta, A., Ohmi, Y., Sato, T., Hozumi, K. and Habu, S. (1997) Involvement of IL-4-producing $V\beta 8.2^+$ $CD4^+$ $CD62L^-$ $CD45RB^-$ T cells in non-MHC gene-controlled predisposition toward skewing into T helper type-2 immunity in BALB/c mice. *J. Immunol.*, **158**, 5698–706.

Noble, A. and Kemeny, D.M. (1995) Interleukin-4 and interferon-γ regulate differentiation of $CD8^+$ T cells into populations with divergent cytokine profiles. *Int. Arch. Allergy Immunol.*, **107**, 186–8.

Noble, A., Macary, P.A. and Kemeny, D.M. (1995) IFN-γ and IL-4 regulate growth and differentiation of $CD8^+$ T cells into subpopulations with distinct cytokine profiles. *J. Immunol.*, **155**, 2928–37.

Ohmen, J.D., Hanifin, J.M., Nickoloff, B.J., Rea, T.H., Wyzykowski, R., Kim, J., Jullien, D., McHugh, T., Nassif, A.S., Chan, S.C. and Modlin, R.L. (1995) Overexpression of IL-10 in atopic dermatitis. Contrasting cytokine patterns with delayed-type hypersensitivity reactions. *J. Immunol.*, **154**, 1956–63.

Ohshima, Y. and Delespesse, G. (1997) T cell-derived IL-4 and dendritic cell-derived IL-12 regulate the lymphokine-producing phenotype of alloantigen-primed naive human CD4 T cells. *J. Immunol.*, **158**, 629–36.

Palmer, E.M. and van Seventer, G.A. (1997) Human T helper cell differentiation is regulated by the combined action of cytokines and accessory cell-dependent costimulatory signals. *J. Immunol.*, **158**, 2654–62.

Park, H.S. and Hong, C.-S. (1991) The significance of specific IgG and IgG4 antibodies to a reactive dye in exposed workers. *Clin. Exp. Allergy*, **21**, 357–62.

Parronchi, P., Macchia, D., Piccinni, M.-P., Biswas, P., Simonelli, C., Maggi, E., Ricci, M., Ansari, A.A. and Romagnani, S. (1991) Allergen and bacterial antigen-specific T-cell clones established from atopic donors show a different profile of cytokine production. *Proc. Natl Acad. Sci. USA*, **88**, 4538–42.

Pearson, C.I., van Ewijk, W. and McDevitt, H.O. (1997) Induction of apoptosis and T helper 2 (Th2) responses correlates with peptide affinity for the major histocompatibility complex in self-reactive T cell receptor transgenic mice. *J. Exp. Med.*, **185**, 583–99.

Pene, J., Rousset, F., Briere, F., Chretien, I., Paliard, X., Banchereau, J., Spits, H. and de Vries, J. (1988) IgE production by normal human B cells induced by alloreactive T cell clones is mediated by IL-4 and suppressed by IFN-γ. *J. Immunol.*, **141**, 1218–24.

Pfeiffer, C., Stein, J., Southwood, S., Ketelaar, H., Sette, A. and Bottomly, K. (1995) Altered peptide ligands can control CD4 T lymphocyte differentiation *in vivo*. *J. Exp. Med.*, **181**, 1569–74.

Pollard, A.M. and Lipscomb, M.F. (1990) Characterization of murine lung dendritic cells: similarities to Langerhans cells and thymic dendritic cells. *J. Exp. Med.*, **172**, 159–67.

Presky, D.H., Yang, H., Minetti, L.., Chua, A.O., Nabavi, N., Wu, C.-Y., Gately, M.R. and Gubler, U. (1996) A functional interleukin 12 receptor complex is composed of two β-type cytokine receptor subunits. *Proc. Natl Acad. Sci. USA*, **93**, 14002–7.

Punnonen, J., Aversa, G., Cocks, B.G., McKenzie, A.N.J., Menon, S., Zurawski, G., de Waal Malefyt, R. and de Vries, J.E. (1993) Interleukin 13 induces interleukin 4-independent IgG4 and IgE synthesis and CD23 expression by human B cells. *Proc. Natl Acad. Sci. USA*, **90**, 3730–4.

Razi-Wolf, Z., Falo, L.D. Jr and Reiser, H. (1994) Expression and function of the costimulatory molecule B7 on murine Langerhans cells: evidence for an alternative CTLA-4 ligand. *Eur. J. Immunol.*, **24**, 805–11.

Reiner, S.L. and Seder, R.A. (1995) T helper cell differentiation in immune response. *Curr. Opin. Immunol.*, **7**, 360–6.

Renz, H., Lack, G., Sologa, J., Schwinzer, R., Bradley, K., Loader, J., Kupfer, A., Larsen, G.L. and Gelfand, E.W. (1994) Inhibition of IgE production and normalization of airways responsiveness by sensitized CD8 T cells in a mouse model of allergen-induced sensitization. *J. Immunol.*, **152**, 351–60.

Rincon, M., Anguita, J., Nakamura, T., Fikrig, E. and Flavell, R.A. (1997) Interleukin (IL)-6 directs the differentiation of IL-4 producing $CD4^+$ T cells. *J. Exp. Med.*, **185**, 461–9.

Robinson, D., Hamid, Q., Bentley, A., Ying, S., Kay, A.B. and Durham, S.R. (1993) Activation of $CD4^+$ T cells, increased Th2-type cytokine mRNA expression and eosinophil recruitment in bronchoalveolar lavage after allergen inhalation challenge in patients with atopic asthma. *J. Allergy Clin. Immunol.*, **92**, 313–24.

Robinson, D.S., Hamid, Q., Ying, S., Tsicopoulos, A., Barkans, J., Bentley, A.M., Corrigan, C., Durham, S.R. and Kay, A.B. (1992) Predominant Th2-like bronchoalveolar T-lymphocyte population in atopic asthma. *N. Engl. J. Med.*, **326**, 298–304.

Rogge, L., Barberis-Maino, L., Biffi Passini, N., Presky, D.H., Gubler, U. and Sinigaglia, F. (1997) Selective expression of an interleukin-12 receptor component by human T helper 1 cells. *J. Exp. Med.*, **185**, 825–31.

Romagnani, S. (1991) Human TH1 and TH2 subsets: doubt no more. *Immunol. Today*, **12**, 256–7.

Romagnani, S. (1992) Induction of TH1 and TH2 responses: a key role for the 'natural' immune response. *Immunol. Today*, **13**, 379–81.

Romagnani, S., Del Prete, G., Maggi, E., Parronchi, P., De Carli, M., Macchia, D., Manetti, R., Sampognaro, S., Piccinni, M.-P., Giudizi, M.G., Biagiotti, R. and Almerigogna, F. (1992) Human TH1 and TH2 subsets. *Int. Arch. Allergy Immunol.*, **99**, 242–5.

Romagnani, S., Del Prete, G., Maggi, E., Parronchi, P., Tiri, A., Macchia, D., Giudizi, M.G., Almerigogna, F. and Ricci, M. (1989) Role of interleukins in induction and regulation of human IgE synthesis. *Clin. Immunol. Immunopathol.*, **50**, S13–S23.

Rook, G.A.W., Hernandez-Pando, R. and Lightman, S.L. (1994) Hormones, peripherally activated prohormones and regulation of the TH1/TH2 balance. *Immunol. Today*, **15**, 301–3.

Rousset, F., Garcia, E. and Banchereau, J. (1991) Cytokine-induced proliferation and immunoglobulin production of human B lymphocytes triggered through their CD40 antigen. *J. Exp. Med.*, **173**, 705–10.

Sad, S., Marcotte, R. and Mosmann, T.R. (1995) Cytokine-induced differentiation of precursor mouse $CD8^+$ T cells into cytotoxic $CD8^+$ T cells secreting Th1 or Th2 cytokines. *Immunity*, **2**, 271–9.

Salerno, A., Dieli, F., Sireci, G., Bellavia, A. and Asherson, G.L. (1995) Interleukin-4 is a critical cytokine in contact sensitivity. *Immunology*, **84**, 404–9.

Salvaggio, J.E., Butcher, B.T. and O'Neil, C.E. (1988) Occupational asthma due to chemical agents. *J. Allergy Clin. Immunol.*, **78**, 1053–7.

Schmitt, E., Hoehn, P., Huels, C., Goedert, S., Palm, N., Rude, E. and German, T. (1994a) T helper type 1 development of naive $CD4^+$ T cells requires the co-ordinate action of interleukin-12 and interferon-γ and is inhibited by transforming growth factor-β. *Eur. J. Immunol.*, **24**, 793–8.

Schmitz, J., Thiel, A., Kuhn, R., Rajewsky, K., Muller, W., Assenmacher, M. and Radbruch, A. (1994b) Induction of interleukin 4 (IL-4) expression in T helper (Th) cells is not dependent on IL-4 from non-Th cells. *J. Exp. Med.*, **179**, 1349–53.

Schon-Hegrad, M.A., Oliver, J., McMenamin, P.G. and Holt, P.G. (1991) Studies on the density, distribution and surface phenotype of intraepithelial class II MHC anti-

gen (Ia)-bearing dendritic cells (DC) in the conducting airways. *J. Exp. Med.*, **173**, 1345–56.
Secrist, H., De Kruyff, R.H. and Umetsu, D.T. (1995) Interleukin 4 production by $CD4^+$ T cells from allergic individuals is modulated by antigen concentration and antigen-presenting cell type. *J. Exp. Med.*, **181**, 1081–9.
Seder, A., Gazzinelli, R., Sher, A. and Paul, W.E. (1993) Interleukin 12 acts directly on $CD4^+$ T cells to enhance priming for interferon γ production and diminishes interleukin 4 inhibition of such priming. *Proc. Natl Acad. Sci. USA*, **90**, 10188–92.
Seder, R.A. and Le Gros, G.G. (1995) The functional role of $CD8^+$ T helper type 2 cells. *J. Exp. Med.* **181**, 5–7.
Selgrade, M.K., Lawrence, D.A., Ullrich, S.E., Gilmour, M.I., Schuyler, M.R. and Kimber I (1997) Modulation of T-helper cell populations: potential mechanisms of respiratory hypersensitivity and immune suppression. *Toxicol. Appl. Pharmacol.*, **145**, 218–29.
Sertl, K., Takemura, T., Tschachler, E., Ferrans, V.J., Kaliner, M.A. and Shevach, E.M. (1986) Dendritic cells with antigen-presenting capability reside in the airway epithelium, lung parenchyma and visceral pleura. *J. Exp. Med.*, **163**, 436–51.
Shimoda, K., van Deursen, J., Sangster, M.Y., Sarawar, S.R., Carson, R.T., Tripp, R.A., Chu, C., Quelle, F.W., Nusaka, T., Vignali, D.A.A., Doherty, P.C., Grosveld, G., Paul, W.E. and Ihle, J.N. (1996) Lack of IL-4-induced Th2 responses and IgE class switching in mice with disrupted Stat6 gene. *Nature*, **380**, 630–3.
Smiley, S.T., Kaplan, M.H. and Grusby, M.J. (1997) Immunoglobulin E production in the absence of interleukin-4-secreting CD1-dependent cells. *Science*, **275**, 977–9.
Steinbrink, K., Sorg, C. and Macher, E. (1996) Low zone tolerance to contact allergens in mice: a functional role for $CD8^+$ T helper type 2 cells. *J. Exp. Med.*, **183**, 759–68.
Stockinger, B., Zal, T., Zal, A. and Gray, D. (1996) B cells solicit their own help from T cells. *J. Exp. Med.*, **183**, 891–9.
Swain, S.L., Bradley, L.M., Croft, M., Tonkonogy, S., Atkins, G., Weinberg, A.D., Duncan, D.D., Hedrick, S.M., Dutten, R.W. and Huston, G. (1991) Helper T cell subsets: phenotype, function and the role of lymphokines in regulating their development. *Immunol. Rev.*, **123**, 115–44.
Szabo, S.L., Jacobson, N.G., Dighe, A.S., Gubler, U. and Murphy, K.M. (1995) Developmental commitment to the Th2 lineage by extinction of IL-12 signalling. *Immunity*, **2**, 665–75.
Szabo, S.J., Dighe, A.S., Gubler, U. and Murphy, K.M. (1997) Regulation of the interleukin (IL)-12R $\beta 2$ subunit expression in developing T helper 1 (Th1) and Th2 cells. *J. Exp. Med.*, **185**, 817–24.
Tepper, R.I., Levinson, D.A., Stanger, B.Z., Compos-Torres, J., Abbas, A.K. and Leder, P. (1990) IL-4 induces allergic-like inflammatory disease and alters T cell development in transgenic mice. *Cell*, **62**, 457–67.
Thepen, T., van Rooijen, N. and Kraal, G. (1989) Alveolar macrophage elimination *in vivo* is associated with an increase in pulmonary immune response of mice. *J. Exp. Med.*, **170**, 499–509.
Thepen, T., McMenamin, C., Oliver, J., Kraal, G. and Holt, P.G. (1991) Regulation of immune responses to inhaled antigen by alveolar macrophages: differential effects of *in vivo* alveolar macrophage elimination on the induction of tolerance vs immunity. *Eur. J. Immunol.*, **21**, 2845–50.
Thepen, T., McMenamin, C., Girn, B., Kraal, G. and Holt, P.G. (1992) Regulation of IgE production in presensitized animals: *in vivo* elimination of alveolar macrophages selectively increases IgE responses to inhaled allergen. *Clin. Exp. Allergy*, **22**, 1107–14.
Thierfelder, W.E., van Deursen, J.M., Yamamoto, K., Tripp, R.A., Sarawar, S.R., Carson, R.T., Sangster, M.Y., Vignali, D.A.A., Doherty, P.C., Grosveld, G.C. and

Ihle, J.N. (1996) Requirement for Stat4 in interleukin-12-mediated response of natural killer and T cells. *Nature*, **382**, 171–4.

Till, S., Dickason, R., Huston, D., Humbert, M., Robinson, D., Larche, M., Durham, S., Kay, A.B. and Corrigan, C. (1997a) IL-5 secretion by allergen-stimulated $CD4^+$ T cells in primary culture. Relationship to expression of allergic disease. *J. Allergy Clin. Immunol.*, **99**, 563–9.

Till, S., Durham, S., Dickason, R., Huston, D., Bungre, J., Walker, S., Robinson, D., Kay, A.B. and Corrigan, C. (1997b) IL-13 production by allergen-stimulated T cells is increased in allergic disease and associated with IL-5 but not IFN-γ expression. *Immunology*, **91**, 53–7.

Topping, M.D., Forster, H.W., Ide, C.W., Kennedy, F.M., Leach, A.M. and Sorkin, S. (1989) Respiratory allergy and specific immunoglobulin E and immunoglobulin G antibodies to reactive dyes used in the wool industry. *J. Occup. Med.*, **31**, 859–62.

Topping, M.D., Venables, K.M., Luczynska, C.M., Howe, W. and Newman Taylor, A.J. (1986) Specificity of human IgE responses to inhaled acid anhydrides. *J. Allergy Clin. Immunol.*, **77**, 834–42.

Trinchieri, G. (1993) Interleukin-12 and its role in the generation of TH1 cells. *Immunol. Today*, **14**, 335–8.

Trinchieri, G. and Scott, P. (1994) The role of interleukin 12 in the immune response, disease and therapy. *Immunol. Today*, **15**, 460–3.

Tsitoura, D.C., Verhoef, A., Gelder, C.M., O'Hehir, R.E. and Lamb, J.R. (1996) Altered T cell ligands derived from a major house dust mite allergen enhance IFN-γ but not IL-4 production by human $CD4^+$ T cells. *J. Immunol.*, **157**, 2160–5.

Tsuyuki, S., Tsuyuki, J., Einsle, K., Kopf, M. and Coyle, A.J. (1997) Costimulation through B7-2 (CD86) is required for the induction of lung mucosal T helper cell 2 (Th2) immune responses and altered airway responsiveness. *J. Exp. Med.*, **185**, 1671–9.

Tunon-de-Lara, J.M., Redington, A.E., Bradding, P., Church, M.K., Hartley, J.A., Semper, A.E. and Holgate, S.T. (1996) Dendritic cells in normal and asthmatic airways: expression of the α subunit of the high affinity immunoglobulin E receptor (FcεRI-α). *Clin. Exp. Allergy*, **26**, 648–55.

Underwood, S.L., Kemeny, D.M., Lee, T.H., Raeburn, D. and Karlsson, J.-A. (1995) IgE production, antigen-induced airway inflammation and airway hyperreactivity in the Brown-Norway rat: the effects of ricin. *Immunology*, **85**, 256–61.

van der Heijden, F.L., Wierenga, E.A., Bos, J.D. and Kapsenberg, M.L. (1991) High frequency of IL-4-producing $CD4^+$ allergen-specific T lymphocytes in atopic dermatitis lesional skin. *J. Invest. Dermatol.*, **97**, 389–94.

Vento, K.L., Dearman, R.J., Kimber, I., Basketter, D.A. and Coleman, J.W. (1996) Selectivity of IgE responses, mast cell sensitization and cytokine expression in the immune response of Brown Norway rats to chemical allergens. *Cell. Immunol.*, **172**, 246–53.

Vicari, A.P. and Zlotnik, A. (1996) Mouse $NK1.1^+$ T cells: a new family of T cells. *Immunol. Today*, **17**, 71–6.

von der Weid, T., Beebe, A.M., Roopenian, D.C. and Coffman, R.L. (1996) Early production of IL-4 and induction of Th2 responses in the lymph node originate from an MHC class I-independent $CD4^+$ $NK1.1^-$ T cell population. *J. Immunol.*, **157**, 4421–7.

Weigmann, B., Schwing, J., Huber, H., Ross, R., Mossmann, H., Knop, J. and Reske-Kunz, A.B. (1997) Diminished contact hypersensitivity response in IL-4 deficient mice at a late phase of the elicitation reaction. *Scand. J. Immunol.*, **45**, 308–14.

Weltzien, H.U., Moulon, C., Martin, S., Padovan, E., Hartmann, U. and Kohler, J. (1996) T cell immune responses to haptens. Structural models for allergic and autoimmune reactions. *Toxicology*, **107**, 141–51.

Wenner, C.A., Guler, M.L., Macatonia, S.E., O'Garra, A. and Murphy, K.M. (1996) Roles of IFN-γ and IFN-α in IL-12-induced T helper cell-1 development. *J. Immunol.*, **156**, 1442–7.

Xia, W., Pinto, C.E. and Kradin, R.L. (1995) The antigen-presenting activities of Ia^+ dendritic cells shift dynamically from lung to lymph node after an airway challenge with soluble antigens. *J. Exp. Med.*, **181**, 1275–83.

Xu, H., DiIulio, N.A. and Fairchild, R.L. (1996) T cell populations primed by hapten sensitization in contact sensitivity are distinguished by polarized patterns of cytokine production. Interferon-γ-producing (Tc1) effector $CD8^+$ T cells and interleukin (IL)4/IL-10 producing (Th2) negative regulatory $CD4^+$ T cells. *J. Exp. Med.*, **183**, 1001–12.

Xu, H., Banerjee, A., DiIulio, N.A. and Fairchild, R.L. (1997) Development of effector $CD8^+$ T cells in contact hypersensitivity occurs independently of $CD4^+$ T cells. *J. Immunol.*, **158**, 4721–8.

Yoshimoto, T. and Paul, W.E. (1994) $CD4^+$ $NK1.1^+$ T cells promptly produce IL-4 in response to *in vivo* challenge with anti-CD3. *J. Exp. Med.*, **179**, 1285–95.

Yoshimoto, T., Bendelac, A., Hu-Li, J. and Paul, W.E. (1995a) Defective IgE production by SJL mice is linked to the absence of $CD4^+$ $NK1.1^+$ T cells that promptly produce interleukin 4. *Proc. Natl Acad. Sci. USA*, **92**, 11931–4.

Yoshimoto, T., Bendelac, A., Watson, C., Hu-Li, J. and Paul, W.E. (1995b) Role of $NK1.1^+$ T cells in a Th2 response and in immunoglobulin E production. *Science*, **270**, 1845–7.

Zeiss, C.R., Patterson, R., Pruzansky, J.J. Miller, M., Rosenberg, M. and Levitz, D. (1977) Trimellitic anhydride-induced airway syndromes: clinical and immunologic studies. *J. Allergy Clin. Immunol.*, **60**, 96–103.

Zhang, Y., Rogers, K.H. and Lewis, D.B. (1996) β2-Microglobulin-dependent T cells are dispensable for allergen-induced T helper 2 responses. *J. Exp. Med.*, **184**, 1507–12.

12 T Lymphocyte Subpopulations in Human Allergic Disease

Mark Schuyler and Julie Wilder

Introduction

Asthma is a common illness in the US (an estimated 5% of the population has or has had asthma) with increasing mortality both in the US and worldwide. There are differences in T helper cell populations which appear to play a role in asthma. This chapter will review the immunological responses associated with asthma, describe immunological responses following local instillation of antigen in human subjects and review the role that T cells play in murine models of allergic asthma. We will not be describing the recently developed models of asthma in the rat, nor the impact of exposure to air pollutants may have on immunological responses associated with asthma, as these will be discussed in Chapter 13.

As the introductory chapters of this volume indicate, there is substantial evidence that $CD4^+$ cells can be functionally separated into those that are important in arming macrophages for defence against intracellular organisms (i.e. Th1), and those that are important in providing help for immunoglobulin (especially IgE and IgG4 in humans and IgE and IgG1 in mice) production (i.e. Th2). Specific antibody isotype and cell-mediated immunity reflect the activity of different types of $CD4^+$ T cells.

Murine $CD4^+$ cells can be divided into Th1 or Th2 subsets by their patterns of cytokine secretion. Th1 cells preferentially secrete interferon-γ (IFN-γ), and tumour necrosis factor-β (TNFβ); activate macrophages; are responsible for cell-mediated immunity reactions and CTL (cytotoxic lymphocytes); and provide help for IgG2a production. Th2 $CD4^+$ cells secrete IL-4, IL-5, IL-9, IL-10 and IL-13; provide help for immunoglobulin (particularly IgE and IgG1) secretion; enhance eosinophil production, survival, and activity; and promote mast cell proliferation and maturation. Development of either an antigen specific predominant Th1 or Th2 response in mice is dependent on many factors,

---

T Lymphocyte Subpopulations in Immunotoxicology. Edited by Ian Kimber and MaryJane K. Selgrade.
 ISBN 0 471 97194 4

including attributes of the antigen; site of delivery; adjuvant used; age, sex and genetic background of the mouse; and nature of the antigen-presenting cell encountered. Cytokines secreted by one $CD4^+$ subset inhibit the development of the reciprocal subset leading to a predominance of one of the subsets and polarization of the antigen specific immune response (Finkelman, 1995). IL-4 is particularly important in inducing differentiation of Th2 cells, and IL-12 and IFN-γ in inducing differentiation of Th1 cells, from Thp (Th precursor) cells. Irreversible commitment to a Th1 or Th2 phenotype is probably mediated by rapid loss of the ability of IL-4, or IL-12 and IFN-γ, to initiate intracellular signalling in differentiated Th1 or Th2 cells, respectively. This lack of responsiveness is caused by changes either in cytokine receptors or in cytoplasmic and/or nuclear signal transducing factors (Pernis *et al.*, 1995; Szabo *et al.*, 1997). IL-12 and IFN-γ are produced by macrophages, dendritic cells and NK cells. The source of the IL-4 present during the initial differentiation of precursor $CD4^+$ cells to the Th2 phenotype is unclear. Candidates include mast cells, eosinophils (Sabin and Pearce, 1995), $CD4^+$ cells (perhaps under the influence of APC produced IL-6) (Rincon *et al.*, 1997), and a subset of $NK1.1^+$ T cells (Coffman and von der Weid, 1997). These concepts are explored in more detail in Chapter 11 of this volume.

The above model developed in mice has not been entirely confirmed in humans in the sense that Th1 and Th2 cells may not exist as exact counterparts in humans. However, the concept that the cytokine milieu (derived from many possible sources, such as macrophages, dendritic cells, $CD4^+$ cells, $CD8^+$ cells, NK cells, mast cells, eosinophils, $\gamma\delta$T cells) at the time of antigen specific $CD4^+$ cell differentiation, determines the later pattern of cytokine secretion and function of the differentiated cells is valid.

Human studies

Asthma is defined as a disease of bronchial mucosal inflammation, variable airflow obstruction (defined as decrease of FEV1 (forced expiratory volume in one second) and airway hyperreactivity (defined as increased reactivity, i.e. decrease of FEV1, in response to non-specific stimuli such as methacholine). The ability safely to perform bronchoscopy, bronchoalveolar lavage and bronchial biopsy in subjects with even moderate asthma (Van Vyve *et al.*, 1992), has allowed repetitive examination of the lung in asthmatics with less than fatal disease. Over the past few years, it has become evident that even mild asthmatics exhibit substantial bronchial pathology indicative of inflammation (Laitinen *et al.*, 1993; Ollerenshaw and Woolcock, 1992), presumably as a result of damage from products of eosinophils and mast cells. Eosinophil products include cytokines, cyclooxygenase products, reactive oxygen species and various intracellular proteins (major basic protein, eosinophil cationic protein, eosinophil peroxidase, eosinophil derived neurotoxin, etc.). Injurious mast cell products (histamine, serotonin, leukotrienes, prostaglandins, platelet activating factor, various proteases, etc.) released after binding of allergen to specific IgE attached to mast cell IgE receptors and subsequent complex intracellular signalling events also damage the bronchial mucosa.

There is evidence of dysfunction of both Th2 and Th1 cells in atopic asthma. Common features of atopy (eosinophilia, eosinophil activation, mast cell hyperplasia, and increased serum levels of IgE and IgG4), are indicative of increased Th2 $CD4^+$ cell activity. Decreased cell mediated immunity and contact hypersensitivity reactions (Hanifin and Lobitz, 1977; Rees *et al.*, 1990) suggest decreased Th1 activity.

Bronchoalveolar lavage fluid (BALF) reflects the milieu of the bronchi and lung. There are increased numbers of eosinophils and amount of eosinophilic cationic protein in BALF of asthmatics (Bousquet *et al.*, 1991; Sedgwick *et al.*, 1991). The level of eosinophilic cationic protein correlates with the severity of symptoms (Bousquet *et al.*, 1991). The proportion of BALF $CD4^+$ cells that are HLA-DR^+ or VLA-1^+ (early and late markers of lymphocyte activation) is increased in asthmatics (Robinson *et al.*, 1993a; Wilson *et al.*, 1992). The number of activated $CD4^+$ cells in BALF correlates with the number of BALF eosinophils (Robinson *et al.*, 1993a), implying that $CD4^+$ cells might be responsible for the BALF eosinophilia. Compared with normal subjects, there are increased amounts of IL-4 and IL-5 in BALF of asthmatics, suggesting a Th2 environment (Sedgwick *et al.*, 1991; Walker *et al.*, 1992). BALF and bronchial mucosal T cells of atopic asthmatics are of the Th2 subset as determined by *in situ* hybridization of cytokine mRNA and RT-PCR (i.e. express mRNA for IL-4, IL-5 and IL-10 and not IFN-γ (Robinson *et al.*, 1992; Ying *et al.*, 1995). The proportion of BALF lymphocytes expressing message for IL-4 correlates with symptoms, airflow obstruction and the degree of bronchial reactivity (Robinson *et al.*, 1993a). $CD4^+$ cell lines derived from BALF cells from atopic asthmatics secrete more IL-5 and GM-CSF (Th2 pattern) than cell lines from both atopic non-asthmatics and non-atopic non-asthmatics (Till *et al.*, 1995).

Experimental interventions in asthmatics have been used to acquire insight into pathophysiological processes. Allergen challenge of atopic asthmatics causes increases of BALF CD25 (IL-2 receptor)$^+$ $CD4^+$ cells, eosinophils, and $CD2^+$ (pan T cell marker) lymphocytes expressing message for IL-4, IL-5, and GM-CSF, but not IL-2, IL-3, or IFN-γ, 24 h after allergen inhalation challenge (Robinson *et al.*, 1993b). Krishnaswamy and colleagues found no difference in baseline cytokine message expression in BALF cells of asthmatics and normal subjects, but did report increase of BALF cell IL-5 message in asthmatics after local allergen bronchial challenge (Krishnaswamy *et al.*, 1993). Treatment of asthmatics with glucocorticosteroids not only decreased methacholine responsiveness and the number of BALF eosinophils, but also decreased the proportion of BALF cells expressing mRNA for IL-4 and IL-5 and increased the proportion expressing mRNA for IFN-γ (Robinson *et al.*,

1993c), strengthening the link between Th2 cytokines, eosinophils and asthma severity.

There are increased numbers of T lymphocytes (especially $CD4^+$ lymphocytes) and eosinophils in the bronchial mucosa of asthmatics (Poston *et al.*, 1992), compared with normal airways. The lymphocytes and eosinophils exhibit markers of activation such as expression of CD25, and EG2 (indicator of cleavage of eosinophil cationic protein), respectively (Azzawi *et al.*, 1990; Beasley *et al.*, 1989; Bentley *et al.*, 1992; Bradley *et al.*, 1991). There are also increased number of cells with IL-5 mRNA (Bentley *et al.*, 1993) in the bronchial mucosa. The presence of activated $CD4^+$ lymphocytes, and eosinophils or their products, such as eosinophil cationic protein and major basic protein in bronchial mucosa, correlates with symptoms of asthma and measures of bronchial hyperreactivity (Bentley *et al.*, 1992; Bradley *et al.*, 1991; Lam *et al.*, 1987; Robinson *et al.*, 1993a; Wardlaw *et al.*, 1988). The bronchial mucosal inflammation is associated with increased expression of ICAM-1 and E-selectin on endothelial cells (Bentley *et al.*, 1994). This increase of adhesion molecules might provide a mechanism of localization of inflammatory cells such as eosinophils within the lung after allergen exposure. VCAM-1 expression may (Ohkawara *et al.*, 1995) or may not (Bentley *et al.*, 1994) be increased in bronchial mucosa of asthmatics. Allergen specific $CD4^+$ clones derived from the bronchial mucosa of asthmatics are predominantly Th2 (i.e. preferentially secrete IL-4 and not IFN-γ) and provide help for IgE production (Del Prete *et al.*, 1993). Treatment with 2 weeks of systemic glucocorticosteroids results in decrease of bronchial mucosal $CD3^+$ cells; $EG2^+$ cells; mast cells; and cells with mRNA for IL-4 and IL-5; but not IFN-γ; and is associated with improvement of pulmonary function tests (Bentley *et al.*, 1996). Therefore, bronchial mucosal cell composition further links activated $CD4^+$ cells, Th2 cytokine profile, eosinophils and bronchial hyperreactivity.

There are even changes of $CD4^+$ cells in peripheral blood of asthmatics. Peripheral blood $CD4^+$ cells from asthmatics with a severe episode express the activation markers CD25, HLA-DR and VLA-1 which tend to revert towards normal as the asthma episode resolves (Corrigan *et al.*, 1988, 1993; Virchow *et al.*, 1994). Serum eosinophil survival promoting activity (presumably IL-5) also inversely correlates with pulmonary function (Virchow *et al.*, 1994). After allergen challenge there is a decrease of the number of peripheral blood $CD4^+$ cells and increase of the number of lymphocytes with activation markers (Gerblich *et al.*, 1984). Peripheral blood $CD4^+$ cells from moderately ill asthmatics compared to atopic non-asthmatic subjects, preferentially transcribe and secrete IL-3, IL-4, IL-5 and GM-CSF. These differences from normal subjects are ablated by 1 week of systemic glucocorticosteroid (Corrigan *et al.*, 1995).

Local challenge of a bronchus with allergen to which a subject is sensitized provides a model of events that occur after allergen exposure. Within 10 min of challenge there is evident local bronchial mucosal oedema, severe enough to narrow the bronchial lumen (Metzger *et al.*, 1987). The number and type of cells in BALF is unchanged. Histamine, tryptase, PGD_2, TxB_2 and multiple chemotactic factors (IL-16 (specific for $CD4^+$ cells), IL-8 and MIP1α) are present in BALF from allergen challenged bronchi of asthmatics (Cruikshank *et al.*, 1995; Liu *et al.*, 1991; Sedgwick *et al.*, 1991), presumably representing a mechanism of subsequent recruitment of lymphocytes to the site of allergen exposure. This is associated with increased permeability to proteins, especially low molecular weight proteins. This increased permeability persists for at least 4 h after allergen exposure (Fick *et al.*, 1987).

Six to 8 h after challenge, there is marked infiltration of the bronchial mucosa with polymorphonuclear leukocytes, eosinophils, tryptase+ cells, and $CD3^+$ lymphocytes and increased expression of E-selectin, ICAM-1, but not VCAM, on vascular endothelium. Histamine is no longer increased in BALF, but the number of lymphocytes and granulocytes is somewhat increased.

Twenty-four to 48 h after challenge, the bronchial mucosa has returned to prechallenge status, but there are greatly increased (up to 100 fold) numbers of eosinophils in the BALF (Metzger *et al.*, 1986; Sedgwick *et al.*, 1992; Frew *et al.*, 1996), with increased amounts of IL-5 (Ohnishi *et al.*, 1993; Sedgwick *et al.*, 1991; Sur *et al.*, 1995). The eosinophils demonstrate characteristics of activation (low density, L-selectin negative, increased expression of CD11b/CD18 and HLA-DR) (Georas *et al.*, 1992; Kroegel *et al.*, 1994; Sedgwick *et al.*, 1992). They also demonstrate greater superoxide generation and adherence to collagen and endothelial cell monolayers after FMLP activation than corresponding blood eosinophils. The BALF contains eosinophil granular proteins, such as major basic protein, eosinophil-derived neurotoxin, eosinophil cationic protein, and eosinophil peroxidase (Sedgwick *et al.*, 1991) and increased amount of soluble ICAM-1 (Takahashi *et al.*, 1994) and E-selectin (Georas *et al.*, 1992). The amount of BALF histamine, tryptase and albumin has returned to control levels (Frew *et al.*, 1996), but there is substantial increase of the number of BAL basophils (Guo *et al.*, 1994) and lymphocytes (Liu *et al.*, 1991).

A previous exposure to an antigen can result in an altered response when a subject is re-exposed to that antigen. This is most clearly demonstrated by the anamnestic immunoglobulin response to a second antigen exposure in which there are alterations of both amount and isotype of immunoglobulin produced (Jones *et al.*, 1993). Previous exposure to an antigen also affects cell-mediated immunity in the lung (Waldman *et al.*, 1973). It is difficult to be assured of precise knowledge of previous exposure of subjects to common aeroantigens, since immunologic evidence of previous exposure typically diminishes over time. Differences in the extent and/or consequences of previous exposure could strikingly alter the results of later exposure. Use of a neoantigen to which the subjects have not been previously exposed avoids the difficulty of distinguishing inherent differences of immune response from differences of the extent and nature of prior exposure. With respect to this chapter, use of a neoantigen provides the means to compare Th responses following initial exposure to asthmatics to antigen.

We have extensively characterized the response of normal humans to intrapulmonary immunization to KLH (Weissman *et al.*, 1991). Keyhole limpet hemocyananin (KLH) is the blood protein that is responsible for oxygen transport of an inedible marine mollusc found in the coastal waters of southern California. It is strongly immunogenic in humans and immunologically distinct without cross-reacting antibodies. Normal humans exhibit no evidence of prior sensitization to this protein (Birdsall and Rossen, 1983; Paul *et al.*, 1974; Zuckerman *et al.*, 1974). Extensive work in animals has demonstrated that response to KLH is a T cell dependent phenomenon and that KLH is a potent agent for inducing both specific antibody and cell-mediated immunity. Response to KLH has been used to assess changes of immune response in aged animals (Kirschmann and Murasko, 1992), animals treated with substances that may alter immune response (Hornquist and Lycke, 1993; Irwin, 1993; Schiffer *et al.*, 1991; Sikorski *et al.*, 1991), normal humans (Bird *et al.*, 1990; Devey *et al.*, 1990; Sikorski *et al.*, 1991; Waldo *et al.*, 1994), humans with HIV infection (Whittle *et al.*, 1993) and humans treated with various agents thought to change immune response (Sidell *et al.*, 1990). KLH has also been used to assess immune response in the gut (Elson, 1992; Vajdy and Lycke, 1992) and the nose (Waldo *et al.*, 1994). Therefore, KLH is well suited for use as an antigen to study the development of a primary immune response after intrapulmonary exposure. Instillation of 500 μg KLH into the lung induces serum antibody to KLH (both IgG and IgA) 10 days after immunization. The amount of serum antibody after 500 μg KLH intrapulmonary deposition is similar to that produced after 100 μg KLH subcutaneously. After intrapulmonary deposition, serum IgG (total) anti-KLH persists for at least 28 days, but serum anti-KLH IgA (total) decreases to baseline by 21 days. Peripheral blood cells producing specific anti-KLH can be detected in most subjects 8–10 days after immunization but disappear by 15 days (IgG) or 12 days (IgA). Anti-KLH antibody can also be detected in bronchoalveolar lavage fluid (BALF) 10–15 days after immunization. The IgA anti-KLH is dimeric in both BALF and serum, implying a pulmonary origin (Weissman *et al.*, 1994). Both smoking and HIV infection decrease the amount of antibody and Ig producing PBMC in blood (Weissman *et al.*, 1991, 1994).

To detect differences in initial responses to antigen in asthmatics, we immunized both allergic asthmatic and non-allergic non-asthmatic subjects with 500 μg KLH in 1 ml sterile normal saline in the lingula (a division of the left upper lobe of the lung). Serum class specific antibody to KLH was measured by ELISA at days 0, 4, 7, 9, 11, 14, 18, 21, and 25 after immunization. Bronchoalveolar lavage (BAL) was performed in three separate sites at the time of and after immunization. The three sites were the right middle lobe at the time of immunization (control lobe); and the superior division of the lingula (immunized lobe) and the right middle lobe (non-immunized lobe) 11 days after immunization. This time was chosen as previous work indicated that this was the time of maximal serum anti-KLH antibody levels (Weissman *et al.*, 1994). We sampled different sites within the lung to determine if there were differences of local responses dependent on antigen exposure.

The subjects in both groups were young (mean ages 23.9 and 22.8 years), with a predominance of female subjects. The atopic group had multiple positive immediate type skin tests (mean 8, range 5–11) to a panel of 11 common aeroallergens, whereas the control group had negative skin tests. Three of the nine asthmatic subjects had baseline FEV_1 < 70% predicted and thus did not undergo methacholine challenge. However, their FEV_1 did increase at least 20% (mean 34%) after inhalation of 200 μg albuterol. The remaining six asthmatic subjects exhibited increased methacholine reactivity (mean $PC_{20}FEV_1$ 5.4, range 0.4–15.8 mg/ml). There were no adverse effects of instillation of KLH, similar to our previous experiments (Weissman *et al.*, 1991, 1992, 1994).

Detectable serum anti-KLH IgG1, IgG2, IgG3, IgG4, IgA1, IgA2, and IgM appeared at 5–9 days after KLH instillation, increased to maximal at 12 days and then decreased (IgA subclasses) or remained the same (IgG subclasses). There was no difference between the groups of the values of anti-KLH antibody for IgG1, IgG2, IgG3, IgA1, IgA2, and IgM, but the values for anti-KLH IgG4 were higher in the atopic asthmatic than in the normal group (Figure 12.1). When only peak values of anti-KLH antibody were considered, the

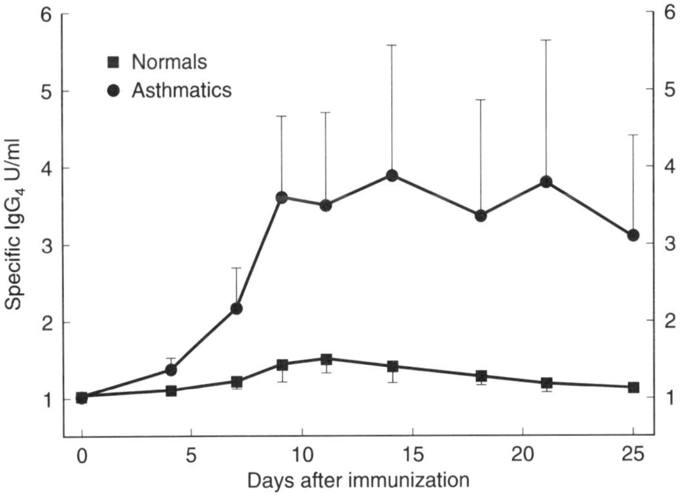

Figure 12.1 Serum IgG4 anti-KLH antibody after pulmonary immunization with 500 μg KLH. There are nine subjects in each group. Using multivariance ANOVA, there was an effect of time and a difference between the groups on the level of IgG4 anti-KLH antibody.

asthmatic group exhibited increased level of anti-KLH IgG4 ($p < 0.05$) compared to the control group.

In bronchoalveolar lavage fluid, there was no difference between asthmatics and normals of the percent return of instilled normal saline, protein concentration, total number or type of cells between the sites of lavage (control, immunized, non-immunized). There was no detectable IgG2, IgG3, IgG4 and IgE anti-KLH antibody in the lavage fluid of both populations, but considerable amounts of IgM, IgA1, IgA2, IgG1 anti-KLH antibody. When both normals and asthmatics were considered together, there was more specific anti-KLH IgA1 and IgM in the lavage from the immunized than from the non-immunized lobe. There were no differences between the asthmatic and normal populations of the amount of IgM, IgA1, IgA2, IgG1 anti-KLH antibody in bronchoalveolar lavage fluid.

Therefore, we found that the serum antibody response of asthmatics following intrapulmonary exposure to a neoantigen can be characterized as a Th2 type response with no evidence of differences between asthmatics and non-asthmatic controls with respect to local pulmonary antibody responses. Further antigen exposure would presumably result in exaggerated Th2 responses (similar to that which occurs after natural exposures to usual allergens, such as pollens and animal allergens).

Animal models of asthma

Modelling a human disease in mice has long been a powerful tool to understanding the pathophysiology of the disease owing to the vast array of tools (monoclonal antibodies, cytokines and drugs) which can be used to dissect the murine immune response. Murine models of atopic asthma have proven to be no exception. The best murine models, of course, are those which mirror the human disease most closely. Although many different experimental models exist, in general, mice have been shown to respond to antigens delivered to the lung by 1) exhibiting eosinophilic pulmonary inflammation, 2) mounting a vigorous antigen-specific IgE response and 3) displaying signs of antigen-specific T cell activation such as shifts of pulmonary T cells to a memory or activated cell phenotype and induction of cytokine secretion. Most importantly, mice have been shown to exhibit airway hyperreactivity (AHR) after antigen challenge via the lung, most often induced by a non-specific bronchoprovocating agent. The above are characteristics of a good animal model of asthma. The critical role of T cells in each of these facets of murine atopic asthma has been well described.

One of the first descriptions of the murine immune response to an aerosolized antigen was from Renz and colleagues (Renz *et al.*, 1992). In these series of studies, BALB/c mice were exposed to nebulized ovalbumin (OVA) daily for a period of 10 days. The mice developed OVA-specific IgE, increased airway responsiveness to intravenously delivered methacholine, and had increased numbers of T lymphocytes recoverable from their peribronchial lymph nodes. Some of these T cells were antigen specific as demonstrated by their ability to proliferate in culture in response to OVA. In addition, passive transfer of peribronchial lymph node cells from OVA-immune BALB/c mice into naive recipients conferred airway responsiveness after a single aerosol exposure to OVA as measured by electrical field stimulation of isolated tracheal smooth muscle segments (Larsen *et al.*, 1992). The donor cells responsible for this increase in airway responsiveness were subsequently shown to be T lymphocytes which specifically expressed the Vβ8.1/8.2 T cell receptor (Renz *et al.*, 1993b). It is unclear in this system whether these T cells directly induce AHR. Their transfer to naive recipients also resulted in OVA-specific IgE production in response to a single OVA challenge, and this IgE could also play a part in the induction of airway responsiveness. The critical role of Vβ8.1/8.2$^+$ T cells in inducing AHR and antigen-specific IgE production was further confirmed when very similar results were found in BALB/c mice exposed to multiple ragweed aerosolizations (Renz *et al.*, 1993a).

One drawback to the use of multiple pulmonary exposures to immunize mice is the absence of any significant pulmonary inflammation, the presence of which is a hallmark of human asthma. Although many other sensitization protocols have been described which do induce pulmonary eosinophilic inflammation (Coyle *et al.*, 1996; Gavett *et al.*, 1994; Lukacs *et al.*, 1994), the most commonly used is the systemic immunization of mice with OVA in the presence of alum adjuvant followed by delivery of the antigen via the pulmonary route (either by aerosolization, intranasally or intratracheally). This type of sensitization also results in increased levels of OVA-specific IgE and AHR. The role of T lymphocytes in these processes has been a topic of intense investigation.

OVA-immune mice subsequently exposed to one or more OVA aerosolizations have been shown to recruit increased numbers of T lymphocytes to bronchoalveolar lavage fluid (BALF) (Garlisi *et al.*, 1995; Kennedy *et al.*, 1995) and peribronchial lymph nodes (Krinzman *et al.*, 1996b). Additionally, flow cytometric evaluation of these T cells showed that there were increased numbers of activated cells represented, expressing increased levels of interleukin-2 receptor (CD25) (Krinzman *et al.*, 1996a) as well as cells of the memory phenotype (CD44$^+$, CD45RB^{lo}) (Garlisi *et al.*, 1995; Kennedy *et al.*, 1995).

The cytokine production profile of pulmonary T cells after OVA immunization has been repeatedly shown to be reflective of an expansion of the Th2 subset with IL-4 and/or IL-5 being detected. Zhang *et al.* found increased levels of IL-4 and IL-5 mRNA in the peribronchial lymph nodes of OVA-alum immunized and intranasally challenged BALB/c mice when analyzed by RT-PCR (Zhang *et al.*, 1997). Similarly, Bell *et al.* found increased mRNA for IL-5 in BALF cells from OVA-immune and intratracheally challenged BALB/c mice, as well as IL-4 and IL-5 protein in BALF as measured by ELISA (Bell *et al.*, 1996). Keane-Myers *et al.* also found increased levels of mRNA for IL-4 and IL-5 in whole lung tissue, as well as the corresponding proteins in BALF, in A/J mice sensitized systemically with OVA in the absence of alum and challenged by OVA aspiration (Keane-Myers *et al.*, 1997). Garlisi *et al.* described increased levels of IL-4 and IL-5 mRNA in whole lung tissue in B6D2F1 mice immunized systemically in the presence of alum and then by aerosolization (Garlisi *et al.*, 1995), whereas Corry *et al.* showed that whole lung cells from OVA immune and aerosolized mice secreted IL-4 as measured by ELISPOT analysis (Corry *et al.*, 1996). Additionally, the production of Th2 type cytokines by pulmonary T cells has been shown using other types of sensitization protocols such as those employing *Schistosoma mansoni* eggs (Lukacs *et al.*, 1994), house dust mite antigen (Coyle *et al.*, 1996) and repeated OVA aerosolizations in the absence of systemic immunization (Renz *et al.*, 1996).

Much is known about the roles that these activated T cells and their products play in the induction of the pulmonary inflammation, airway hyperresponsiveness and increased levels of IgE that characterize the models of atopic asthma. Complete depletion of T lymphocytes from antigen sensitized mice before pulmonary challenge inhibited pulmonary eosinophilia, and Th2 cytokine mRNA expression in the lung (Garlisi *et al.*, 1995) as well as AHR (Gavett *et al.*, 1994). These studies suggest that eosinophil recruitment, activation and local accumulation cannot proceed in the absence of pulmonary Th2 cell production of IL-5, which may in turn permit the development of AHR. These studies do not rule out, however, that production of IL-4 by pulmonary Th2 cells also contributes to eosinophil recruitment and AHR.

Rather than simply depleting immune mice of T cells, two different groups of investigators chose to examine the effect of inhibiting T cell activation using CTLA4-Ig, a reagent which blocks the interaction of T cells with antigen-presenting cells (Krinzman *et al.*, 1996b; Keane-Myers *et al.*, 1997). Both Krinzman *et al.* and Keane-Myers *et al.* found that blockade of T cell activation in OVA immune mice during the aerosol challenge phase prohibited the development of AHR, pulmonary inflammation and increases in serum IgE levels. Additionally, the latter group also detected less IL-4 in the BALF of mice treated with CTLA4-Ig. Both groups also reported that blockade of T cell activation during the sensitization phase with CTLA4-Ig similarly ablated IgE production, pulmonary inflammation and AHR. These studies demonstrated, therefore, that activation of T cells is required not only for generating the primary immunity to an antigen but also during the effector stage of the pulmonary immune response. Again, whether the T cells themselves or their cytokine products are ultimately responsible for manifesting pulmonary inflammation and AHR remains unclear.

Many studies addressing the specific role of IL-4 and IL-5 in inducing the pathophysiology of murine atopic asthma have been conducted, either by depletion of the cytokines by monoclonal antibody treatment, by exogenous addition of the cytokines or by immunizing mice which have been genetically altered such that production of the cytokines cannot proceed (cytokine knockout mice). These investigations have led to some conflicting results, probably due to slight differences in immunization protocols and, in the case of treating the mice with exogenous reagents, the timing and route of delivery of the immunomodulating agents. In addition, murine strain differences may play a role. Corry *et al.* administered anti-IL-4 antibody once weekly to BALB/c mice during both the systemic immunization phase (OVA given subcutaneously for 4 weeks, once weekly) and the aerosolization phase (two or three aerosolizations of OVA delivered 5 days apart), or during the aerosolization phase only (Corry *et al.*, 1996). When mice were treated with anti-IL-4 throughout both the sensitization and challenge periods, no airway hyperreactivity to acetylcholine developed despite the presence of pulmonary eosinophilia. In contrast, anti-IL-4 antibody given during the aerosolization phase only had no effect on AHR. In these studies, treating mice with antibodies to neutralize IL-5 abrogated the pulmonary eosinophilia induced in control antibody treated animals, but failed to inhibit the induction of AHR. Nagai *et al.* also reported on the failure of an anti-IL-5 antibody to inhibit the induction of AHR in BALB/c mice when it was administered intratracheally to OVA-immune mice 2 h prior to aerosol challenge (Nagai *et al.*, 1996). Pulmonary eosinophilia, however, was abolished. These studies, therefore, would not support a role for eosinophils in contributing to antigen-induced non-specific airway hyperreactivity. In contrast, Hamelmann *et al.*, employing multiple aerosolizations of OVA in the absence of systemic immunization, demonstrated that intravenous administration of anti-IL-5 antibody to BALB/c mice (every other day during the 10 day aerosolization protocol) abrogated the development of AHR and pulmonary eosinophilia without affecting IgE production or T cell activation (Hamelmann *et al.*, 1997). However, in a study conducted by Hessel and colleagues, neither antibodies specific for IL-4 nor IL-5 were able to inhibit AHR in OVA-immune BALB/c mice when administered intraperitoneally 1 day before a series of aerosol challenges with OVA (Hessel *et al.*, 1997). Not unexpectedly, anti-IL-5 did inhibit the pulmonary eosinophilia induced by OVA aerosolization. Interestingly, these authors showed that neutralization of IFN-γ prior to the aerosolization challenge was effective in abrogating AHR, suggesting a role for this Th1 type cytokine in manifesting bronchoconstriction.

An alternative to depleting cytokines with a monoclonal antibody has been the use of knock-out mice. Brusselle *et al.* used C57BL/6 IL-4-knock-out mice to show that systemic immunization with OVA in alum followed by multiple OVA aerosolizations failed to generate the pulmonary eosinophilia, increased OVA-specific IgE levels and airway hyperresponsiveness observed in wild-type (IL-4 sufficient) C57BL/6 mice (Brusselle *et al.*, 1995). Although this study supports the findings of Corry *et al.* (1996), these authors failed to demonstrate any T cell reactivity to OVA (even of the Th1 type) suggesting that perhaps IL-4-deficient mice fail to mount any systemic immune response to OVA and therefore fail to respond to the pulmonary challenge. IL-5-knock-out mice have also been shown to be deficient when compared with their wild-type C57BL/6 counterparts in mounting pulmonary eosinophilia and AHR in response to systemic immunization with OVA followed by multiple OVA aerosolizations (Foster *et al.*, 1996). These IL-5 knock-out mice were, however, shown to be capable of mounting a systemic immune response to OVA by virtue of their ability to produce equivalent levels of OVA-specific IgE. In addition, AHR and airway eosinophilia was restored in these mice by pulmonary delivery of a recombinant vaccinia virus expressing IL-5. This study, therefore, strongly supports a role for IL-5 in directly inducing not only pulmonary eosinophilia, but AHR as well. In addition, it suggests that OVA-specific IgE, which is presumably bound to pulmonary mast cells, is not sufficient to induce AHR when crosslinked by a pulmonary delivery of OVA in the absence of eosinophilia.

Further evidence of the important role of IL-5 in inducing pulmonary eosinophilia and AHR has come from a recent study by Kaminuma *et al.* (1997). These authors showed that intravenous transfer of IL-5 producing T cell clones to naive BALB/c mice could confer pulmonary eosinophilia and AHR when the mice were exposed to a single OVA aerosol challenge. No OVA-specific IgE was present. Both the eosinophilia and AHR could be blocked by anti-IL-5 administration immediately before and after the aerosol challenge. Further, the extent of eosinophilia correlated well with the amount of IL-5 produced by the various T cell clones. Transfer of Th1 type clones, producers of IL-2 and IFN-γ, were unable to produce either eosinophilia or AHR.

As the production of Th2 type cells (and their cytokines IL-5 and IL-4) has been so closely linked with the pathophysiology of atopic asthma, many investigations into blocking their development have been conducted. By treating mice with IFN-γ or IL-12, T cell development can be skewed away from the Th2 type and toward Th1 cell expansion and the consequent production of more IFN-γ. Delivery of IFN-γ to the lung by nebulization has been shown to inhibit AHR and decrease OVA-specific IgE levels when administered either before or after a series of OVA aerosolizations to BALB/c mice (Lack *et al.*, 1994, 1996). Similarly, if the production of IFN-γ is stimulated in the lung by mucosal gene expression (Li *et al.*, 1996), antigen or Th2 cell induced increases in pulmonary eosinophilia and AHR are blocked. Additionally, treatment of mice with IL-12, a potent inducer of IFN-γ production, has been shown to inhibit antigen-induced airway eosinophilia, AHR and, if administered early in the immunization protocol, antigen-specific IgE production (Gavett *et al.*, 1995; Kips *et al.*, 1996; Sur *et al.*, 1996). Interestingly, Kips *et al.* (1996) found that intraperitoneal administration of IL-12 to actively sensitized animals during the aerosol challenge phase inhibited the development of AHR to intravenous delivered carbachol, without affecting the OVA-specific IgE levels in serum. These data again support the theory that mechanisms other than IgE mediated degranulation of mast cells can induce AHR.

Although the precise mechanisms of bronchoconstriction mediated by the components of the immune system remain somewhat murky, it is clear that T helper cells, primarily of the Th2 type, play very important roles in the induction and pathophysiology of murine atopic asthma. Many elegant studies have richly contributed to our knowledge of how these T cells and their cytokines can mediate the pulmonary eosinophilia, atopy and AHR observed in these murine models. Our own data, using two different sensitization protocols, support the concept that neither the absolute extent of eosinophilia nor antigen-specific IgE levels predict whether pulmonary delivery of antigen will mediate AHR (Lyons *et al.*, 1996; Wilder *et al.*, 1997). Instead, it is most likely a combination of genetic background and inherent triggerability of bronchial smooth muscle which determines whether antigen-induced non-specific AHR will be manifested. It seems intriguing that inhibiting the development of Th2 cells, by administration of IFN-γ or IL-12, may provide an effective therapy for allergen-induced eosinophilia and bronchoconstriction. The efficacy of IFN-γ or IL-12 on allergen-specific IgE, however, may depend upon the timing and route of intervention as IgE producing plasma cells are most likely not affected by these agents.

Figure 12.2 illustrates the postulated immunopathogenesis of asthma.

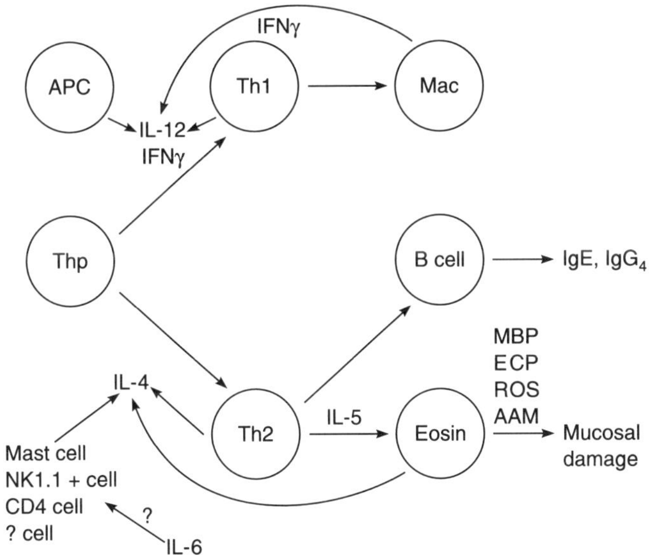

Figure 12.2 Differentiation of either Th1 or Th2 cells. Asthma is postulated to be a Th2 predominant disease. MBP = major basic protein, ECP = eosinophil cationic protein, ROS = reactive oxygen species, AAM = arachidonic acid metabolites, ? cell = unknown cell.

References

Azzawi, M., Bradley, B., Jeffery, P.K., Frew, A.J., Wardlaw, A.J., Knowles, G., Assoufi, B., Collins, J.V., Durham, S. and Kay, A.B. (1990) Identification of activated T lymphocytes and eosinophils in bronchial biopsies in stable atopic asthma. *Am. Rev. Respir. Dis.*, **142**, 1407–13.

Beasley, R., Roche, W.R., Roberts, J.A. and Holgate, S.T. (1989) Cellular events in the bronchi in mild asthma and after bronchial provocation. *Am. Rev. Respir. Dis.*, **139**, 806–17.

Bell, S.J., Metzger, W.J., Welch, C.A. and Gilmour, M.I. (1996) A role for Th2 T-memory cells in early airway obstruction. *Cell. Immunol.*, **170**, 185–94.

Bentley, A.M., Menz, G., Storz, C., Robinson, D.S., Bradley, B., Jeffery, P.K., Durham, S.R. and Kay, A.B. (1992) Identification of T lymphocytes, macrophages, and activated eosinophils in the bronchial mucosa in intrinsic asthma. Relationship to symptoms and bronchial responsiveness. *Am. Rev. Respir. Dis.*, **146**, 500–6.

Bentley, A.M., Meng, Q., Robinson, D.S., Hamid, Q., Kay, A.B. and Durham, S.R. (1993) Increases in activated T lymphocytes, eosinophils, and cytokine mRNA expression for interleukin-5 and granulocyte/macrophage colony-stimulating factor in bronchial biopsies after allergen inhalation challenge in atopic asthmatics. *Am. J. Respir. Cell. Mol. Biol.*, **8**, 35–42.

Bentley, A.M., Durham, S.R. and Kay, A.B. (1994) Comparison of the immunopathology of extrinsic, intrinsic and occupational asthma. *J. Invest. Allerg. Clin. Immunol.*, **4**, 222–32.

Bentley, A.M., Hamid, Q., Robinson, D.S., Schotman, E., Meng, Q., Assoufi, B., Kay, A.B. and Durham, S.R. (1996) Prednisolone treatment in asthma. Reduction in the numbers of eosinophils, T cells, tryptase-only positive mast cells, and modulation of IL-4, IL-5, and interferon-gamma cytokine gene expression within the bronchial mucosa. *Am J. Respir. Crit. Care Med.*, **153**, 551–6.

Bird, P., Calvert, J.E. and Amlot, P.L. (1990) Distinctive development of IgG4 subclass antibodies in the primary and secondary responses to keyhole limpet haemocyanin in man. *Immunology*, **69**, 355–60.

Birdsall, H.H. and Rossen, R.D. (1983) Characterization of anti-Fab′ antibodies in human sera: identification of soluble immune complexes that contain hidden anti-KLH and blocking anti-immunoglobulins following immunization with keyhole limpet haemocyanin. *Clin. Exp. Immunol.*, **53**, 497–504.

Bousquet, J., Chanez, P., Lacoste, J.Y., Enander, I., Venge, P., Peterson, C., Ahlstedt, S., Michel, F.B. and Godard, P. (1991) Indirect evidence of bronchial inflammation assessed by titration of inflammatory mediators in BAL fluid of patients with asthma. *J. Allergy Clin. Immunol.*, **88**, 649–60.

Bradley, B.L., Azzawi, M., Jacobson, M., Assoufi, B., Collins, J.V., Irani, A.M., Schwartz, L.B., Durham, S.R., Jeffery, P.K. and Kay, A.B. (1991) Eosinophils, T-lymphocytes, mast cells, neutrophils, and macrophages in bronchial biopsy specimens from atopic subjects with asthma: comparison with biopsy specimens from atopic subjects without asthma and normal control subjects and relationship to bronchial hyperresponsiveness. *J. Allergy Clin. Immunol.*, **88**, 661–74.

Brusselle, G., Kips, J., Joos, G., Bluethmann, H. and Pauwels, R. (1995) Allergen-induced airway inflammation and bronchial responsiveness in wild-type and interleukin-4-deficient mice. *Am. J. Respir. Cell. Mol. Biol.*, **12**, 254–9.

Coffman, R.L. and von der Weid, T. (1997) Multiple pathways for the initiation of T helper 2 (Th2) responses. *J. Exp. Med.*, **185**, 373–5.

Corrigan, C.J., Hartnell, A. and Kay, A.B. (1988) T lymphocyte activation in acute severe asthma. *Lancet*, **1**, 1129–32.

Corrigan, C.J., Haczku, A., Gemou-Engesaeth, V., Doi, S., Kikuchi, Y., Takatsu, K., Durham, S.R. and Kay, A.B. (1993) CD4 T-lymphocyte activation in asthma is accompanied by increased serum concentrations of interleukin-5. Effect of glucocorticoid therapy. *Am. Rev. Respir. Dis.*, **147**, 540–7.

Corrigan, C.J., Hamid, Q., North, J., Barkans, J., Moqbel, R., Durham, S., Gemou-Engesaeth, V. and Kay, A.B. (1995) Peripheral blood CD4 but not CD8 T-lymphocytes in patients with exacerbation of asthma transcribe and translate messenger RNA encoding cytokines which prolong eosinophil survival in the context of a Th2-type pattern: effect of glucocorticoid therapy. *Am. J. Respir. Cell. Mol. Biol.*, **12**, 567–78.

Corry, D.B., Folkesson, H.G., Warnock, M.L., Erle, D.J., Matthay, M.A., Wiener-Kronish, J.P. and Locksley, R.M. (1996) Interleukin 4, but not interleukin 5 or eosinophils, is required in a murine model of acute airway hyperreactivity. *J. Exp. Med.*, **183**, 109–17.

Coyle, A.J., Wagner, K., Bertrand, C., Tsuyuki, S., Bews, J. and Heusser, C. (1996) Central role of immunoglobulin (Ig) E in the induction of lung eosinophil infiltration and T helper 2 cell cytokine production: inhibition by a non-anaphylactogenic anti-IgE antibody. *J. Exp. Med.*, **183**, 1303–10.

Cruikshank, W.W., Long, A., Tarpy, R.E., Kornfeld, H., Carroll, M.P., Teran, L., Holgate, S.T. and Center, D.M. (1995) Early identification of interleukin-16 (lymphocyte chemoattractant factor) and macrophage inflammatory protein 1 alpha (MIP1 alpha) in bronchoalveolar lavage fluid of antigen-challenged asthmatics. *Am. J. Respir. Cell. Mol. Biol.*, **13**, 738–47.

Del Prete, G.F., De Carli, M., Elios, M.M., Maestrelli, P., Ricci, M., Fabbri, L. and Romagnani, S. (1993) Allergen exposure induces the activation of allergen-specific Th2 cells in the airway mucosa of patients with allergic respiratory disorders. *Eur. J. Immunol.*, **23**, 1445–9.

Devey, M.E., Bleasdale-Barr, K.M., Bird, P. and Amlot, P.L. (1990) Antibodies of different human IgG subclasses show distinct patterns of affinity maturation after immunization with keyhole limpet haemocyanin. *Immunology*, **70**, 168–74.

Elson, C.O. (1992) Cholera toxin as a mucosal adjuvant: effects of H-2 major histocompatibility complex and lps genes. *Infect. Immun.*, **60**, 2874–9.

Fick, R.B. Jr., Metzger, W.J., Richerson, H.B., Zavala, D.C., Moseley, P.L., Schoderbek, W.E. and Hunninghake, G.W. (1987) Increased bronchovascular permeability after allergen exposure in sensitive asthmatics. *J. Appl. Physiol.*, **63**, 1147–55.

Finkelman, F.D. (1995) Relationships among antigen presentation, cytokines, immune deviation, and autoimmune disease. *J. Exp. Med.*, **182**, 279–82.

Foster, P.S., Hogan, S.P., Ramsay, A.J., Matthaei, K.I. and Young, I.G. (1996) Interleukin 5 deficiency abolishes eosinophilia, airways hyperreactivity, and lung damage in a mouse asthma model. *J. Exp. Med.*, **183**, 195–201.

Frew, A.J., St-Pierre, J., Teran, L.M., Trefilieff, A., Madden, J., Peroni, D., Bodey, K.M., Walls, A.F., Howarth, P.H., Carroll, M.P. and Holgate, S.T. (1996) Cellular and mediator responses twenty-four hours after local endobronchial allergen challenge of asthmatic airways. *J. Allergy Clin. Immunol.*, **98**, 133–43.

Garlisi, C.G., Falcone, A., Kung, T.T., Stelts, D., Pennline, K.J., Beavis, A.J., Smith, S.R., Egan, R.W. and Umland, S.P. (1995) T cells are necessary for Th2 cytokine production and eosinophil accumulation in airways of antigen-challenged allergic mice. *Clin. Immunol. Immunopathol.*, **75**, 75–83.

Gavett, S.H., Chen, X., Finkelman, F. and Wills-Karp, M. (1994) Depletion of murine CD4+ T lymphocytes prevents antigen-induced airway hyperreactivity and pulmonary eosinophilia. *Am. J. Respir. Cell. Mol. Biol.*, **10**, 587–93.

Gavett, S.H., O'Hearn, D.J., Li, X, Huang, S.K., Finkelman, F.D. and Wills-Karp, M. (1995) Interleukin 12 inhibits antigen-induced airway hyperresponsiveness, inflammation, and Th2 cytokine expression in mice. *J. Exp. Med.*, **182**, 1527–36.

Georas, S.N., Liu, M.C., Newman, W., Beall, L.D., Stealey, B.A. and Bochner, B.S. (1992) Altered adhesion molecule expression and endothelial cell activation accompany the recruitment of human granulocytes to the lung after segmental antigen challenge. *Am. J. Respir. Cell. Mol. Biol.*, **7**, 261–9.

Gerblich, A.A., Campbell, A.E. and Schuyler, M.R. (1984) Changes in T-lymphocyte subpopulations after antigenic bronchial provocation in asthmatics. *N. Engl. J. Med.*, **310**, 1349–52.

Guo, C.B., Liu, M.C., Galli, S.J., Bochner, B.S., Kagey-Sobotka, A. and Lichtenstein, L.M. (1994) Identification of IgE-bearing cells in the late-phase response to antigen in the lung as basophils. *Am. J. Respir. Cell. Mol. Biol.*, **10**, 384–90.

Hamelmann, E., Oshiba, A., Loader, J., Larsen, G.L., Gleich, G., Lee, J. and Gelfand, E.W. (1997) Antiinterleukin-5 antibody prevents airway hyperresponsiveness in a murine model of airway sensitization. *Am. J. Respir. Crit. Care Med.*, **155**, 819–25.

Hanifin, J. and Lobitz, W. (1977) Newer concepts of atopic dermatitis. *J. Invest. Dermatol.*, **47**, 484–90.

Hessel, E.M., Van Oosterhout, A.J., Van Ark, I., Van Esch, B., Hofman, G., Van Loveren, H., Savelkoul, H.F. and Nijkamp, F.P. (1997) Development of airway hyperresponsiveness is dependent on interferon-gamma and independent of eosinophil infiltration. *Am. J. Respir. Cell. Mol. Biol.*, **16**, 325–34.

Hornquist, E. and Lycke, N. (1993) Cholera toxin adjuvant greatly promotes antigen priming of T cells. *Eur. J. Immunol.*, **23**, 2136–43.

Irwin, M. (1993) Brain corticotropin-releasing hormone- and interleukin-1 beta-induced suppression of specific antibody production. *Endocrinology*, **133**, 1352–60.

Jones, S.E., Davila, D.R., Haley, P.J. and Bice, D.E. (1993) The effects of age on immune responses in the antigen-instilled dog lung. Antibody responses in the lung and lymphoid tissues following primary and secondary antigen instillation. *Mech. Ageing Dev.*, **68**, 191–207.

Kaminuma, O., Mori, A., Ogawa, K., Nakata, A., Kikkawa, H., Naito, K., Suko, M. and Okudaira, H. (1997) Successful transfer of late phase eosinophil infiltration in the lung by infusion of helper T cell clones. *Am. J. Respir. Cell. Mol. Biol.*, **16**, 448–54.

Keane-Myers, A., Gause, W.C., Linsley, P.S., Chen, S.J. and Wills-Karp, M. (1997) B7-CD28/CTLA-4 costimulatory pathways are required for the development of T helper cell 2-mediated allergic airway responses to inhaled antigens. *J. Immunol.*, **158**, 2042–9.

Kennedy, J.D., Hatfield, C.A., Fidler, S.F., Winterrowd, G.E., Haas, J.V., Chin, J.E. and Richards, I.M. (1995) Phenotypic characterization of T lymphocytes emigrating into lung tissue and the airway lumen after antigen inhalation in sensitized mice. *Am. J. Resp. Cell. Mol. Biol.*, **12**, 613–23.

Kips, J.C., Brusselle, G.J., Joos, G.F., Peleman, R.A., Tavernier, J.H., Devos, R.R. and Pauwels, R.A. (1996) Interleukin-12 inhibits antigen-induced airway hyperresponsiveness in mice. *Am. J. Respir. Crit. Care Med.*, **153**, 535–9.

Kirschmann, D.A. and Murasko, D.M. (1992) Splenic and inguinal lymph node T cells of aged mice respond differently to polyclonal and antigen-specific stimuli. *Cell. Immunol.*, **139**, 426–37.

Krinzman, S.J., De Sanctis, G.T., Cernadas, M., Kobzik, L., Listman, J.A., Christiani, D.C, Perkins, D.L. and Finn, P.W. (1996) T cell activation in a murine model of asthma. *Am. J. Physiol.* **271**, L476–83.

Krinzman, S.J., De Sanctis, G.T., Cernadas, M., Mark, D., Wang, Y., Listman, J., Kobzik, L., Donovan, C., Nassr, K., Katona, I., Christiani, D.C., Perkins, D.L.

and Finn, P.W. (1996) Inhibition of T cell costimulation abrogates airway hyperresponsiveness in a murine model. *J. Clin. Invest.*, **98**, 2693–9.
Krishnaswamy, G., Liu, M.C., Su, S.N., Kumai, M., Xiao, H.Q., Marsh, D.G. and Huang, S.K. (1993) Analysis of cytokine transcripts in the bronchoalveolar lavage cells of patients with asthma. *Am. J. Respir. Cell. Mol. Biol.*, **9**, 279–86.
Kroegel, C., Liu, M.C., Hubbard, W.C., Lichtenstein, L.M. and Bochner, B.S. (1994) Blood and bronchoalveolar eosinophils in allergic subjects after segmental antigen challenge: surface phenotype, density heterogeneity, and prostanoid production. *J. Allergy Clin. Immunol.*, **93**, 725–34.
Lack, G., Renz, H., Saloga, J., Bradley, K.L., Loader, J., Leung, D.Y., Larsen, G. and Gelfand, E.W. (1994) Nebulized but not parenteral IFN-gamma decreases IgE production and normalizes airways function in a murine model of allergen sensitization. *J. Immunol.*, **152**, 2546–54.
Lack, G., Bradley, K.L., Hamelmann, E., Renz, H., Loader, J., Leung, D.Y., Larsen, G. and Gelfand, E.W. (1996) Nebulized IFN-gamma inhibits the development of secondary allergic responses in mice. *J. Immunol.*, **157**, 1432–9.
Laitinen, L.A., Laitinen, A. and Haahtela, T. (1993) Airway mucosal inflammation even in patients with newly diagnosed asthma. *Am. Rev. Respir. Dis.*, **147**, 697–704.
Lam, S., LeRiche, J., Phillips, D. and Chan-Yeung, M. (1987) Cellular and protein changes in bronchial lavage fluid after late asthmatic reaction in patients with red cedar asthma. *J. Allergy Clin. Immunol.*, **80**, 44–50.
Larsen, G.L., Renz, H., Loader, J.E., Bradley, K.L. and Gelfand, E.W. (1992) Airway response to electrical field stimulation in sensitized inbred mice. Passive transfer of increased responsiveness with peribronchial lymph nodes. *J. Clin. Invest.*, **89**, 747–52.
Li, X.M., Chopra, R.K., Chou, T.Y., Schofield, B.H., Wills-Karp, M. and Huang, S.K. (1996) Mucosal IFN-gamma gene transfer inhibits pulmonary allergic responses in mice. *J. Immunol.*, **15**, 3216–19.
Liu, M.C., Hubbard, W.C., Proud, D., Stealey, B.A., Galli, S.J., Kagey-Sobotka, A., Bleecker, E.R. and Lichtenstein, L.M. (1991) Immediate and late inflammatory responses to ragweed antigen challenge of the peripheral airways in allergic asthmatics. Cellular, mediator, and permeability changes. *Am. Rev. Respir. Dis.*, **144**, 51–8.
Lukacs, N.W., Strieter, R.M., Chensue, S.W. and Kunkel, S.L. (1994) Interleukin-4-dependent pulmonary eosinophil infiltration in a murine model of asthma. *Am. J. Respir. Cell. Mol. Biol.*, **10**, 526–32.
Lyons, C.R., Wilder, J.A., Collie, D.D.S., Wilson, B.S., Bice, D.E. and Lipscomb, M.F. (1996) Dissociation of airway hyperreactivity from eosinophilia and IgE levels in alum immunized and aerosolized mice. *Am. J. Resp. Crit. Care Med.*, **153**, A139.
Metzger, W.J., Richerson, H.B., Worden, K., Monick, M. and Hunninghake, G.W. (1986) Bronchoalveolar lavage of allergic asthmatic patients following allergen bronchoprovocation. *Chest*, **89**, 477–83.
Metzger, W.J., Zavala, D., Richerson, H.B., Moseley, P., Iwamota, P., Monick, M., Sjoerdsma, K. and Hunninghake, G.W. (1987) Local allergen challenge and bronchoalveolar lavage of allergic asthmatic lungs. Description of the model and local airway inflammation. *Am. Rev. Respir. Dis.*, **135**, 433–40.
Nagai, H., Yamaguchi, S., Maeda, Y. and Tanaka, H. (1996) Role of mast cells, eosinophils and IL-5 in the development of airway hyperresponsiveness in sensitized mice [see comments]. *Clin. Exp. Allergy*, **26**, 642–7.
Ohkawara, Y., Yamauchi, K., Maruyama, N., Hoshi, H., Ohno, I., Honma, M., Tanno, Y., Tamura, G., Shirato, K. and Ohtani, H. (1995) *In situ* expression of the cell adhesion molecules in bronchial tissues from asthmatics with air flow limitation: *in*

vivo evidence of VCAM-1/VLA-4 interaction in selective eosinophil infiltration. *Am. J. Respir. Cell. Mol. Biol.*, **12**, 4–12.

Ohnishi, T., Kita, H., Weiler, D., Sur, S., Sedgwick, J.B., Calhoun, W.J., Busse, W.W., Abrams, J.S. and Gleich, G.J. (1993) IL-5 is the predominant eosinophil-active cytokine in the antigen-induced pulmonary late-phase reaction. *Am. Rev. Respir. Dis.*, **147**, 901–7.

Ollerenshaw, S.L. and Woolcock, A.J. (1992) Characteristics of the inflammation in biopsies from large airways of subjects with asthma and subjects with chronic airflow limitation. *Am. Rev. Respir. Dis.*, **145**, 922–7.

Paul, S., Kenny, A. and Hitzig, W. (1974) Immune response to keyhole limpet hemocyanin in the human. *Int. Arch. Allergy*, **47**, 155–65.

Pernis, A., Gupta, S., Gollob, K.J., Garfein, E., Coffman, R.L., Schindler, C. and Rothman, P. (1995) Lack of interferon gamma receptor beta chain and the prevention of interferon gamma signaling in TH1 cells. *Science*, **269**, 245–7.

Poston, R.N., Chanez, P., Lacoste, J.Y., Litchfield, T., Lee, T.H. and Bousquet, J. (1992) Immunohistochemical characterization of the cellular infiltration in asthmatic bronchi. *Am. Rev Respir Dis* **145**, 918–21.

Rees, J., Friedmann, P.S. and Matthews, J.N. (1990) Contact sensitivity to dinitrochlorobenzene is impaired in atopic subjects. Controversy revisited. *Arch. Dermatol.*, **126**, 1173–5.

Renz, H., Smith, H.R., Henson, J.E., Ray, B.S., Irvin, C.G. and Gelfand, E.W. (1992) Aerosolized antigen exposure without adjuvant causes increased IgE production and increased airway responsiveness in the mouse. *J. Allergy Clin. Immunol.*, **89**, 1127–38.

Renz, H., Bradley, K., Saloga, J., Loader, J., Larsen, G.L. and Gelfand, E.W. (1993a) T cells expressing specific V beta elements regulate immunoglobulin E production and airways responsiveness *in vivo*. *J. Exp. Med.*, **177**, 1175–80.

Renz, H., Saloga, J., Bradley, K.L., Loader, J.E., Greenstein, J.L., Larsen, G. and Gelfand, E.W. (1993b) Specific V beta T cell subsets mediate the immediate hypersensitivity response to ragweed allergen. *J. Immunol.*, **151**, 1907–17.

Renz, H., Bradley, K. and Gelfand, E.W. (1996) Production of interleukin-4 and interferon-gamma by TCR-V beta-expressing T-cell subsets in allergen-sensitized mice. *Am. J. Respir. Cell. Mol. Biol.*, **14**, 36–43.

Rincon, I., Anguita, J., Nakamura, T., Fikrug, E. and Flavell, R.A. (1997) Interleukin (IL)-6 directs the differentiation of IL-4-producing CD4+ T cells. *J. Exp. Med.*, **185**, 461–9.

Robinson, D.S., Hamid, Q., Ying, S., Tsicopoulos, A., Barkans, J., Bentley, A.M., Corrigan, C., Durham, S.R. and Kay, A.B. (1992) Predominant TH2-like bronchoalveolar T-lymphocyte population in atopic asthma. *N. Engl. J. Med.*, **326**, 298–304.

Robinson, D.S., Bentley, A.M., Hartnell, A., Kay, A.B. and Durham, S.R. (1993a) Activated memory T helper cells in bronchoalveolar lavage fluid from patients with atopic asthma: relation to asthma symptoms, lung function, and bronchial responsiveness. *Thorax*, **48**, 26–32.

Robinson, D.S., Hamid, Q., Bentley, A., Ying, S., Kay, A.B. and Durham, S.R. (1993b) Activation of $CD4^+$ T cells, increased TH2-type cytokine mRNA expression, and eosinophil recruitment in bronchoalveolar lavage after allergen inhalation challenge in patients with atopic asthma. *J. Allergy Clin. Immunol.*, **92**, 313–24.

Robinson, D.S., Hamid, Q., Ying, S., Bentley, A., Assoufi, B., Durham, S. and Kay, A.B. (1993c) Prednisolone treatment in asthma is associated with modulation of bronchoalveolar lavage cell interleukin-4, interleukin-5, and interferon-gamma cytokine gene expression. *Am. Rev. Respir. Dis.*, **148**, 401–6.

Robinson, D.S., Ying, S., Bentley, A.M., Meng, Q., North, J., Durham, S.R., Kay, A.B. and Hamid, Q. (1993d) Relationships among numbers of bronchoalveolar lavage cells expressing messenger ribonucleic acid for cytokines, asthma symptoms, and airway methacholine responsiveness in atopic asthma. *J. Allergy Clin. Immunol.*, **92**, 397–403.

Sabin, E.A. and Pearce, E.J. (1995) Early IL-4 production by non-CD4$^+$ cells at the site of antigen deposition predicts the development of a T helper 2 cell response to Schistosoma mansoni eggs. *J. Immunol.*, **155**, 4844–53.

Schiffer, R.B., Sunderman, F.W. Jr, Baggs, R.B. and Moynihan, J.A. (1991) The effects of exposure to dietary nickel and zinc upon humoral and cellular immunity in SJL mice. *J. Neuroimmunol.*, **34**, 229–39.

Sedgwick, J.B., Calhoun, W.J., Gleich, G.J., Kita, H., Abrams, J.S., Schwartz, L.B., Volovitz, B., Ben-Yaakov, M. and Busse, W.W. (1991) Immediate and late airway response of allergic rhinitis patients to segmental antigen challenge. Characterization of eosinophil and mast cell mediators. *Am. Rev. Respir. Dis.*, **144**, 1274–81.

Sedgwick, J.B., Calhoun, W.J., Vrtis, R.F., Bates, M.E., McAllister, P.K. and Busse, W.W. (1992) Comparison of airway and blood eosinophil function after *in vivo* antigen challenge. *J. Immunol.*, **149**, 3710–18.

Sidell, N., Connor, M.J., Chang, B., Lowe, N.J. and Borok, M. (1990) Effects of 13-cis retinoic acid therapy on human antibody responses to defined protein antigens. *J. Invest. Dermatol.*, **95**, 597–602.

Sikorski, E.E., Burns, L.A., McCoy, K.L., Stern, M. and Munson, A.E. (1991) Suppression of splenic accessory cell function in mice exposed to gallium arsenide. *Toxicol. Appl. Pharmacol.*, **110**, 143–56.

Sur, S., Gleich, G.J., Offord, K.P., Swanson, M.C., Ohnishi, T., Martin, L.B., Wagner, J.M., Weiler, D.A. and Hunt, L.W. (1995) Allergen challenge in asthma: association of eosinophils and lymphocytes with interleukin-5. *Allergy* **50**, 891–8.

Sur, S., Lam, J., Bouchard, P., Sigounas, A., Holbert, D. and Metzger, W.J. (1996) Immunomodulatory effects of IL-12 on allergic lung inflammation depend on timing of doses. *J. Immunol.*, **157**, 4173–80.

Szabo, S.J., Dighe, A.S., Gubler, U. and Murphy, K.M. (1997) Regulation of the interleukin (IL)-12R beta 2 subunit expression in developing T helper 1 (Th1) and Th2 cells. *J. Exp. Med.*, **185**, 817–24.

Takahashi, N., Liu, M.C., Proud, D., Yu, X.Y., Hasegawa, S. and Spannhake, E.W. (1994) Soluble intracellular adhesion molecule 1 in bronchoalveolar lavage fluid of allergic subjects following segmental antigen challenge. *Am. J. Respir. Crit. Care Med.*, **150**, 704–9.

Till, S., Li, B., Durham, S., Humbert, M., Assoufi, B., Huston, D., Dickason, R., Jeannin, P., Kay, A.B. and Corrigan, C. (1995) Secretion of the eosinophil-active cytokines interleukin-5, granulocyte/macrophage colony-stimulating factor and interleukin-3 by bronchoalveolar lavage CD4$^+$ and CD8$^+$ T cell lines in atopic asthmatics, and atopic and non-atopic controls. *Eur. J. Immunol.*, **25**, 2727–31.

Vajdy, M. and Lycke, N.Y. (1992) Cholera toxin adjuvant promotes long-term immunological memory in the gut mucosa to unrelated immunogens after oral immunization. *Immunology*, **75**, 488–92.

Van Vyve, T., Chanez, P., Bousquet, J., Lacoste, J.Y., Michel, F.B. and Godard, P. (1992) Safety of bronchoalveolar lavage and bronchial biopsies in patients with asthma of variable severity. *Am. Rev. Respir. Dis.*, **146**, 116–21.

Virchow, J.C., Oehling, A., Boer, L., Hansel, T.T., Werner, P., Matthys, H., Blaser, K. and Walker, C. (1994) Pulmonary function, activated T cells, peripheral blood eosinophilia, and serum activity for eosinophil survival *in vitro*: a longitudinal study in bronchial asthma. *J. Allergy Clin. Immunol.*, **94**, 240–9.

Waldman, R.H., Jurgensen, P.F., Olsen, G.N., Ganguly, R. and Johnson, J.E. (1973) Immune response of the human respiratory tract. I. Immunoglobulin levels and influenza virus vaccine antibody response. *J. Immunol.*, **111**, 38–41.

Waldo, F.B., van den Wall Bake, A.W., Mestecky, J. and Husby, S. (1994) Suppression of the immune response by nasal immunization. *Clin. Immunol. Immunopathol.*, **72**, 30–4.

Walker, C., Bode, E., Boer, L., Hansel, T.T., Blaser, K. and Virchow, J.C. Jr (1992) Allergic and nonallergic asthmatics have distinct patterns of T-cell activation and cytokine production in peripheral blood and bronchoalveolar lavage. *Am. Rev. Respir. Dis.*, **146**, 109–15.

Wardlaw, A.J., Dunnette, S., Gleich, G.J., Collins, J.V. and Kay, A.B. (1988) Eosinophils and mast cells in bronchoalveolar lavage in subjects with mild asthma. Relationship to bronchial hyperreactivity. *Am. Rev. Respir. Dis.*, **137**, 62–9.

Weissman, D.N., Bice, D., Crowell, R. and Schuyler, M. (1991) Human lung immunization: Impairment of humoral responses in smokers. *Am. Rev. Resp. Dis.*, **143**, A227.

Weissman, D.N., Bice, D.E., Muggenburg, B.A., Haley, P.J., Shopp, G.M. and Schuyler, M.R. (1992) Primary immunization in the canine lung. Soluble antigen induces a localized response. *Am. Rev. Resp. Dis.*, **145**, 6–12.

Weissman, D.N., Bice, D.E., Crowell, R.E. and Schuyler, M.R. (1994) Intrapulmonary antigen deposition in the human lung: local responses. *Am. J. Resp. Cell. Mol. Biol.*, **11**, 607–14.

Whittle, H., Egboga, A., Todd, J., Morgan, G., Rolfe, M., Sabally, S., Wilkins, A. and Corrah, T. (1993) Immunological responses of Gambians in relation to clinical stage of HIV-2 disease. *Clin. Exp. Immunol.*, **93**, 45–50.

Wilder, J.A., Collier, D.D., Forrister, B.J., Izzo, A.A., Bice, D.E., Lipscomb, M.F. and Lyons, C.R. (1997) Antigen-induced airway hyperreactivity in ovalbumin T cell receptor transgenic mice. *Am. J. Resp. Crit. Care Med.*, **155**, A11.

Wilson, J.W., Djukanovic, R., Howarth, P.H. and Holgate, S.T. (1992) Lymphocyte activation in bronchoalveolar lavage and peripheral blood in atopic asthma. *Am. Rev. Respir. Dis.*, **145**, 958–60.

Ying, S., Durham, S.R., Corrigan, C.J., Hamid, Q. and Kay, A.B. (1995) Phenotype of cells expressing mRNA for TH2-type (interleukin 4 and interleukin 5) and TH1-type (interleukin 2 and interferon gamma) cytokines in bronchoalveolar lavage and bronchial biopsies from atopic asthmatic and normal control subjects. *Am. J. Respir. Cell. Mol. Biol.*, **12**, 477–87.

Zhang, Y., Lamm, W.J., Albert, R.K., Chi, E.Y., Henderson, W.R. Jr and Lewis, D.B. (1997) Influence of the route of allergen administration and genetic background on the murine allergic pulmonary response. *Am. J. Respir. Crit. Care Med.*, **155**, 661–9.

Zuckerman, K.S., Neidhart, J.A., Balcerzak, S.P. and LoBuglio, A.F. (1974) Immunologic specificity of transfer factor. *J. Clin. Invest.*, **54**, 997–1000.

13 Modulation of T Lymphocyte Responses by Air Pollutants

Matthew I. Gilmour and MaryJane K. Selgrade

Introduction

A number of air pollution episodes documented over the last 7 decades has been associated with significant increases in respiratory disease and mortality in the exposed population (reviewed by Pope *et al.*, 1995). Recent epidemiological studies in the United States and Europe have found that hospital admissions for pneumonia, asthma and bronchitis increase as air pollution levels rise (Bates and Sitzo, 1987; Cody *et al.*, 1992; Waldbott, 1972). While many of these increases in respiratory illness are thought to be associated with pollutant-induced irritation of the airways, particularly in individuals with pre-existing disease such as asthmatics and cardio-pulmonary patients, there is emerging evidence that air pollutant exposure can both suppress host resistance to pathogens, and increase allergic immune responses. Experiments in animals have shown that a wide array of inhaled substances ranging from gaseous air pollutants such as ozone (O_3) and nitrogen dioxide (NO_2) to cigarette smoke, acid aerosols, metals, organic compounds, combustion products and a variety of occupational materials can, depending on the agent and the exposure regime, either impair or stimulate T cell function (reviewed in Gardner, 1982; Selgrade and Gilmour, 1994). It is now known that immune responses directed against certain pathogens and allergens are regulated by different subpopulations of $CD4^+$ T cells, and several animal models have demonstrated that alterations in T cell function can lead to decreases in protective immune responses against infectious agents, and increased development of allergic immune responses (Mosmann and Coffman, 1987). Thus, while the causative agents and mechanisms of tissue damage differ between infectious and allergic diseases, modulation of T lymphocyte function by air pollutant exposure can result in a concomitant disruption of host defences to pathogens and increased immunity to allergens.

T Lymphocyte Subpopulations in Immunotoxicology. Edited by Ian Kimber and MaryJane K. Selgrade. Published in 1998 by John Wiley & Sons Ltd. ISBN 0 471 97194 4

As indicated in Chapter 2 of this volume, knowledge of immune regulation has increased greatly over the last decade with the identification of two types of $CD4^+$ T helper lymphocytes, termed Th1 and Th2, that produce contrasting cytokine profiles and have opposing functions in the generation of specific immune responses and hypersensitivity diseases (reviewed in Mosmann and Coffman, 1987; Mosmann and Sad, 1996). In general, Th1 cells produce interleukin 2 (IL-2), and interferon γ which evoke cell-mediated reactions to combat bacterial and viral infections (Gajewski and Fitch, 1988). Th2 cells secrete IL-4, which stimulates allergic IgE antibody production, IL-5, a chemotactic and maturation factor for eosinophils, and IL-10, which promotes the maturation and development of Th2 responses and suppresses Th1 activity (Paul and Seder, 1994). The Th2 cytokine profile dominates the immune response to some protozoal infections (Kalyanasundanam *et al*., 1996), and is also increased in lung tissue and lymphocytes of allergic asthmatics (Robinson *et al*., 1992).

This chapter will focus on O_3, NO_2 and inhaled particles. These air pollutants are ubiquitous and result primarily from the combustion of fossil fuel. Among the most common methods used to evaluate the effects of air pollutants (or for that matter any toxicant) on T lymphocytes is enumeration of T cell populations circulating in peripheral blood, in lymph nodes, and in systemic immune sites such as the thymus and spleen using fluorescent tagged antibodies to various T cell markers. While these studies (reviewed in the next section) do not provide direct information on Th1 or Th2 populations (for which cell surface markers have not been available), they do indicate that T cells are affected by exposure to air pollutants. General information on T cell responses to pollutant exposure have also been provided by assessing maximal T cell proliferative responses in the presence of mitogenic lectins, or after stimulation of their receptors such as CD3.

The third section of this chapter will review the effects of air pollutants on innate immunity and related cytokine responses in the local microenvironment. Innate (non-specific) immune responses are usually regarded as the first line of defence against pathogens but are also thought to play an important role in the polarization of T lymphocytes to either the Th1 or Th2 phenotype (Fearon and Locksley, 1996; Swain, 1995). Cytokines produced by macrophages, natural killer cells, basophils, mast cells, as well as non-myeloid cells such as the respiratory epithelium influence the local microenvironment during the period of primary antigen contact and have a crucial role in determining whether Th1 or Th2 responses dominate (Swain, 1995). For example, macrophage derived IL-12 upregulates Th1 cell populations and is a potent suppresser of Th2 activity (Trichieri and Gerosa, 1996). In contrast, IL-6 and IL-10 which are produced by a number of different cell types promote the induction and expression of IL-4 in T cells resulting in increased IgE production and Th2 activity (Rincon *et al*., 1997). Modulation of these responses by air pollutants has the potential to polarize Th and Tc responses.

The fourth section of this chapter will review studies on the effects of air pollutants on the functional (proliferative) capacity of T cells to respond to specific antigen after immunization or infection. The fifth and sixth sections will describe the impact that air pollutants have on infectious and allergic disease. There is a large data base on the effects of air pollutants on infectious and allergic disease models in laboratory rodents as well as some data in humans. These studies provide suggestive evidence that Th responses (and perhaps Tc responses) are modulated by pollutant exposure.

Most of the studies reviewed in subsequent sections provide only indirect evidence for modulation of Th (and possibly Tc) responses. Recent advances in cytokine biology have led to the development of numerous antibodies and molecular probes that can be used to measure the genetic expression, transcription and synthesis of a substantial number of immune mediators. In addition, flow cytometric techniques are being developed to enable simultaneous intracellular detection of individual cytokines while also staining for specific phenotypic markers of different T lymphocyte subpopulations (Schauer *et al*., 1996). Application of these newer technologies to study air pollution modulation of immune responses has only just begun, and is the subject of the last section of this chapter. Future research using these tools is likely to provide more concrete evidence for the modulation of type 1 and type 2 Th and Tc responses by exposure to air pollutants.

General changes in T lymphocytes by air pollutants

A number of investigators have observed changes in T cell numbers and function as a consequence of air pollutant exposure. While the majority of these findings have been made in experimental animals, a few limited studies have reported similar effects in humans.

A review of the immunotoxicology of ozone (O_3) by Jakab *et al.* (1995) identified a general pattern of changes in the immune response depending upon the O_3 exposure regime and the endpoint under study. In mice exposed daily to relatively high concentrations of O_3 (0.5–0.8 ppm), a general suppression of immune function occurred in the first few days which was followed by re-establishment of lymphocyte activity. More chronic exposure led to suppressed immunity in systemic tissue such as the spleen and blood, while heightened immunity could be seen in the lung. Experiments using NO_2 are less detailed but have shown that the CD4/CD8 ratio of T lymphocyte sub-populations in the spleen is decreased by acute exposure to 2 ppm NO_2 (Richters and Damji, 1988), and that lymphocyte proliferation to mitogens is decreased following more chronic exposure (Holt *et al*., 1979).

In a series of studies in mice, Dziedzic and White (1986a,b) found that exposure to O_3 (0.3–0.7 ppm, 20 h per day) caused a four-phase change in cellularity of the bronchus associated lymph nodes over a 28-day period which was dose dependent (depicted in Figure 13.1). During the first 2 days,

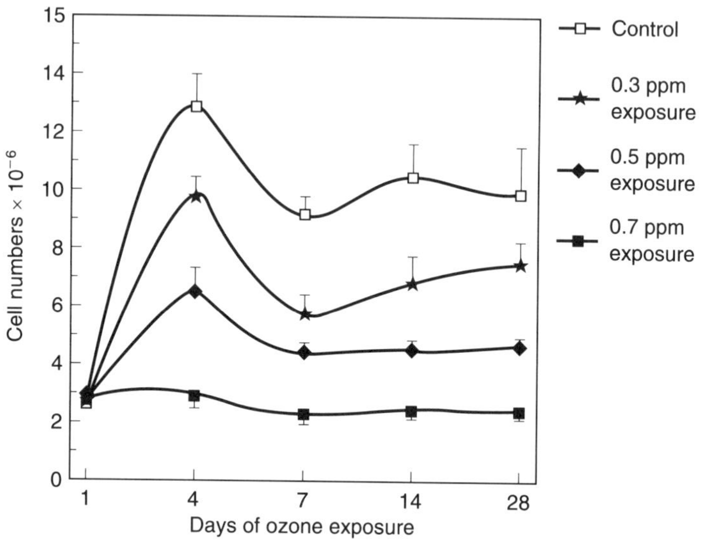

Figure 13.1 Effect of ozone exposure on cell numbers in the bronchial lymph nodes of mice exposed for 20 h per day for up to 28 days (From Dziedzic and White, 1986. Reproduced by permission from Academic Press).

they observed cellular depletion in the lymph node which was followed on the third and fourth day by cellular hyperplasia and increased mitotic activity in the paracortical T-cell areas. Between 5 and 7 days of exposure, the cell number decreased incrementally before levelling out at a steady state elevated baseline value for the remaining 20 days of exposure. At this later stage, cells retrieved from the lymph node displayed increased proliferative responses to the mitogen, concanavalin A. Between 4 and 14 days the lungs showed marked T cell infiltration around the centriacinar region where the O_3-induced lesions were most prominent (Bleavins and Dziedzic, 1990). Similarly, rats exposed to 0.5 ppm O_3, 20 h/day for 1–14 days displayed increased lymphocyte proliferation in both bronchus associated lymphoid tissue (BALT) and mediastinal lymph nodes which peaked on day 3 of exposure and was still significantly elevated at day 7 (Dziedzic *et al.*, 1990). O_3 exposure also caused thymic atrophy in mice which was abrogated by approximately 50% with prior adrenalectomy (Dziedzic and White, 1986a), suggesting that the decrease in thymus weight was partially due to stress. The involvement of the thymus in lymph node cellularity was also established, since the hyperplastic response to O_3 was reduced by 40% in thymectomized animals. These changes in T lymphocyte populations in various tissues could be due to altered cell trafficking following acute or sub-chronic exposures to O_3; however, experiments designed to monitor T cell migration after pollutant exposure have yet to be reported.

T lymphocytes were studied in another series of experiments using athymic, T cell deficient (nu/nu) mice. These animals experienced much less lymphoid hyperplasia in response to O_3 exposure compared with heterozygote (nu/t) O_3-exposed controls, but exhibited markedly increased lung wet weights, alveolar lesions and extent of PMN infiltration (Dziedzic and White, 1987a). Furthermore, normal CD1 mice pre-treated with the interferon inducer, poly I:C experienced fewer lung lesions and neutrophilic infiltrates compared with O_3-exposed control animals, while animals treated with anti-interferon (presumably α/β) antiserum showed more severe lesions (Dziedzic and White, 1987b). These findings led to the hypothesis that lymphoid hyperplasia induced by O_3 exposure conferred protection against further oxidative damage and suggested that the most likely mechanisms for this effect rested either with a T cell product which induced favorable alterations in the resistance of target (epithelial) cells to oxidant exposure, or in changes in the regulation of neutrophil accumulation at the site of damage. Overall conclusions from these studies were that O_3 exposure caused profound changes in the number and function of pulmonary T cells and that some of these effects may have been via cell trafficking from the thymus.

Experiments investigating T lymphocyte function in humans exposed to O_3 have been more limited, but nevertheless have shown that T cell function can be affected by oxidant gases. Several studies have found that humans exposed to 0.4 ppm O_3 experience decrements in lung function, increased pulmonary inflammation and changes in circulating lymphocyte function. Orlando *et al.* (1988) found that while exercise increased T cell proliferation to phytohaemagglutinin (PHA) 24 h later, concomitant O_3 exposure ameliorated this effect. They reported that the O_3 exposure resulted in increased cortisol and beta-endorphin levels and suggested that these stress hormones downregulated the exercise-enhanced immune function. The role that stress hormones play in altering Th1/Th2 responses is considered in Chapter 10. Using an *in vitro* system, Becker *et al.* (1989) found that human blood cells exposed to O_3 *in vitro* had less proliferative activity toward mitogens compared with the same cells exposed to air.

Observations of altered blood lymphocyte function in humans, as well as reduced spleen and thymus weights and changes in lymph node cellularity in mice indicate clearly that the effects of O_3 exposure are not limited to the lung but also occur systemically. In another animal study, plasma from mice exposed to 0.8 ppm O_3 contained lipid peroxidation products which were toxic to cultured thymocytes (Li and Richters, 1991). Other recent studies in humans and animals have identified a number of mediators such as the pro-inflammatory cytokines IL-6, IL-8 and the monokine macrophage inflammatory protein-1 (MIP-1) as well as arachidonic acid metabolites such as prostaglandin E_2 (PGE_2), which are produced in the lung following ozone exposure

(Koren *et al.*, 1989), but may exert effects elsewhere (Canning *et al.*, 1991). At this point, the role which these mediators play in influencing mucosal and systemic immune responses is poorly understood, although results from *in vitro* experiments indicate a wide range of effects depending upon the cells and endpoints under study. The following section will summarize how products from cells other than lymphocytes might affect subsequent T cell function, particularly polarization toward type 2 responses.

Changes in innate immune function with potential impact on T cell polarization

The respiratory tract is lined by several types of ciliated and non-ciliated epithelial cells which transport particles and cell debris in a mucous blanket to the oropharynx and act as a semi-permeable protective barrier to maintain tissue homeostasis (Farmer and Hay, 1990). It is now appreciated that epithelial cells also express a variety of receptors important in cell signalling and adhesion, and can synthesize a number of mediators which regulate inflammatory and immune processes in the lung (Figure 13.2). In addition, other cell types such as mast cells, goblet cells and dendritic cells are interspersed

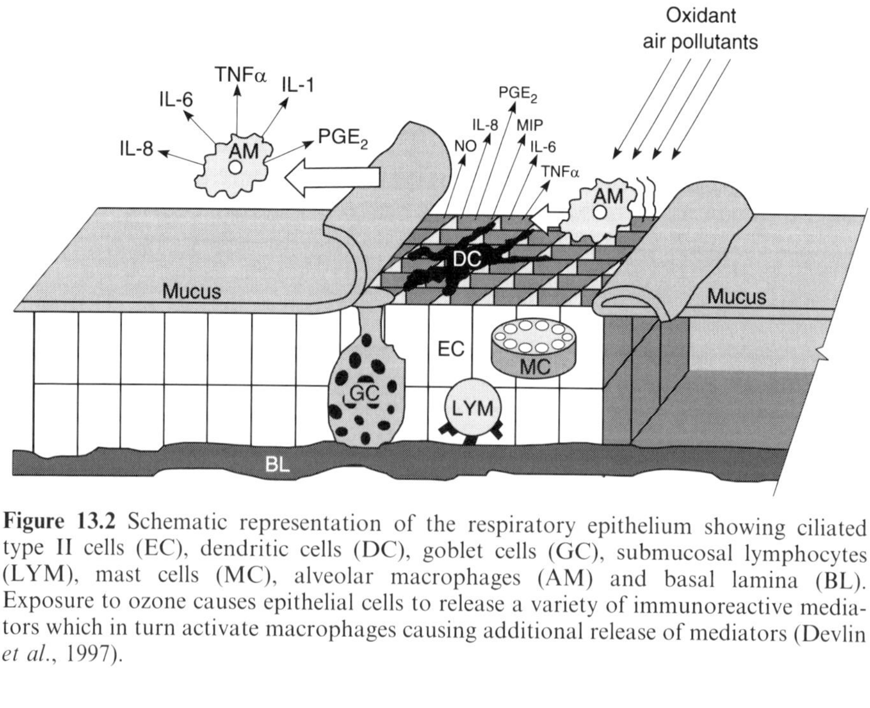

Figure 13.2 Schematic representation of the respiratory epithelium showing ciliated type II cells (EC), dendritic cells (DC), goblet cells (GC), submucosal lymphocytes (LYM), mast cells (MC), alveolar macrophages (AM) and basal lamina (BL). Exposure to ozone causes epithelial cells to release a variety of immunoreactive mediators which in turn activate macrophages causing additional release of mediators (Devlin *et al.*, 1997).

throughout the respiratory epithelium and are available for cell/cell interaction via direct contact and/or signalling through soluble mediators. Experiments have thus far shown that epithelial cells exposed to air pollutants *in vitro* can produce mediators which may influence the development of Th1 and Th2 responses. For example, type II epithelial cells produce NO (Punjabi *et al.*, 1994), PGE_2 (Leikauf *et al.*, 1988; McKinnon *et al.*, 1992) IL-6 and IL-8 (Koren *et al.*, 1991) following exposure to ozone. PGE_2 inhibits production of Th1 lymphokines, IL-2 and IFN-γ, but does not affect the Th2 cytokines, IL-4 and IL-5 (Betz and Fox, 1991; Hilkens *et al.*, 1996; Snijdewint *et al.*, 1993). NO inhibits Th1 but not Th2 cells, while IL6 is capable of stimulating IL-4 production (Barnes and Liew, 1995).

Another important cell in the regulation of inflammatory and immune processes in the lung is the alveolar macrophage (AM). These cells patrol the epithelial surface, engulfing inhaled materials and cell debris, and are a rich source of cytokines and mediators. Several investigators have demonstrated that AM phagocytosis is impaired following exposure to oxidant air pollutants such as O_3 or NO_2 in mice (Goldstein *et al.*, 1971; Gilmour *et al.*, 1991), rats (Oosting *et al.*, 1991), and rabbits (Driscoll *et al.*, 1987), as well as in human volunteers (Devlin *et al.*, 1991a,b). Decreased AM function has been associated with delayed clearance of inhaled particles and enhanced susceptibility to Gram positive bacterial infections (reviewed by Morrow, 1992; Selgrade and Gilmour, 1994). This effect has been partially attributed to increased levels of PGE_2 in bronchoalveolar lavage (BAL) fluid and serum since treatment with the cyclo-oxygenase inhibitor indomethacin blocked decreased phagocytosis by O_3 exposure (Canning *et al.*, 1991), and abrogated O_3-enhanced susceptibility to infection in mice (Gilmour *et al.*, 1993). Recent clinical studies have confirmed the role of PGE_2 in O_3 related lung injury in humans, since pretreatment with the anti-inflammatory drug ibuprofen, reduced PGE_2 levels in the BAL of ozone-exposed human subjects and resulted in smaller changes in lung function than would be expected with O_3 exposure (Hazucha *et al.*, 1996). *In vitro* experiments have shown that both AMs and epithelial cells produce PGE_2 in response to O_3 (Leikauf *et al.*, 1988; McKinnon *et al.*, 1992); however, the relative contribution by each cell *in vivo* is currently not known. In terms of lymphocyte regulation, AM produce a number of inflammatory mediators in response to O_3 including, PGE_2, NO, TNF-α, IL-1, IL-6 and IL-8 (Pendino *et al.*, 1993, 1994), whose effects on T lymphocyte subpopulations were described above. To our knowledge, the effects of air pollutants on AM production of IL-12, another important cytokine in the development of Th1 responses have not been explored.

Less information is available on the effects of air pollutants on pulmonary natural killer (NK) cell activity, thought to be important in the production of interferon and development of Th1 responses. Two studies demonstrated that higher concentrations (1 ppm) of O_3 suppressed pulmonary NK cell activity whereas lower concentrations (0.2 and 0.4 ppm enhanced this activity

(Burleson *et al.*, 1989; Van Loveren *et al.*, 1990). More work is needed to determine whether air pollutant modulations of NK cell activity and the production of interferon occurs and influences T cell responses.

Clearly, a number of mediators are produced in the local milieu as a result of exposure to air pollution. These have the potential not only to affect local immune responses, but also to enter into the blood and influence systemic responses. Depending upon the concentration and half-life of the mediator in the circulation, a gradient effect will arise such that the micro-environment surrounding exposed tissue will be affected to a greater extent than cells peripheral to the site of damage. Future research using animals and cell lines which lack or over express these mediators, as well as neutralization experiments with monoclonal antibodies and receptor antagonists, will be useful for defining cell/cell signalling after exposure to air pollutants and their subsequent effects on immune regulation.

Changes in antigen-specific T cell responses by air pollutants

During the course of a primary infection or after vaccination, antigen-specific T and B lymphocyte populations are selected for and expanded into effector and memory sub-populations which circulate throughout the body and can home to particular mucosal sites such as the respiratory and gastrointestinal tracts. The local specific immune response (antibodies and cytotoxic T cells) then eliminates the pathogen or neutralizes its toxic products. Reinfection with the same agent is usually more limited because memory T and B cells are available for rapid expansion into an effector population.

Several immune function assays measuring lymphocyte proliferation, delayed type hypersensitivity reactions (DTH) and antibody production to experimental antigens have been developed as surrogates for measuring these protective immune responses. As in the previous sections, most of the published literature pertaining to air pollutant immunotoxicity has focused on O_3. In a series of studies, Fujimaki (1989) and Fujimaki *et al.* (1984, 1987) reported that mice exposed to 0.8 ppm O_3 for 1, 3, 7, and 14 days had decreased ability to mount a primary antibody response to sheep red blood cells. In addition, O_3 exposure over the 7-day exposure period gradually depressed DTH reactions in the footpad, although by 14 days there was no difference from controls. The suppressive effects were in part dependent upon the experimental protocol, since mice exposed to O_3 for 3 days prior to immunization mounted an effective DTH response, while animals exposed after immunization had suppressed DTH reactions. Similarly, rats exposed to 0.75 ppm O_3 for 7 days prior to infection with Listeria, had a normal DTH response to the organisms, while exposure to O_3 for 7 days after the infection resulted in significant suppression of immunity.

A few studies have indicated that NO_2 exposure suppresses systemic immune responses in mice including spleen cell responses to T and B-cell mitogens

(Azoulay-Dupuis *et al.*, 1987; Holt *et al.*, 1979; Maigetter *et al.*, 1978), antibody-forming cell responses in spleen or serum antibody levels (Azoulay-Dupuis *et al.*, 1985; Ehrlich *et al.*, 1975; Fujimaki *et al.*, 1981, 1982; Holt *et al.*, 1979; Lefkowitz *et al.*, 1986), and percent T lymphocytes and natural killer (NK) cells present in spleen (Richters and Damji, 1988). In general, it seems that chronic exposure to air pollutants results in suppression of Th1 type immunity in systemic sites while local immunity may be suppressed or enhanced depending upon the exposure regime and the endpoint under study. These effects have been shown in several experimental systems to translate into increased infectious and allergic lung disease.

Increased susceptibility to infectious agents

The T lymphocyte plays a central role in regulating specific immune responses toward pathogens and maintains the balance between health and disease. Once an individual has become immune through previous infection, vaccination or by passive transmission of maternal antibody, these specific defences effectively control most viral and bacterial pathogens. Cell-mediated immune responses consist of T cell recognition and activation which ultimately (via cytokines) directs AMs to kill obligate intracellular microbes such as tubercle bacilli and *Listeria monocytogenes.* Humoral antibody production by T cell-directed B lymphocytes results in opsonization and toxin neutralization as occurs with *S. pneumoniae* and *Bordetella pertussis.* Viruses and obligate intracellular organisms are typically inactivated by the dual mechanisms of neutralizing antibody and cytotoxic T lymphocytes. The most striking example of the importance of T cells can be seen in HIV-infected individuals, who gradually experience a depletion of their $CD4^+$ T cells and become increasingly susceptible to a variety of opportunistic infections (Edelman and Zolla-Pazner, 1989).

Most experimental studies of pollutant-enhanced infection have focused on extracellular pathogens such as *Streptococcus* and *Staphylococcus* spp. which, at least in animal models, are combated by innate aspects of the immune response such as phagocytes and bactericidal enzymes (reviewed by Selgrade and Gilmour, 1994). In one early study, however, Campbell and Hilsenroth (1976) reported that mice exposed to O_3 during tetanus vaccination, were less resistant to subsequent toxin challenge than controls exposed to air, suggesting a defect in antibody responses to the toxin. Van Loveren *et al.* (1988) investigated the effect of a 7-day exposure to up to 1 ppm O_3 on resistance to intratracheal infection with *L. monocytogenes.* They found that O_3 exposure suppressed the development of cellular immune responses to the organism, which was associated with reduced clearance of the bacteria from the lung and the spleen. In addition, the histological changes (inflammatory responses) associated with the infection were markedly enhanced in O_3-exposed animals, and all the effects were most apparent when O_3 exposure was conducted during a primary infection. Mice infected with influenza virus after, or during O_3

exposure, also developed increased lesions compared to controls (Jakab and Basset, 1990), and experienced greater mortality to virus infection (Selgrade *et al.*, 1988). Th2 responses appear to have a primary role in the pulmonary pathology associated with influenza infection, whereas Th1 responses are thought to be protective (Graham *et al.*, 1994).

These studies suggest that air pollutant exposure can reduce specific immune responses to pathogens and enhance susceptibility to infection. At present, however, it is not clear which point in the development of protective immunity is most affected. The generation of specific antibody, for example, only arises after antigen processing and presentation to T lymphocytes by accessory cells, notably macrophages and dendritic cells. Since it is known that air pollutant exposure suppresses AM phagocytosis, this could result in decreased antigen processing. Furthermore, dendritic cells exist in a contiguous network throughout the airway epithelium in both rats and humans (Holt *et al.*, 1990) and may be directly affected by oxidant exposure, or indirectly, via immunomodulatory mediators produced by epithelial cells in the microenvironment. Analysis of these primary signals and their effect on the subsequent generation of immunity will provide valuable mechanistic information on how air pollutant exposure can increase susceptibility to infectious agents.

Air pollutants as adjuvants for allergic lung disease

A number of epidemiological studies have found that the incidence and severity of asthma attacks increase as air quality deteriorates. While many of these effects are thought to be a result of direct irritation of already diseased tissue (e.g. in asthmatic airways), a number of animal studies have also demonstrated that air pollutants can act as adjuvants to enhance allergic IgE antibody production and bronchial reactivity to pharmacological agents. Although many of these animal models are criticized because of their lack of chronic disease, they do exhibit many features of acute exacerbations seen in allergic asthmatics such as antigen-induced bronchoconstriction, mucus production and eosinophilic inflammation of the airways.

Pulmonary allergy is dependent upon a number of host susceptibility factors, and interactions between antigens, accessory (antigen presenting) cells, immune cells and mediators. Immediate type hypersensitivity is caused by allergen specific IgE antibodies, which bind to $Fc\epsilon$ receptors on the surface of mast cells in the airways. Upon re-exposure to allergen, antigenic molecules crosslink immunoglobulin receptor sites, causing mast cell degranulation and release of vasoactive amines and other mediators responsible for inflammation and bronchoconstriction (reviewed by Daniele, 1988). These allergens are also recognized as being common triggers in asthma attacks (Carswell, 1988; Schou and Lind, 1991). Allergic alveolitis and delayed type hypersensitivity reactions are caused by IgG/antigen complexes and T lymphocytes respectively, and result in cellular activation, inflammation, lung remodelling and granuloma formation. A central component of all forms of pulmonary allergy is that they are immune-mediated and acquired following natural exposure to antigen. The reaction times following antigen exposure differ however, such that hay fever type reactions (type I hypersensitivity responses) occur almost immediately after allergen contact while later immune-complex and cell-mediated reactions occur between 4 and 48 h post-allergen exposure and have a more chronic nature. Allergic (extrinsic) asthmatics usually display immediate hypersensitivity reactions as well as late phase responses which are associated with pulmonary inflammation and lung remodelling (Danielle, 1988).

Most experimental evidence for pollutant-enhanced allergy has come from studies of O_3, sulphur dioxide (SO_2) and NO_2 and all have been conducted using type I (immediate) allergic reactions in animals and humans (reviewed in Gilmour, 1995). In 1970, Matsumura reported that guinea pigs exposed to very high levels of O_3 (5–10 ppm), NO_2 (70 ppm) or SO_2 (330 ppm) for 30 min prior to airway sensitization with ovalbumin antigen had elevated specific antibody titres and increased anaphylactic responses following challenge with aerosolized antigen. Later studies in mice (Osebold *et al.*, 1979, 1980), rats (Gilmour *et al.*, 1996), guinea pigs (Reidel, 1988) and in one case, monkeys (Biagini, 1986), confirmed these results and were then extended to show that diesel exhaust particles (DEP) have an adjuvant effect on IgE production. In these latter experiments, mice immunized intraperitoneally (Muranaka, 1986) or intranasally (Takafuji *et al.*, 1987) with ovalbumin mixed with DEP produced elevated levels of antigen-specific IgE compared with controls. Lymphocytes from lung lymph nodes of mice given intratracheal DEP and ovalbumin antigen had increased proliferative responses to antigen which were associated with increased IL4 production and IgE antibody in culture supernates and sera, respectively.

In a model of immune-mediated pulmonary hypersensitivity developed in our laboratory, Brown Norway rats immunized with house dust mite allergen produce high levels of specific IgG and IgE antibody compared with controls (Gilmour and Selgrade, 1996). Intratracheal antigen challenge causes a rapid bronchoconstriction, a large eosinophilic inflammatory response, and increases in mucous producing goblet cells in the airways. At the peak of inflammation (2 days post-challenge), the rats have increased airway responses to acetylcholine, indicating airway hyperresponsiveness. In order to determine whether air pollutant exposure could alter immune function and subsequent clinical symptoms and pathology, rats were exposed to 5 ppm NO_2 prior to both the sensitization and challenge procedures (Gilmour *et al.*, 1996). Pollutant exposure resulted in significantly higher levels of dust mite specific serum IgE and BAL IgA, IgE and IgG antibody, increased numbers of inflammatory cells in the lungs, and elevated lymphocyte responsiveness to antigen, compared to air-exposed controls. In later experiments, we examined whether intratracheal instillation of 1 mg residual oil fly ash (ROFA) 3 days before local sensitization and/or challenge could alter specific antibody levels, immune-mediated inflammation and allergen-induced airway hyperresponsiveness. ROFA instillation 3 days prior to sensitization resulted in increased levels of IgG and IgE antibody (Figure 13.3) but no differences in immune-mediated inflammation or airway hyperresponsiveness 1 day post-antigen challenge (Figures 13.4 and 13.5). ROFA treatment 3 days prior to challenge did not affect antibody levels (Figure 13.3) but enhanced pulmonary eosinophil numbers (Figure 13.4) and airway hyperresponsiveness (Figure 13.5) to intravenous acetylcholine compared to ROFA instilled rats which were not challenged with antigen. Exposure to ROFA prior to both sensitization and challenge procedures resulted in significant increases in all the immune, inflammatory and physiological parameters compared to saline instilled controls which received HDM or ROFA alone. The effects of ROFA have also been studied in a mouse model using ovalalbumin as the allergen. Gavett *et al.* (1996, 1997) demonstrated that intratracheal instillation of ROFA in ovalbumin-sensitized and challenged mice enhanced eosinophil recruitment and increased airway reactivity to methacholine challenge. IL-4 levels in BAL fluid from ROFA treated mice were three-fold greater than that of OVA-sensitized animals treated with vehicle. The data from these experiments show that exposure to NO_2 or ROFA at the sensitization period enhances the development of allergic immune responses, while exposure to pollutants at the challenge phase augments immune-mediated inflammation and airway hyperresponsiveness.

Controlled clinical studies have confirmed that exposure to O_3 and NO_2 increased pulmonary inflammation and airway responses after allergen challenge in allergic individuals. Molfino *et al.* (1995) and Tunnicliffe *et al.* (1996) reported increased responsiveness to allergen in allergic asthmatics exposed to O_3 and NO_2, respectively. Peden *et al.* (1996) demonstrated that allergic asthmatics exposed to O_3 produced a greater nasal eosinophil response than if they were exposed to clean air. Recent human experiments have also demonstrated that intranasal challenge with DEP increased expression and production of IgE antibody in allergic asthmatics (Diaz-Sanchez *et al.*, 1994). Peripheral monocytes and B lymphocytes from non-allergic human donors could also be stimulated *in vitro* by extracts of DEP to produce more IgE than vehicle-exposed control cells (Takenaka *et al.*, 1995).

Taken together, these animal and human studies demonstrate that a number of gaseous and particulate air pollutants can act as adjuvants to promote allergic (Th2-type) immune responses in the lung. Furthermore, air pollutants can act as irritants to exacerbate allergic reactions in individuals with pre-existing disease such as asthma.

Conclusions and future research needs

In summary, it is clear that oxidant air pollutants modulate immune responses both in the lung and systemically. These effects include modulation of T lymphocyte numbers and function in BALT and mediastinal lymph nodes and elevation of specific IgE antibody, but suppression of primary antibody

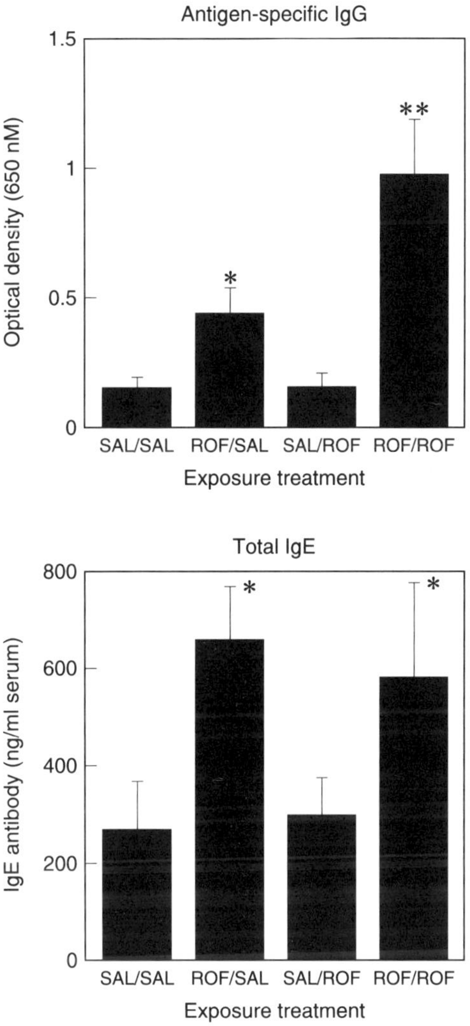

Figure 13.3 Antigen specific IgG and total IgE levels in the serum of rats exposed to 1 mg residual oil fly ash (ROFA) 3 days before allergic sensitization (ROF/SAL) with house dust mite (HDM), 3 days before allergen challenge (SAL/ROF), or 3 days before both sensitization and challenge (ROF/ROF). * and ** indicate significant ($p < 0.05$) differences from saline controls (SAL/SAL) and between treatments, respectively.

responses to sheep red blood cells and DTH responses. These changes have been attributed to alterations in lymphocyte trafficking, stress hormones, and the production of cytokines and other mediators which can influence immune regulation and function. Whereas many of the early studies showed that air pollutants decreased immune responses, these endpoints are now recognized as

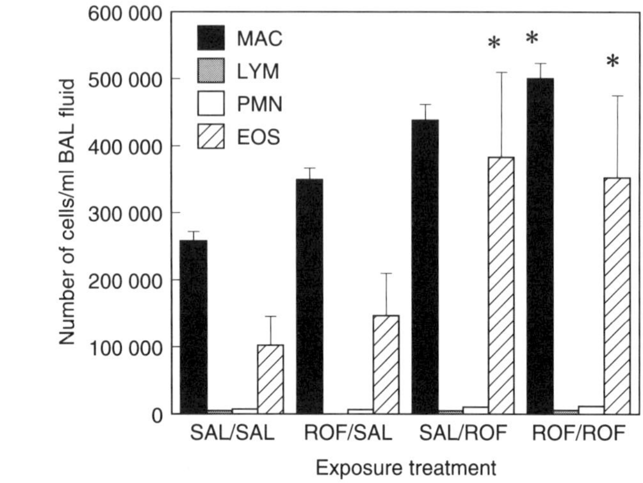

Figure 13.4 Number of pulmonary inflammatory cells (alveolar macrophages [MAC], lymphocytes [LYM], neutrophils [PMN] and eosinophils [EOS]) in the bronchoalveolar lavage fluid (BAL) of rats exposed to 1 mg residual oil fly ash (ROFA) 3 days before allergic sensitization (ROF/SAL) with house dust mite (HDM), 3 days before allergen challenge (SAL/ROF), or 3 days before both sensitization and challenge (ROF/ROF). * indicates significant ($p < 0.05$) differences from saline controls (SAL/SAL).

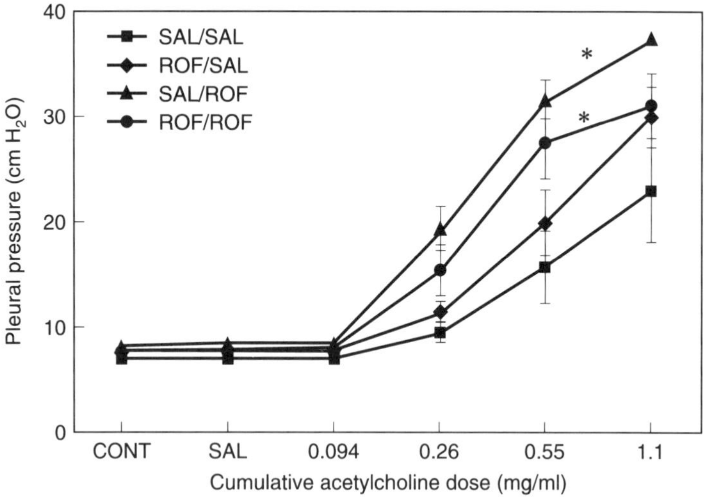

Figure 13.5 Pleural pressure measurements following stepwise increases of intravenous acetylcholine in rats exposed to 1 mg residual oil fly ash (ROFA) 3 days before allergic sensitization (ROF/SAL) with house dust mite (HDM), 3 days before allergen challenge (SAL/ROF), or 3 days before both sensitization and challenge (ROF/ROF). * indicates significant ($p < 0.05$) differences from saline controls (SAL/SAL).

being Th1 in nature. It is now becoming clear that this suppression may coincide with an upswing in Th2 cell function with subsequent increased IgE responses and enhanced immediate hypersensitivity reactions. This paradigm provides a rationale by which air pollutants can simultaneously enhance both allergic and infectious disease despite the apparent contradiction of immunosuppression and immuno-enhancement. The hypothesis that O_3, NO_2 and other lower respiratory tract irritants create an environment that favours the development of Th2 responses requires further study, perhaps using intracellular cytokine markers directly to identify Th cell sub-populations. Future studies should examine the effect of pollutant exposure on local (pulmonary) and systemic cytokine responses and changes in T cell subpopulations in response to antigenic challenge. Attention should be paid to the exposure regime and the route and nature of antigen administration. Analysis of mediators released by the respiratory epithelium will provide insights into the mechanisms associated with pollutant induced changes in immune function, and improve our understanding of the effects of air pollutants on pulmonary and systemic disease. For now, there is certainly a great deal of suggestive and consistent evidence to indicate that oxidant air pollutants tend to suppress Th1 responses and enhance Th2 responses.

Disclaimer

This chapter has been reviewed by the Health Effects Research Laboratory, US Environmental Protection Agency and approved for publication. Approval does not signify that the contents necessarily reflect the views and policies of the agency, nor does mention of trade names or commercial products constitute endorsement or recommendation for use. M.I. Gilmour is supported by Cooperative Agreement CR817643 with the US Environmental Protection Agency.

References

Azoulay-Dupuis, E., Levacher, M., Muffat-Joly, M. and Pocidalo, J.J. (1985) Humoral immunodepression following acute NO_2 exposure in normal and adrenalectomized mice. *J. Toxicol. Environ. Health*, **15**, 149–62.

Azoulay-Dupuis, E., Gougerot-Picidalo, M.A., Kraus, L. and Moreau, J. (1987) *In vivo* nitrogen dioxide exposure depresses spleen cell *in vitro* mitogenic responses: Effects of sulfur compounds. *Environ. Res.*, **42**, 130–9.

Barnes, P.J. and Liew, F.Y. (1995) Nitric oxide and asthmatic inflammation. *Immunol. Today*, **16**, 128–30.

Bates, D.V. and Sixto, (1987) Hospital admissions and air pollutants in Southern Ontario: The acid summer haze effect. *Environ. Res.*, **43**, 317–31.

Becker, S., Jordan, R.L., Orlando, G.S. and Koren, H.S. (1989) *In vitro* ozone exposure inhibits mitogen-induced lymphocyte proliferation and IL2 production. *J. Toxicol. Environ. Health*, **24**, 469–83.

Betz, M. and Fox, B.S. (1991) Prostaglandin E2 inhibits production of Th1 lymphokines but not of Th2 lymphokines. *J. Immunol.*, **146**, 108–13.

Biagini, R.E., Moorman, W.J., Lewis, T.R. and Bernstein, I.L. (1986) Ozone enhancement of platinum asthma in a primate model. *Am. Rev. Respir. Dis.*, **134**, 719–25.

Bleavins, M.R. and Dziedzic, D. (1990) An immunofluorescence study of T and B lymphocytes in ozone-induced pulmonary lesions in the mouse. *Toxicol. Appl. Pharmacol.*, **105**, 93–102.

Burleson, G.R., Keyes, L.L. and Stutzman, J.D. (1989) Immunosuppression of pulmonary natural killer activity by exposure to ozone. *Immunopharmacol. Immunotoxicol.*, **11**, 715–35.

Campbell, K.I. and Hilsenroth, R.H. (1976) Impaired resistance to toxin in toxoid-immunized mice exposed to ozone or nitrogen dioxide. *Clin. Toxicol.*, **9**, 943–54.

Canning, B.J., Hmieleski, R.R., Spannhake, E.W. and Jakab, G.J. (1991) Ozone reduces murine alveolar and peritoneal macrophage phagocytosis: the role of prostanoids. *Am. J. Physiol.*, 261 (*Lung Cell. Mol. Physiol.*, **5**): L227–l282.

Carswell, F. (1988) State of the art: mites and human allergy. *Immunology*, **65**, 497–500.

Cody, R.P., Weisel, C.P., Birnbaum, G. and Lioy, P.J. (1992) The effect of ozone associated with summertime photochemical smog and the frequency of asthma visits to hospital emergency departments. *Environ. Res.*, **58**, 184–94.

Daniele, R.P. (1988) *Immunology and Immunologic Diseases of the Lung*. Blackwell Scientific, Boston.

Devlin, R.B., McDonnell, W.F., Mann, R., Becker, S., House, D.E., Schreinemachers, D. and Koren, H.S. (1991a) Exposure of humans to ambient levels of ozone for 6.6 hours causes cellular and biochemical changes in the lung. *Am. J. Resp. Cell. Mol. Biol.*, **4**, 72–81.

Devlin, R.B., McDonnell, W.F., Perez, R., Becker, S., House, D.E. and Koren, H.S. (1991b) Prolonged exposure of humans to ambient levels of ozone causes cellular and biochemical changes in the lung. *Am. Rev. Resp. Dis.*, **4**, 72–81.

Devlin, R.B., Raub, J.A. and Folinsbee, L.J. (1997) Health effects of ozone. *Sci. Med.*, **4**, 8–17.

Diaz-Sanchez, D., Dotson, A.R., Takeaka, H. and Saxon, A. (1994) Diesel exhaust particles induce local IgE production *in vivo* and alter the pattern of messenger RNA isoforms. *J. Clin. Invest.*, **94**, 1417–25.

Driscoll, K.E., Vollmuth, T.A. and Schlesinger, R.B. (1987) Acute and subchronic ozone inhalation in the rabbit: response of the alveolar macrophages. *J. Toxicol. Environ. Health*, **21**, 27–43.

Dzeidzic, D. and White, H.J. (1986a) Thymus and pulmonary lymph node response to acute and subchronic ozone exposure in the mouse. *Environ. Res.*, **41**, 598–609.

Dzeidzic, D. and White, H.J. (1986b) T cell activation in pulmonary lymph nodes of mice exposed to ozone. *Environ. Res.*, **41**, 610–22.

Dzeidzic, D. and White, H.J. (1987a) Response of T-cell deficient mice to ozone exposure. *J. Toxicol. Environ. Health*, **21**, 57–71.

Dzeidzic, D. and White, H.J. (1987b) Quantitation of ozone-induced lung lesion density after treatment with an interferon inducer or an anti-interferon antibody. *Toxicol. Lett.*, **39**, 51–62.

Dzeidzic, D., Wright, E.S. and Sargen, N.E. (1990) Pulmonary response to ozone: reaction of bronchus associated lymphoid tissue and lymph node lymphocytes in the rat. *Environ. Res.*, **51**, 194–208.

Edelman, A.S. and Zolla-Pazner, S. (1989) AIDS: a syndrome of immune dysregulation, dysfunction, and deficiency. *FASEB J.*, **3**, 22–30.

Ehrlich, R., Silverstein, E., Maigetter, R. and Fenters, J.F. (1975) Immunologic response in vaccinated mice during long-term exposure to nitrogen dioxide. *Environ. Res.*, **10**, 217–23.

Farmer, S.G. and Hay, D.W.P. (1991) *The Airway Epithelium: Physiology, Pathophysiology and Pharmacology*. Marcel Dekker, New York.

Fearon, D.L. and Locksley, R.M. (1996) The instructive role of innate immunity in the acquired immune response. *Science*, **272**, 50–6.

Fujimaki, H. (1989) Impairment of humoral immune responses in mice exposed to nitrogen dioxide and ozone mixtures. *Environ. Res.*, **48**, 211–17.

Fujimaki, H., Shimizu, F. and Jubota, K. (1981) Suppression of antibody response in mice by acute exposure to nitrogen dioxide: *in vitro* study. *Environ. Res.*, **26**, 490–6.

Fujimaki, H., Shimizu, F. and Jubota, K. (1982) Effect of subacute exposure to NO_2 on lymphocytes required for antibody responses. *Environ. Res.* **29**, 280–286.

Fujimaki, H., Oawa, M., Imai, T. and Shimizu, F. (1984) Effect of short-term exposure to ozone on antibody response in mice. *Environ. Res.* **35**, 490–96.

Fujimaki, H., Shiraishi, F., Ashikawa, T. and Murikami, M. (1987) Changes in delayed type hypersensitivity reaction in mice exposed to ozone. *Environ. Res.*, **43**, 186–90.

Gajewski, F.F. and Fitch, F.W. (1988) Anti-proliferative effect of IFN γ in immune regulation. *J. Immunol.*, **140**, 4245–51.

Gardner, D.E. (1982) Effects of airborne gases and airborne particles on lung infections, in *Air Pollution: Physiological Effects* (eds J.J. McGrath and C.D. Barnes), Academic, New York, pp. 48–79.

Gavett, S.H., Madison, S.L. and Costa, D.L. (1996) Allergen challenge amplifies pulmonary responses to residual oil fly ash and ambient air particulate matter samples in BALB/cj mice. *Am. J. Crit. Care Med.*, **153**, A15.

Gavett, S.H., Madison, S.L., Ledbetter, A.D. and Costa, D.L. (1997) Increased levels of interleukin-4 are associated with enhanced pulmonary responses to inhaled residual oil fly ash in allergen challenged BALB/cj mice. *Am. J. Crit. Care Med.*, **155**, A244.

Gilmour, M.I. (1995) Interaction of air pollutants and pulmonary allergic responses in experimental animals. *Toxicology*, **105**, 335–42.

Gilmour, M.I. and Selgrade, M.J.K. (1996) A model of immune-mediated lung disease in rats sensitized to house dust mite. *Chest*, **109**, 69S.

Gilmour, M.I., Hmieleski, R.R., Stafford, E.A. and Jakab, G.J. (1991) Suppression and recovery of the alveolar macrophage phagocytic system during continuous exposure to 0.5 ppm ozone. *Exp. Lung Res.*, **17**, 547–58.

Gilmour, M.I., Park, P., Doefler, D. and Selgrade, M.J.K. (1993) Factors that influence the suppression of pulmonary anti-bacterial defenses in mice exposed to ozone. *Exp. Lung Res.*, **19**, 299–314.

Gilmour, M.I., Park, P. and Selgrade, M.J.K. (1996) Increased immune and inflammatory responses to dust mite antigen in rats exposed to 5 ppm nitrogen dioxide. *Toxicol. Appl. Pharmacol.*, **135**, 65–70.

Goldstein, E., Tyler, W.S., Hoeprich, P.D. and Eagle, C. (1971) Ozone and the antibacterial defense mechanisms of the murine lung. *Arch. Int. Med.*, **127**, 1099–102.

Graham, M.B., Braciale, V.L. and Braciale, T.J. (1994) Influenza virus-specific $CD4^+$ T helper type 2 T lymphocytes do not promote recovery from experimental virus infection. *J. Exp. Med.*, **180**, 1273–82.

Hazucha, M.J., Madden, M., Pape, G., Becker, S., Devlin, R.B., Koren, H.S., Kehrl, H. and Bromberg, P.A. (1996) Effects of cyclo-oxygenase inhibition on ozone-induced respiratory inflammation and lung function changes. *Eur. J. Appl. Physiol.*, **73**, 17–27.

Hilkens, C.M., Snijders, A., Vermeulen, H., van der Meide, P.H., Wierenga, E.A. and Kapsenberg, M.L. (1996) Accessory cell-derived IL-12 and prostaglandin E2 determine the IFN-γ level of activated human $CD4^+$ T cells. *J. Immunol.*, **156**, 1722–7.

Holt, P.G., Finlay-Jones, L.M., Keast, D. and Papadimitrou, J.M. (1979) Immunological function in mice chronically exposed to nitrogen oxides (NOx). *Environ. Res.*, **19**, 154–62.

Holt, P.G., Schon-Hegrad, M.A., Oliver, J., Holt, B.J. and McMenamin, P.G. (1990) A contiguous network of dendritic antigen presenting cells within the respiratory epithelium. *Int. Arch. Allergy Appl. Immunol.*, **91**, 155–9.

Jakab, G.J. and Basset, D.J.P. (1990). Influenza virus infection, ozone exposure and fibrogenesis. *Am. Rev. Respir. Dis.*, **141**, 1307–15.

Jakab, G.J., Spannhake, E.W., Kleeberger, S.R., Canning, B.J. and Gilmour, M.I. (1995) The effects of ozone on immune function. *Environ. Health Perspect.*, **103**, 77–89.

Kalyanasunundanam, R., Negrao-Correa, D. and Bel, R. (1996) Local intestinal immune responses to infections with *Trichinella spiralis*. *J. Immunol.*, **156**, 4328–37.

Koren, H.S., Devlin, R.B., Graham, D.E., Mann, R., McGee, M.P., Horstman, D.H., Kozumbo, W.S., Becker, S., House, D.E., McDonnell, W.F. and Bromberg, P.A. (1989) Ozone induced inflammation in the lower airways of human subjects *Am. Rev. Resp. Dis.*, **139**, 404–15.

Koren, H.S., Devlin, R.B. and Becker, S. (1991) Ozone-induced inflammatory response in pulmonary cells, in *Xenobiotics and Inflammation: Roles of Cytokines and Growth Factors* (eds L.B. Schook and D.L. Laskin), Academic Press, New York, pp. 249–81.

Lefkowitz, S.S., McGrath, J.J. and Lefkowitz, D.L. (1986) Effects of NO_2 on immune responses. *J. Toxicol. Environ. Health*, **17**, 241–8.

Leikauf, G.D., Driscoll, K.E. and Wey, H.E. (1988) Ozone-induced augmentation of eicosanoid metabolism in epithelial cells from bovine trachea. *Am. Rev. Resp. Dis.*, **137**, 435–42.

Li, A.F. and Richters, A. (1991) Effects of 0.7 ppm ozone exposure on thymocytes *in vivo* and *in vitro* studies. *Inhalation Toxicology*, **3**, 61–71.

Maigetter, R.Z., Fenters, J.D., Findlay, J.C., Ehrlich, R. and Gardner, D.E. (1978) Effect of exposure to nitrogen dioxide on T and B cells in mouse spleens. *Toxicol. Lett.*, **2**, 157–61.

Matsumara, Y. (1970) The effects of ozone, nitrogen dioxide and sulfur dioxide on the experimentally induced allergic respiratory disorder in guinea pigs. I. The effect of sensitization with albumin through the airway. *Am. Rev. Respir. Dis.*, **102**, 430–7.

McKinnon, K., Madden, M.C., Noah, T.L. and Devlin, R.B. (1992) *In vitro* ozone exposure increases release of arachidonic acid products from a human bronchial epithelial cell line. *J. Toxicol. Appl. Pharmacol.*, **118**, 215–23.

Molfino, N.A., Wright, S.C., Katz, I., Tarlo, S., Silverman, F., McClean, P.A., Szalai, J.P., Raizenne, M., Slutsky, A.S. and Zamel, N. (1991) Effect of low concentrations of ozone on inhaled allergen responses in asthmatic subjects. *Lancet*, **338**, 199–203.

Morrow, P.E. (1992) Dust overloading of the lungs: update and appraisal. *Toxicol. Appl. Pharmacol.*, **113**, 1–12

Mosmann, T.R. and Coffman, R.L. (1987) Two types of mouse T helper clone: implication for immune regulation. *Immunol. Today*, **8**, 233–6.

Mosmann, T.R. and Sad, S. (1996) The expanding universe of T-cell subsets: Th1, Th2 and more. *Immunol. Today*, **17**, 138–46.

Muranaka, M., Suzuki, S., Koizumi, K., Takafuji, S., Miuamoto, T., Ikemori, R. and Tokiwa, H. (1986) Adjuvant activity of diesel exhaust particulates for the production of IgE antibody in mice. *J. Allergy Clin. Immunol.*, **77**, 616–23.

Oosting, R.S., Van Golde, L.M.G., Verhoef, J. and Van Bree, L. (1991) Species differences in impairment of alveolar macrophage functions following single and repeated ozone exposures. *Toxicol. Appl. Pharmacol.*, **110**, 170–8.

Orlando, G.S., House, D., Daniel, E.G., Koren, H.S. and Becker, S. (1988) Effect of ozone on T cell proliferation and serum levels of beta-endorphins in exercising males. *Inhal. Toxicol.*, **1**, 53–63.
Osebold, J.W., Owens, S.L., Zee, Y.C., Dotson, W.M. and LaBarre, D.D. (1979) Immunological alterations in the lungs of mice following ozone exposure. *Arch. Environ. Health*, **34**, 258–65.
Osebold, J.W., Gershwin, L.J. and Zee, Y.C. (1980) Studies on the enhancement of allergic lung sensitization by inhalation of ozone and sulfuric acid aerosol. *J. Environ. Pathol. Toxicol.*, **3**, 221–34.
Paul, W.E. and Seder, R.A. (1994) Lymphocyte response and cytokines. *Cell*, **76**, 241–51.
Peden, D.B. (1996). Effect of air pollution in asthma and respiratory allergy. *Otolaryngol. Head Neck Surg.*, **114**, 242–7.
Pendino, K.J., Laskin, J.S., Shuler, R.L., Punjabi, C.J. and Laskin, D.L. (1993) Enhanced production of nitric oxide by rat alveolar macrophages after inhalation of a pulmonary irritant is associated with increased expression of nitric oxide synthase. *J. Immunol.*, **151**, 7196–205.
Pendino, K.J., Shuler, R.L., Laskin, J.D. and Laskin, D.L. (1994) Enhanced production of interleukin-1, tumor necrosis factor-α, and fibronectin by rat lung phagocytes following inhalation of a pulmonary irritant. *Am. J. Respir. Cell Mol. Biol.*, **11**, 279–86.
Pope, C.A., Dockery, D.W. and Schwartz, J. (1995) Review of epidemiological evidence of health effects of particulate air pollution. *Inhal. Toxicol.*, **7**, 1–18.
Punjabi, C.J., Laskin, J.D., Pendino, K.J., Goller, N.L., Durham, S.K. and Laskin, D.L. (1994) Production of nitric oxide by rat type II pneumocytes: increased expression of inducible nitric oxide synthase following inhalation of pulmonary irritant. *Am. J. Respir. Cell. Mol. Biol.* **11**, 165–72.
Reidel, F., Kramer, M., Scheibenbogen, C. and Reiger, C.H.L. (1988) Effects of SO_2 exposure on allergic sensitization in the guinea pig. *J. Allergy Clin. Immunol.*, **82**, 527–34.
Richters, A. and Damji, K.S. (1988) Changes in T-lymphocyte subpopulations and natural killer cells following exposure to ambient levels of nitrogen dioxide. *J. Toxicol. Environ. Health*, **25**, 247–56.
Rincon, M., Anguita, J., Nakumara, T., Fikrig, E. and Flavell, R.A. (1997) Interleukin 6 directs the differentiation of $CD4^+$ T cells. *J. Exp. Med.*, **185**, 461–9.
Robinson, D.S., Hamid, Q., Ying, S., Tsicopoulos, A., Barkans, J., Bentley, A.M., Corrigan, C., Durham, S.R. and Kay, A.B. (1992) Predominant Th2-like bronchoalveolar T lymphocyte population in atopic asthma. *N. Engl. J. Med.*, **326**, 298–304.
Schauer, U., Jung, T., Krug, N. and Frew, A. (1996) Measurement of intracellular cytokines. *Immunol. Today*, **17**, 305–6.
Schou, C. and Lind, P. (1991) The antigenicity of house dust mites. *Allergy*, **46**, 10–13.
Selgrade, M.J.K. and Gilmour, M.I. (1994) Effects of gaseous air pollutants on immune responses and susceptibility to infectious and allergic disease, in *Immunotoxicology and Immunopharmacology* (eds J.H. Dean, M.I. Luster, A.E. Munson and I. Kimber), Raven Press, New York, pp. 395–411.
Selgrade, M.J.K., Illing, J.W., Starnes, D.M., Stead, A.G., Menache, M.G. and Stevens, M.A. (1988) Evaluation of effects of ozone exposure on influenza infection in mice using several indicators of susceptibility. *Fundam. Appl. Toxicol.*, **11**, 169–80.
Snijdewint, F.G.M., Kalinski, P., Wierenga, E.A., Bos, J.D. and Kapsenberg, M.L. (1993) Prostaglandin E2 differentially modulates cytokine secretion profiles of human T helper lymphocytes. *J. Immunol.*, **150**, 5321–9.
Swain, S.L. (1995) Who does the polarizing? *Curr. Biol.*, **5**, 849–51.

Takafuji, S., Suzuki, S., Koizumi, K., Tadokoro, K., Miyamoto, T., Ikemori, R. and Muranaka, M. (1987) Diesel exhaust particulates inoculated by the intranasal route have an adjuvant activity for IgE production in mice. *J. Allergy Clin. Immunol.*, **79**, 639–45.

Takenaka, H., Zhang, K., Diaz-Sanchez, D., Tsien, A. and Saxon, A. (1995) Enhanced human IgE production results from exposure to the aromatic hydrocarbons from diesel exhaust: direct effects on B cell IgE production. *J. Allergy Clin. Immunol.*, **95**, 103–15.

Trichieri, G. and Gerosa, F. (1996) Immunoregulation by interleukin-12. *J. Leukocyte Biol.*, **59**, 505–11.

Tunnicliffe, W.S., Burge, P.S. and Ayres, P.S. (1994) Effect of domestic concentrations of nitrogen dioxide on airway responses to inhaled allergen in asthmatic patients. *Lancet*, **344**, 1668–71.

Van Loveren, H., Rombout, P.J.A., Wagenaar, S.S., Valvoort, H.C. and Vos, J.G. (1988) Effects of ozone on the defense to a respiratory Listeria monocytogenes infection in the rat. *Toxicol. Appl. Pharmacol.*, **94**, 374–93.

Van Loveren, H., Krajnc, E.I., Rombout, P.J.A., Blommaert, F.A. and Vos, J.G. (1990) Effects of ozone, hexachorobenzene, and bis (tri-*n*-butyltin) oxide on natural killer activity in the rat lung. *Toxicol. Appl. Pharmaco*l., **102**, 21–33.

Waldbott, G.L. (1972) Air pollutants as a cause of conditions which simulate allergic respiratory disease. *Ann. Allergy*, **30**, 460–3.

14 T Lymphocyte Subpopulations in Autoimmune Responses to Chemicals and Drugs

David B.G. Oliveira and J.W. Coleman

Drug-induced autoimmunity – introduction

Autoimmune disease results from breakdown of immunological tolerance to self. As a consequence of this disruption, the immune system mounts an attack against the body's own cells, tissues and plasma proteins. This can in turn lead to inflammation, tissue damage and associated pathologies. The causes of autoimmunity are complex and various, but it is becoming increasingly recognized that exposure to drugs and environmental chemicals is a major risk factor in genetically predisposed individuals and certain strains of animals. In fact current theories of the causes of autoimmunity place much emphasis on the concept that self antigens may become modified, for example by chemicals, such that they then appear to the immune system as foreign or 'neo-self'. The maintenance of immunological tolerance is largely achieved via clonal deletion of lymphocytes that recognize self antigens. This process operates during early T cell differentiation in the thymus (Kappler *et al.*, 1987) and B cell differentiation in germinal centres of lymphoid tissues (Joshua *et al.*, 1989). However, deletion of T and B lymphocytes that recognize self is not absolute and these cells are present at low numbers in the blood and lymphoid tissues of healthy individuals. They can undergo clonal expansion and differentiation following appropriate stimulation, and this culminates in autoantibody production resulting possibly in symptoms of autoimmune disease.

Autoimmune reactions vary considerably in their aetiology and in the target tissues that are affected. This is true also for chemical-induced autoimmunity, in which reactions may be tissue specific or systemic and different types of reaction are associated with different inducing compounds. Tissue specific reactions include quinidine-induced thrombocytopenia, penicillin or methyl-

T Lymphocyte Subpopulations in Immunotoxicology. Edited by Ian Kimber and MaryJane K. Selgrade.
 ISBN 0 471 97194 4

dopa-induced haemolytic anaemia and severe halothane-induced hepatitis (Kenna, 1991; Lerner *et al.*, 1985; Petz and Branch, 1973; Pfueller *et al.*, 1988; Vistentin *et al.*, 1991; Worlledge, 1973). Systemic reactions that often resemble idiopathic systemic lupus erythematosus (SLE), can be induced by the drugs procainamide, hydralazine, phenytoin, methimazole, succinimides, carbamazapine, chlorpromazine, sulphonamides and D-penicillamine (Kitteringham *et al.*, 1988; Uetrecht, 1989; Rubin, 1994) and heavy metals such as gold and mercury (Druet *et al.*, 1978; Hirsch *et al.*, 1982; Schuhmann *et al.*, 1990; Tournade *et al.*, 1990). The distinction between tissue-specific and systemic reactions is not absolute, since in the latter, certain organs and tissues may be affected selectively.

How do drugs and chemicals break T cell-dependent immunological tolerance?

There is little evidence to suggest that drugs and chemicals that cause autoimmunity do so by acting directly on B cells. Rather, a substantial body of evidence implicates strongly the T cell as the primary target. Such T cell activation could occur through specific recognition of neo-antigen, or by direct non-specific pharmacological activation. Before considering in more detail the various theories that have been proposed to explain chemical-induced autoimmunity, it is worth considering how T helper cells (Th cells) are activated by antigen. This process requires two signals: the first is provided via cognate recognition of the antigen presented to the T cell in association with MHC class II molecules on an antigen-presenting cell (e.g. dendritic cell or B cell). Following this cell–cell interaction, the CD28 protein molecule is expressed on the surface of the T cell and this binds to its counterstructure, B7.1 or B7.2, on the antigen presenting cell. This interaction provides the second signal to the Th cell which is now activated to secrete interleukins such as IL-2, IL-4, IL-12 and interferon-γ, the pattern of which determines the further direction and expansion of the immune response, for example towards cell-mediated or antibody-mediated immunity. Of particular relevance to chemical-induced autoimmunity is the fact that the second signal may be provided by alternative mechanisms, for example by direct action of the chemical on an interleukin (e.g. IL-4) producing cell. In this chapter, the mechanisms whereby chemicals may generate neo-antigens to provide the first signal and also stimulate production of interleukins to provide the second signal for T cell activation will be considered. Attention will focus on the importance of selective Th cell subset activation in determining the direction of the autoimmune response, and in particular on the Brown Norway rat as a model of mercury-induced autoimmunity. It is important to emphasize that within this conceptual framework the detailed mechanisms may vary considerably between different causative compounds. This should not be considered surprising, bearing in mind that the classes of drugs and chemicals that can cause autoimmune reactions are chemically and pharmacologically extremely heterogeneous, as is the pathology and immunological specificity of the disease activity.

T CELL RECOGNITION OF MODIFIED SELF LEADING TO SPECIFIC AUTOANTIBODY PRODUCTION

Allison and colleagues (Allison, 1989; Allison *et al.*, 1971) suggested that protein-reactive drugs and chemicals might break tolerance by binding to self macromolecules such as to generate a haptenic determinant (neo-antigen or neo-epitope) that could be recognized by T cells. T cells recognizing the neo-antigen would clonally expand and proceed to provide help for B cells that recognized adjacent autoantigens (unmodified self) on the same drug-self conjugate. These B cells in turn would clonally expand and differentiate into autoantibody producing plasma cells. In this way, the normal process of tolerance that operates through clonal or functional deletion of Th cells is effectively bypassed through clonal expansion of a hapten-specific population of Th cells. This process is illustrated in Figure 14.1. A considerable body of experimental evidence shows that haptenation of specific cells or proteins can selectively break tolerance to these targets (Weigle, 1965; Yamashita *et al.*, 1976). Thus, the Allison model is particularly attractive as an explanation for chemical-induced target-specific autoimmunity.

The Allison model is consistent with our understanding of the process of T cell-B cell co-operation in which T and B cells recognize adjacent epitopes on the same protein molecule (Mitchison, 1971). Conceptually, there is a shift in emphasis: in Mitchison's experiments the B cell epitope was a chemical hapten,

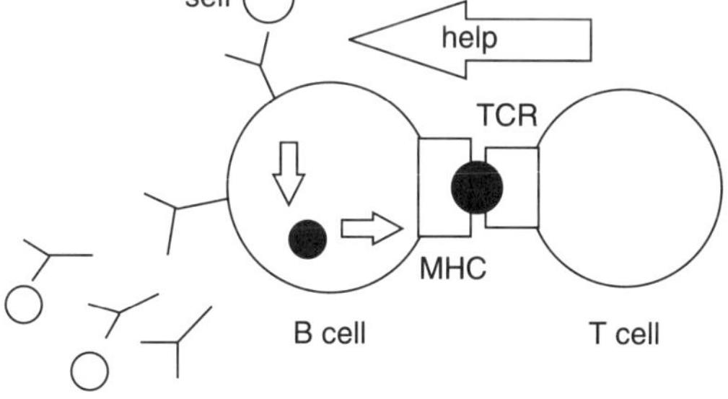

Figure 14.1 Diagram showing how haptenation of self protein can lead to induction of Th cells that then provide help for synthesis of antibodies directed at the self determinant. The open circles represent the unmodified self B cell determinant and the filled circles a haptenated (neo-self) T cell determinant. Small open arrows represent intracellular processing of the antigen. Based on ideas from Allison *et al.* (1971) and Allison (1989).

while the T cell epitope was a foreign protein molecule; in the Allison model the T cell determinant is the hapten, while the B cell determinant is self protein. Both models fit comfortably with current knowledge that B cells capture antigen via cell surface immunoglobulin receptors, process it intracellularly, and present it in conjunction with self MHC class II molecules to T cells (Lanzevecchia, 1985). As a result of this cognate recognition of antigen, the specific T cells and B cells provide costimulatory signals to one another that promote expansion of each population and lead ultimately to B cell antibody synthesis. Presumably, the generation of hapten sensitized Th cells during the primary immune response is dependent on processing of haptenated self by professional antigen presenting cells, such as dendritic cells.

T CELL RECOGNITION OF MODIFIED SELF MHC CLASS II – GVH DISEASE ANALOGY

A further theory of chemical-induced autoimmunity implicating T cell recognition of modified self was proposed by Gleichmann *et al.* (1984) by analogy with graft-versus-host (GVH) disease. In experimentally induced GVH disease in mice, in which grafted T lymphocytes mount a response against allogeneic histocompatibility antigens of the host, a spectrum of immunologically mediated pathological changes is seen. Generally, a Th cell response to MHC class II predominates, leading to production of autoantibodies against nuclear antigens, double stranded DNA, erythrocytes and thymocytes, and elevated IgE production (Gleichmann *et al.*, 1984; Rolink and Gleichmann, 1983; Rolink *et al.*, 1983; van Rappard-van der Veen *et al.*, 1982). In this system, recognition by donor Th cells of host B cells carrying 'foreign' MHC class II leads to polyclonal host B cell activation and uncontrolled antibody production. These observations led Gleichmann *et al.* (1984) to postulate that in chemical-induced autoimmunity the chemical interacts with self MHC class II molecules in such a way as to make them appear as non-self. Subsequent T cell activation would lead to indiscriminate help to all B cells carrying the drug-modified self MHC molecule. Assuming that the drug modifies MHC molecules without regard for the antigen specificity of the B cell, the resulting cognate T-B cell interaction would lead to polyclonal B cell activation. Synthesis of antibodies against antigens that provide a strong B cell stimulus is favoured – these are often antigens with repeating determinants, for example nucleic acids. Figure 14.2 shows schematically how this mechanism would operate.

Studies in rodents support the analogy drawn by Gleichmann and colleagues between GVH-induced and chemical-induced autoimmunity. Repeated injections of low doses of mercuric chloride (Druet *et al.*, 1978; Hirsch *et al.*, 1982), D-penicillamine (Tournade *et al.*, 1990) or gold salts (Schuhmann *et al.*, 1990) into the susceptible Brown Norway strain of rats leads to multiple T cell mediated immune disruptive changes resembling those seen in experimental

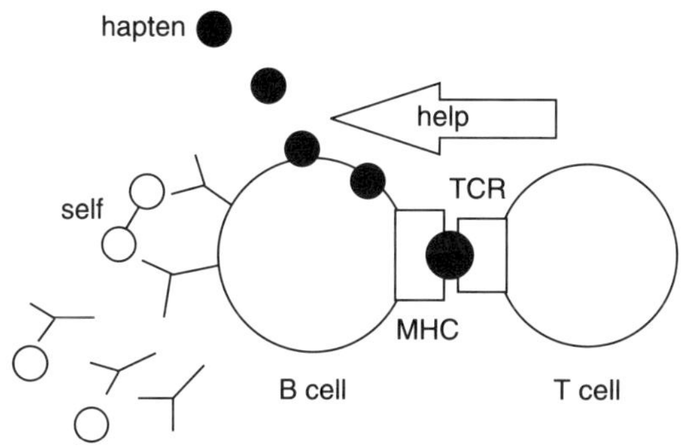

Figure 14.2 Diagram showing how haptenation of self MHC class II molecule or self MHC-associated peptide can lead to T cell activation and autoantibody production by B cells. Open circles represent self determinant and filled circles the modifying chemical. Based on ideas of Gleichmann *et al.* (1984).

GVH disease (see later in this chapter for full description). Furthermore, autoantibodies of shared idiotype can be detected in Brown Norway rats in which autoimmunity was induced by either GVH reactions, or by drugs (D-penicillamine or gold) or mercuric chloride (Guery *et al.*, 1990), suggesting that the same B cell clones are activated in chemical and GVH-induced disease. This supports the concept that both forms of autoimmunity have a common underlying aetiology.

INDUCTION OF T CELLS TO UNMODIFIED SELF

The models of Allison (Allison, 1989; Allison *et al.*, 1971) and Gleichmann *et al.* (1984) postulate T cell recognition of chemically modified self. However, T cells that respond to unmodified self can be induced experimentally *in vitro* by stimulation with autologous MHC class II antigen presenting cells, and are also present in high numbers in autoimmune-susceptible rat strains (Pelletier *et al.*, 1994). In the Brown Norway rat model of D-penicillamine, gold and mercury-induced autoimmunity, T cells that recognize unmodified self MHC class II molecules on normal B cells have been reported, rather than T cells that recognize chemically modified self (Pelletier *et al.*, 1986, 1988, 1991; Rossert *et al.*, 1988). This finding led Druet and colleagues to suggest that that mercury and other autoimmunogenic chemicals might induce directly T cells that recognize self MHC class II (Pelletier *et al.*, 1991, 1994). Thus, in contrast to the Gleichmann model, the induced T cells are anti-native self MHC rather than anti-modified self MHC. Nevertheless, induction of T cells by either mechanism would have exactly the same consequences, namely indiscriminate activation of MHC class II-positive B cells, accompanied by polyclonal antibody production, and again the response is analogous to that seen in GVH disease.

Anti-self T cell lines and clones have been derived from both human and rodent sources; these cells have diverse cytokine secreting profiles (Th1 or Th2 type) and helper/suppressor functions with regard to B cell activation (Clayberger *et al.*, 1984; Pelletier *et al.*, 1991, 1994; Quintans *et al.*, 1986). Figure 14.3 shows schematically how T cells that recognize self MHC class II might be induced by chemicals leading to B cell activation.

HOW CAN CHEMICALS INDUCE T CELL RESPONSES TO UNMODIFIED SELF?

Although it is not known with any certainty how chemicals might induce T cells that recognize self MHC class II, three theories are currently receiving interest:

(i) There could be some kind of pharmacological effect of the chemical, such as induction of supra-normal levels of self MHC class II expression or some other undefined change leading to disruption of the T cell network (Pelletier *et al.*, 1991). More specifically, certain drugs such as 5-azacytidine, hydralazine and procainamide induce methylation of DNA, and this could lead to derepression of gene expression, for example of MHC class II, leading to induction of autoimmunity (Cornacchia *et al.*, 1988; Richardson, 1986).

(ii) Induction of anti-self T cells might occur by the process of 'antigenic shift' or 'epitope spreading'. According to Sercarz and colleagues, immunological tolerance develops to dominant self determinants that are presented to T cells by normal processing but not to 'cryptic' self determinants that are not normally presented (Lehmann *et al.*,

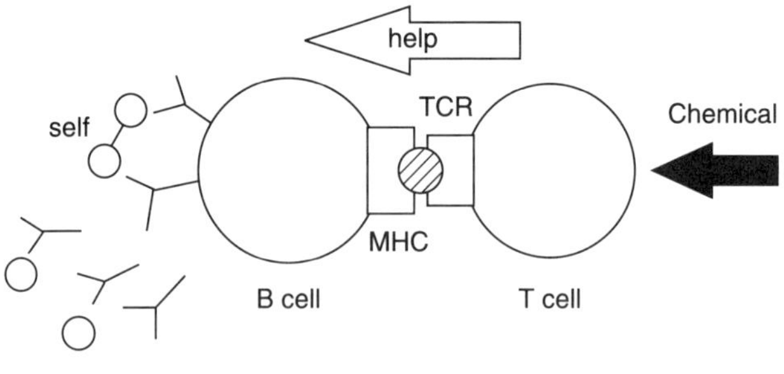

Figure 14.3 Diagram illustrating induction of autoreactive T cells. The chemical induces Th cells that recognize self MHC class II, perhaps by modification of self peptide. The T cells then provide help for anti-self B cells. Open circles represent self determinant. Hatched circle represents chemically modified self peptide occupying MHC molecule. Based on ideas of Pelletier *et al.* (1988, 1991) and Martin and Weltzien (1994).

1992, 1993; Moudgil and Sercarz, 1994). It is suggested that during infection or other perturbation of the immune system the processing of self proteins may be altered, for example by the presence of cytokines in the tissue microenvironment, such that normally cryptic peptides become processed and presented to non-deleted anti-self T cells that then mount a vigorous response. This type of epitope spreading from a dominant to a cryptic self determinant may be favoured when dominant epitopes on the foreign antigen are structurally homologous to cryptic self determinants (Moudgil and Sercarz, 1994). Evidence for epitope spreading in the progression of autoimmunity comes from the mouse model of experimental allergic encephalomyelitis in which immunization with dominant immunogenic peptides of myelin basic protein leads to progressive shifting of the T cell response towards adjacent cryptic peptide sequences (Lehmann *et al.*, 1992). Similarly the response may shift from anti-foreign to anti-self: immunization of mice to hen eggwhite lysozyme leads to an anti-self lysozyme response (Moudgil and Sercarz, 1993).

This concept of epitope spreading has profound significance to chemical-induced autoimmunity. Many metal ions such as Hg(II), Pt(II) and Au(I) form tightly bound complexes with proteins and such interactions could induce abnormal polypeptide unfolding and antigen processing leading to presentation of cryptic self sequences. In this way, an initial T cell response against chemically modified self could develop into an anti-self response (Griem and Gleichmann, 1995). Au(I) can be oxidized by activated macrophages to Au(III) that can induce Au(III)-specific T cells as well as T cells reactive against unmodified self peptides (Griem and Gleichmann, 1995). Such an evolution of immune specificity from anti-modified self to anti-unmodified self has been demonstrated in rodent models of autoimmunity induced not only by gold (III) salts (Griem *et al.*, 1996) but also by mercuric chloride (Kubicka-Muranyi *et al.*, 1996). Presumably other protein reactive autoimmunogenic chemicals may turn out to act via a similar mechanism.

(iii) Autoimmunogenic chemicals may react with MHC-associated self peptides to increase the binding of autoreactive T cells to self peptide/MHC (Martin and Weltzien, 1994). It is established that the surface 'groove' of MHC molecules is occupied by foreign or self peptides. Normally, T cells with high affinity receptors for self peptide + MHC are deleted during development in the thymus, but populations of T cells with low affinity receptors for self peptides are retained. These low affinity anti-self T cells are usually harmless but recent studies suggest that haptenation of peptides may increase their affinity for T cell receptors (Martin and Weltzien, 1994).

It has been known for many years that anti-hapten T cells can be generated *in vitro* by culture with haptenated autologous MHC class II positive cells (Shearer, 1974). It is now known that the chemical may haptenate directly the MHC-associated peptide. This is the case with the T cell sensitizing agents nickel (Romagnoli *et al.*, 1991), and trinitrophenyl (TNP) (Ortmann *et al.*, 1992). Design and synthesis of TNP haptenated peptides has produced molecules with high affinity for the mouse MHC class I molecule H-2K^b. On these peptides, the TNP hapten is not involved in MHC binding but is surface exposed and available for recognition by the T cell antigen receptor (Martin *et al.*, 1992). A detailed analysis revealed that cytotoxic T cell clones recognized two independent 'sub-epitopes' on haptenated octapeptides. One of these sub-epitopes was the 'natural' amino acid side chain in position 3 or 4, while the second epitope was the TNP hapten on the side chain of the amino acid in position 7. Interestingly a proportion of the anti-TNP clones recognized cells carrying the unhaptenated peptide, i.e. the natural sub-epitope only (Martin and Weltzien, 1994). It was these findings that led Weltzien and co-workers to suggest that in chemical-induced autoimmunity haptenation of the second site of an MHC-associated peptide may increase the affinity of its recognition by T cells. Thus T cells that have low affinity recognition of a self peptide, because they recognize only one sub-epitope, acquire high affinity for the modified peptide and are triggered to expand clonally. The expanded population of T cells retains low affinity reactivity for the unhaptenated self peptide and can therefore provide help for or kill (depending on whether they are CD4 or CD8 positive cells) MHC positive autologous cells carrying self peptide (Martin *et al.*, 1995). Furthermore, DNA analyses of T cell receptors from different cytotoxic T cell clones revealed similar sequences of the antigen binding regions suggesting restricted selection of T cell receptor variable elements. All of these clones responded to unhaptenated peptide suggesting that autoreactivity is a common feature of T cells induced by haptenated self peptide (Martin *et al.*, 1995).

IMPORTANCE OF PROTEIN REACTIVITY IN CHEMICAL MODIFICATION OF SELF

Most of the above theories to explain chemical-induced breakdown of tolerance require conjugation of the chemical to self protein or peptide. Many drugs and chemicals have this potential. Some are are inherently protein-reactive, such as the beta-lactam antibiotics including penicillins (Ahlstedt and Kristofferson, 1982; Levine and Redmond, 1967), the sulphydryl drugs D-penicillamine (Coleman *et al.*, 1988; Nagata *et al.*, 1986; O'Donnell and Coleman, 1992) and captopril (Coleman *et al.*, 1986), and sensitizing chemicals such as the dinitrohalobenzenes (Park *et al.*, 1987b). Immunogenic metals bind to amino acids such as cysteine (mercury and gold) or histidine (copper): although the association is non-covalent it is sufficiently tight for the metal to be seen by

T cells as a neo-epitope (Gleichmann *et al.*, 1984; Romagnoli *et al.*, 1991, 1992; Schuhmann *et al.*, 1990; Sinigaglia *et al.*, 1985). The majority of drugs and chemicals that induce autoimmunity, such as phenytoin, mianserin and sulphanilamide, are not spontaneously reactive with peptides but will conjugate only after oxidative metabolism to reactive species (Park and Kitteringham, 1990; Park *et al.*, 1987a; Uetrecht, 1989).

THE SECOND SIGNAL – DIRECT INDUCTION OF CYTOKINE PRODUCTION BY CHEMICALS

As discussed above, not only is some specific antigen-related stimulus required to activate T cells, but a second signal is required too. This signal, which is provided by cytokines, may result from further T/B cell contact, e.g. through CD28/B7, or, in the case of chemical-induced autoimmunity may result from direct induction of cytokine production, possibly bypassing or acting as an adjunct to the CD28/B7 signal. A number of recent studies suggest that autoimmunogenic chemicals do indeed have the potential to induce cytokine expression. For example, Oliveira *et al.* (1995) showed that mercury can induce interleukin-4 expression by peritoneal mast cells from the autoimmune susceptible Brown Norway rat strain. At about the same time, Prigent *et al.* (1995) reported that mercuric chloride induces interleukin-4 mRNA in splenic T cells from Brown Norway rats. Furthermore, MHC class II and CD23 (a B cell surface molecule involved in IgE synthesis) were induced *in vitro* and this effect was abrogated by anti-interleukin-4 antibody. The direct activation of cells by mercuric chloride may be explained by its bivalent thiol-reactive properties. Indeed, it has been shown that mercuric chloride can co-aggregate CD3, CD4 and CD45 on T cells, and this is associated with receptor aggregation and activation of the signal transducing protein kinase $p56^{lck}$ (Nakashima *et al.*, 1994). While it is unlikely that all autoimmunogenic compounds activate cells of the immune system in the same way as mercuric chloride, this chemical nevertheless provides a useful model with which to explore the mechanisms involved in chemical-induced autoimmunity.

The Brown Norway rat model – insights into Th2 cell mediated autoimmunity induced by chemicals or drugs

The administration of mercuric chloride, D-penicillamine or gold compounds to susceptible strains of rat or mouse induces an autoimmune syndrome that appears to be dominated by Th2 cell activity (Goldman *et al.*, 1991). It seems clear that this is an immunotoxicological response because administration of the same compounds to non-susceptible strains does not induce autoimmunity (although it does cause some very early tissue injury, which appears to be mediated by a different mechanism; see below). The best characterized syndrome, mercuric chloride induced autoimmunity in the Brown Norway rat, was first described by Druet and associates in 1977 (Sapin *et al.*, 1977).

MERCURIC CHLORIDE INDUCED AUTOIMMUNITY IN THE BROWN NORWAY RAT

The usual experimental protocol used to induce this syndrome consists of five subcutaneous injections of mercuric chloride, each of 1 mg/kg as a 0.1% solution, given over a period of 10 days. Inhalation of mercury vapour is also capable of producing the autoimmune response (Hua *et al.*, 1993). The resulting syndrome can be divided up into three stages: initiation, autoregulation, and resistance.

The initiation phase peaks approximately 2 weeks after the start of treatment with mercuric chloride. Serologically the response is characterized by a massive rise in the concentration of serum IgE, with an increase from baseline concentrations of less than 10 μg/ml to a peak of several mg/ml (Sapin *et al.*, 1984). Simultaneously, there is a polyclonal increase in serum IgG associated with the production of a wide variety of autoantibodies. Reactivity has been detected to glomerular basement membrane (Sapin *et al.*, 1977), DNA (Hirsch *et al.*, 1982), myeloperoxidase (Esnault *et al.*, 1992), cardiolipin (Marriott *et al.*, 1994), collagen and thyroglobulin (Pusey *et al.*, 1990). The non-specific and polyclonal nature of this response is reflected by the production of antibodies to exogenous antigens such as sheep red blood cells, and haptens such as trinitrophenol (Hirsch *et al.*, 1982). Simultaneously with this peak of serological activity, a variety of tissue manifestations is seen. There is marked lymphoproliferation, with splenomegaly and lymphadenopathy (Hirsch *et al.*, 1982). In the kidney, the anti-glomerular basement membrane antibodies deposit in a linear fashion along the glomerular basement membrane (Sapin *et al.*, 1977). This is associated with the onset of proteinuria, although there is rather little in the way of inflammatory activity within the glomerulus (Sapin *et al.*, 1977). The animals also develop a leucocytoclastic necrotizing vasculitis which affects principally the caecum, although changes are also noted in other tissues (Mathieson *et al.*, 1992). There is an inflammatory arthritis affecting the feet (Kiely *et al.*, 1995a), and inflammation at mucocutaneous junctions (Mathieson *et al.*, 1992).

Following the peak of activity, the syndrome autoregulates; the concentration of IgE and the various IgG autoantibodies (with the exception of antibodies against DNA, Pusey *et al.*, 1990) decrease (Prouvost-Danon *et al.*, 1981) and the tissue injury resolves, even if exposure to mercuric chloride is continued. Following this regulation, the animals are relatively resistant to a re-challenge with mercuric chloride; although a small serological response may be produced, this is of far lesser magnitude than the primary response (Mathieson *et al.*, 1991). In parallel, the animals are resistant to the re-induction of arthritis (Kiely *et al.*, 1995a), and, although this has not been studied formally, it is probable that this applies to the vasculitis as well. The initially linear deposits of immunoglobulin in the glomeruli reorganize to a more granular pattern during the resolution and resistant phases (Druet *et al.*, 1978) and the appearances are then analogous to those seen in human membranous nephropathy.

RELATED MODELS

In addition to mercuric chloride, a number of other agents produce a very similar but less severe syndrome in the Brown Norway rat. These include D-penicillamine (Donker *et al.*, 1991) and a variety of gold compounds including auranofin, gold chloride and sodium aurothiopropanolsulphonate (Tournade *et al.*, 1991). In mouse strains bearing the $H2^S$ haplotype, such as B10.S, administration of mercuric chloride or gold compounds produces a serological response characterised by elevation of IgE and autoantibodies, most notably antinucleolar antibodies (Pietsch *et al.*, 1989). The tempo of this syndrome is rather slower than that seen in the Brown Norway rat, with the titre of the autoantibodies rising over a period of a number of weeks. There has been no reported instance of tissue damage in these mice. D-Penicillamine and gold compounds can also induce autoantibodies in the susceptible strains of mice (Robinson *et al.*, 1986). A similar state of autoimmunity with the production of anti-glomerular basement membrane antibodies can also be induced in rabbits by treatment with mercuric chloride (Roman-Franco *et al.*, 1978).

In humans, it is of interest that the same substances, namely, mercury compounds (Oliveira *et al.*, 1987), D-penicillamine (Hall *et al.*, 1988) and gold compounds (Hall *et al.*, 1987), can also produce immunopathology in the form of a membranous nephropathy. Whether this involves Th2 activation in humans is unknown, but it is of interest that the predominant subclass of immunoglobulins deposited in membranous nephropathy is IgG4 (Doi *et al.*, 1984). This is unexpected because IgG4 is the rarest subclass found in the serum (French and Harrison, 1984); it may be significant that this subclass is strongly dependent on interleukin-4 (Lundgren *et al.*, 1989). There is some supportive evidence in man that gold compounds cause Th2 cell activation, as shown by increased concentrations of IgE (interleukin-4) and eosinophilia (interleukin-5) (Davis *et al.*, 1973).

STRAIN SPECIFICITY AND GENETICS

Work in the rat has shown that the development of mercuric chloride-induced autoimmunity is determined by two or three genes, one of which is linked to the MHC (Aten *et al.*, 1991). Resistant strains such as the Lewis do not develop any autoimmune manifestations following the administration of mercury, but instead develop a relative state of immunosuppression, which is also linked to the MHC (Aten *et al.*, 1991). In mice, the development of an autoimmune response to mercuric chloride is again linked to the MHC, with mice of the haplotype $H2^s$ being susceptible; other strains are either unresponsive to mercuric chloride or develop a state of immunosuppression (Mirtcheva *et al.*, 1989).

It has been observed recently that mercuric chloride (Gillespie *et al.*, 1995; Qasim *et al.*, 1995), D-penicillamine and the gold compound auranofin (latter two unpublished observations) can induce a phase of tissue injury much earlier than that described above, not only in Brown Norway rats, but also in a number of other rat strains, both inbred and outbred (unpublished observations). This injury takes the form of a leucocytoclastic necrotizing vasculitis which, as with the later phase of tissue injury, also affects the gut. However, it occurs within a few hours of the initiation of treatment with mercuric chloride and peaks after 2–3 days. Very different mechanisms appear to be involved in this early phase, as it is not strain specific and is also α/β T cell independent, which contrasts with the later manifestations (Kiely *et al.*, 1995b). There is some evidence to suggest that this early vasculitis is mediated, at least in part, by mast cell degranulation (unpublished observations); this may be an early event favouring the subsequent Th2 cell mediated response in Brown Norway rats (see below). A recent systematic approach to the identification of non-MHC genes contributing to the susceptibility of Brown Norway rats to chemically induced autoimmunity has identified a locus on chromosome 10 close to the cluster of genes coding for Th2 cytokines (Kermarrec *et al.*, 1996). The region containing this locus is still rather large and more precise characterization requires further study.

EVIDENCE FOR SELECTIVE T CELL SUBSET INVOLVEMENT

It has been known for some time that chemically induced autoimmunity in the Brown Norway rat is T cell dependent. T cell depleted animals are resistant to the syndrome (Pelletier *et al.*, 1987) and certain of the manifestations can be transferred with T cells (Pelletier *et al.*, 1988). *In vitro* analysis has shown the appearance of autoreactive T cells with specificity for MHC class II antigens (Rossert *et al.*, 1988).

There are a number of pieces of evidence showing specifically that Th2 cells are involved. The marked rise in IgE concentration clearly implies a role for interleukin-4 as isotype switching to IgE, at least in the mouse, is dependent on this cytokine (Finkelman *et al.*, 1988). As compared with the mouse (see below), less is known in the rat about the dependency of IgG subclasses on interleukin-4. Further evidence of Th2 cell involvement is provided by the exacerbation of vasculitis which is induced by the administration of the monoclonal antibody OX22 (Mathieson *et al.*, 1993). This antibody is directed against the CD45RC molecule and identifies a subset of CD4 cells which have Th1 like properties (Powrie and Mason, 1989). The exacerbation of tissue injury is interpreted as being due to increased Th2 activity following removal of Th1 control. The genetic evidence linking a possible susceptibility gene in the

Brown Norway rat to the cluster of Th2 cytokine genes is intriguing but clearly further analysis is required. Tissue analysis in the Brown Norway rat has shown upregulation of interleukin-4 within a few days of the initiation of treatment with mercuric chloride (Gillespie *et al.*, 1995). This is accompanied by a more modest up-regulation of mRNA for interferon-γ and (although these were only measured semi-quantitatively) mRNAs for interleukin-2, interleukin-6 and interleukin-10 (Gillespie *et al.*, 1996). Although the cellular source of the interleukin-4 has not been identified, this is clearly compatible with the early initiation of a Th2 type response. *In vitro* data have confirmed that exposure to mercuric chloride induces interleukin-4 mRNA in both T cells (Prigent *et al.*, 1995) and mast cells (Oliveira *et al.*, 1995) from the Brown Norway rat.

In the mouse, more direct evidence has been obtained for involvement of interleukin-4 by administration of neutralizing antibodies to this cytokine. This modulates the mercuric chloride-induced syndrome in B10.S mice in that the rise in IgE and IgG1 (both dependent on interleukin-4 in the mouse) is significantly decreased (Ochel *et al.*, 1991).

Weening and associates (van der Meide *et al.*, 1993) have shown that exposure of Brown Norway but not Lewis rats to mercuric chloride reduced numbers of concanavalin A-induced interferon-γ producing spleen cells *in vitro* (van der Meide *et al.*, 1993). Because the Th1 cell product interferon-γ down-regulates interleukin-4 production by Th2 cells (Krishnan and Mossman, Chapter 2 of this book) the Th2 cell-inducing effect of mercuric chloride could be via inhibition of interferon-γ production. Alternatively, the metal could target both subsets directly to induce interleukin-4 and inhibit interferon-γ.

In the Brown Norway rat, there is some evidence for recovery of Th1-type function during the regulation phase of mercuric chloride-induced autoimmunity (Castedo *et al.*, 1994). Whether this is simply an epiphenomenon or whether it plays a direct role in autoregulation is unclear. A CD8+ subset appears to be involved in the subsequent resistance to rechallenge (see below) rather than any CD4+ subset.

In strains of both rat (e.g. Lewis) and mouse (e.g. B10.D2) that are resistant to this type of chemically induced autoimmunity, administration of mercuric chloride instead appears to activate regulatory populations. The Lewis rat develops a state of relative immunosuppression, mediated by CD8+ T cells which are induced by $CD4^+$ cells (Castedo *et al.*, 1993). These CD4+ cells are autoreactive with specificity for self-class II, secrete interleukin-2, and, in contrast to the autoreactive cells induced in the Brown Norway rat, do not provide help for B cells.

In addition to the role of particular CD4+ T cells in chemically induced autoimmunity, a number of studies have examined CD8+ T cells. There are hints that different subsets of CD8+ cells, which do exist in the rat (Noble *et al.*, 1995), might be involved. Depletion of CD8+ cells facilitates the adoptive transfer of autoimmunity by CD4+ cells exposed to mercuric chloride (Pelletier *et al.*, 1988) and results in an increase in serum IgE concentration, both at baseline (Pelletier *et al.*, 1988) and in the early phases of the response to mercuric chloride (Kiely *et al.*, 1996). These observations suggest that there is a CD8+ population capable of playing a regulatory role. On the other hand, depletion of CD8+ cells leads to a significant amelioration of the inflammatory arthritis (but not the vasculitis) found at peak disease (Kiely *et al.*, 1996), suggesting that there is also a proinflammatory population of CD8+ cells. The CD8+ subset does not appear to play any part in the spontaneous autoregulation of the peak response (Mathieson *et al.*, 1991; Pelletier *et al.*, 1990) but does seem to be responsible, at least in part, for the subsequent resistance to rechallenge (Bowman *et al.*, 1984; Mathieson *et al.*, 1991).

MECHANISMS INVOLVED IN SELECTIVE T LYMPHOCYTE SUBPOPULATION ACTIVATION

The mechanisms involved in selective Th2 cell activation in the various models discussed in this section are unclear, but, as discussed earlier in this chapter, could involve some kind of selective pharmacological targeting of Th cell subsets or other interleukin-producing cell types. This is supported by the observation that mercuric chloride is able to induce interleukin-4 mRNA in both T cells and mast cells from Brown Norway rats. The pathways involved in the initial upregulation of interleukin-4 mRNA are unknown. Work with the mast cell has shown that mercuric chloride, D-penicillamine and auranofin can all induce the production of reactive oxygen species within the cell, and that modulation of intracellular redox potential will also modulate the subsequent sensitization for mediator release produced by these agents (Wolfreys and Oliveira, 1997). It is possible that intracellular redox balance via oxidation sensitive transcription factors such as NFκB (Meyer *et al.*, 1993) could directly influence transcription from the interleukin-4 gene in either T cells and/or mast cells, and thereby provide the initial source of interleukin-4 required for the subsequent Th2 cell response (Le Gros *et al.*, 1990; Swain *et al.*, 1990).

Van der Meide *et al.* (1993) have shown that injection of mercuric chloride into Brown Norway rats decreases concanavalin A-stimulated interferon-γ production from isolated spleen cells. Since the interferon-γ response is dependent on reduced glutathione in the culture medium, and because the bivalent mercuric ion binds to and inactivates thiol groups on glutathione, it is suggested that mercury inhibits interferon-γ production by depletion of cellular free glutathione (van der Meide *et al.*, 1993). Overall, it appears that mercury-induced changes in intracellular redox potential via reduction of glutathione may have consequences in terms of induction of IL-4 and inhibition of interferon-γ. The net effect of these changes would be to direct the immune response strongly towards activation of Th2 cells – whether the target for the metal was naive Th cells themselves or other cell types such as mast cells that might be able to influence Th cell differentiation.

A scheme is illustrated in Figure 14.4 that attempts to bring together many of the processes discussed in this chapter whereby mercury breaks tolerance to self and induces expansion of autoreactive Th2 cells. The initial breakdown of tolerance may be by one or more of several mechanisms: it could pharmacological; it could result from chemical-induced epitope spreading or it could be via haptenation of self MHC class II-associated peptides to increase their affinity for T cell antigen receptors. Alongside these processes, it is postulated that the promotion of a Th2 type response is by chemical regulation of cytokine gene expression (up-regulation of IL-4 or down-regulation of IFN-γ) either in uncommitted T (Th0) cells or in bystander cytokine producing cells such as mast cells. This gene regulation may be via depletion of cellular glutathione and/or induction of reactive oxygen species. Clearly, these latter processes are intimately linked but the precise physico-chemical and genetic processes in operation are yet to be worked out fully. This will undoubtedly be an area of active interest in the near future.

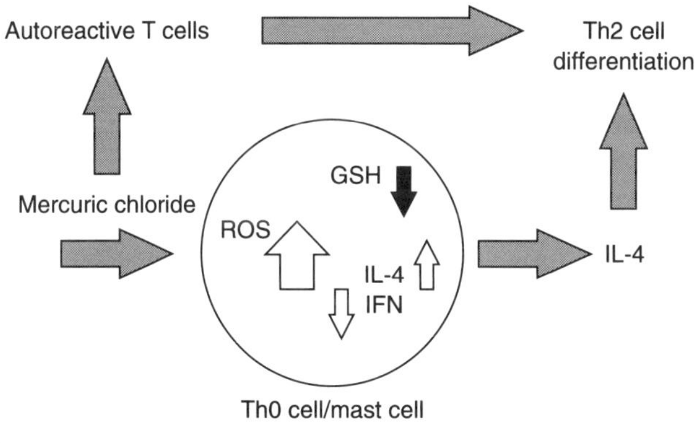

Figure 14.4 Scheme showing dual action of mercuric chloride to generate autoreactive Th2 cells. Initially T cell tolerance is broken by one of several mechanisms as discussed in the text. Alongside this, it is postulated that mercury promotes Th2 type responses by regulation of cytokine gene expression either in uncommitted T (Th0) cells or in bystander cytokine producing cells such as mast cells. This gene regulation may be via depletion of cellular reduced glutathione (GSH) and/or induction of reactive oxygen species (ROS). These intracellular physico-chemical changes lead to induction of interleukin-4 (IL-4) expression and inhibition of interferon-γ (IFN) expression, thereby leading to Th2 cell differentiation.

References

Ahlstedt, S. and Kristofferson, A. (1982) Immune mechanisms for induction of penicillin allergy. *Prog. Allergy*, **30**, 67–134.

Allison, A.C. (1989) Theories of self tolerance and autoimmunity, in *Autoimmunity and Toxicology* (eds M.E. Kammuller, N. Bloksma and W. Seinen), Elsevier, Amsterdam, pp. 67–115.

Allison, A.C., Denman, A.M. and Barnes, R.D. (1971) Co-operating and controlling functions of thymus-derived lymphocytes in relation to autoimmunity. *Lancet*, **ii**, 135–40.

Aten, J., Veninga, A., de Heer, E., Rozing, J., Nieuwenhuis, P., Hoedemaeker, P.J. and Weening, J.J. (1991) Susceptibility to the induction of either autoimmunity or immunosuppression by mercuric chloride is related to the major histocompatibility complex class II haplotype. *Eur. J. Immunol.*, **21**, 611–16.

Bowman, C., Mason, D.W., Pusey, C.D., Lockwood, C.M. (1984) Autoregulation of autoantibody synthesis in mercuric chloride nephritis in the Brown Norway rat. I. A role for T suppressor cells. *Eur. J. Immunol.*, **14**, 464–70.

Castedo, M., Pelletier, L., Rossert, J., Pasquier, R., Villarroya, H., and Druet, P. (1993) Mercury-induced autoreactive anti-class II T cell line protects from experimental autoimmune encephalomyelitis by the bias of CD8+ antiergotypic cells in Lewis rats. *J. Exp. Med.*, **177**, 881–9.

Castedo, M., Pelletier, L., Pasquier, R. and Druet, P. (1994) Improvement of Th1 functions during the regulation phase of mercury disease in Brown Norway rats. *Scand. J. Immunol.*, **39**,144–50.

Clayberger, C., DeKruyff, R.H. and Cantor, H. (1984) Immunoregulatory activities of autoreactive T cells: an Ia-specific T cell clone mediates both help and suppression of antibody responses. *J. Immunol.*, **132**, 2237–43.

Coleman, J.W., Yeung, J.H.K., Roberts, D.H., Breckenridge, A.M. and Park, B.K. (1986) Drug-specific antibodies in patients receiving captopril. *Br. J. Clin. Pharmacol.*, **22**,161–5.

Coleman, J.W., Foster, A.L., Yeung, J.H.K. and Park, B.K. (1988) Drug-protein conjugates XV. A study of the disposition of D-penicillamine in the rat and its relationship to immunogenicity. *Biochem. Pharmacol.*, **3**, 737–42.

Cornacchia, E., Golbus, J., Maybaum, J., Strahler, J., Hanash, S. and Richardson, B. (1988) Hydralazine and procainamide inhibit T cell DNA methylation and induce autoreactivity. *J. Immunol.*, **140**, 2197–200.

Davis, P., Ezeoke, A., Munro, J., Hobbs, J.R., and Hughes, G.R.V. (1973) Immunological studies on the mechanism of gold hypersensitivity. *Br. Med. J.*, **III**, 676–8.

Doi, T., Mayumi, M., Kanatsu, K., Suehiro, F. and Hamashima, Y. (1984) Distribution of IgG subclasses in membranous nephropathy. *Clin. Exp. Immunol.*, **58**, 57–62.

Donker, A.J., Venuto, R.C., Vladutiu, A.O., Brentjens, J.R. and Andres, G.A. (1991) Effects of prolonged administration of D-penicillamine or captopril in various strains of rats. Brown Norway rats treated with D-penicillamine develop autoantibodies, circulating immune complexes, and disseminated intravascular coagulation. *Clin. Immunol. Immunopathol.*, **30**, 142–55.

Druet, P., Druet, E., Potdevin, F. and Sapin, C. (1978) Immune type glomerulonephritis induced by HgCl2 in the Brown Norway rat. *Ann. Immunol.* (Paris), **129C**, 777–92.

Esnault, V.L.M., Mathieson, P.W., Thiru, S., Oliveira, D.B.G. and Lockwood, C.M. (1992) Autoantibodies to myeloperoxidase in Brown Norway rats treated with mercuric chloride. *Lab. Invest.*, **67**, 114–20.

Finkelman, F.D., Katona, I.M., Urban, J.F., Holmes, J., Ohara, J., Tung, A.S., Sample, J.V. and Paul, W.E. (1988) IL-4 is required to generate and sustain *in vivo* IgE responses. *J. Immunol.*, **141**, 2335–41.

French, M.A. and Harrison, G. (1984) Serum IgG subclass concentrations in healthy adults: a study using monoclonal antisera. *Clin. Exp. Immunol.*, **56**, 473–5.

Gillespie, K., Saoudi, A., Kuhn, J., Whittle, C.J., Druet, P., Bellon, B. and Mathieson, P.W. (1996) Th1/Th2 cytokine gene expression after mercuric chloride in susceptible and resistant rat strains. *Eur. J. Immunol.*, **26**, 2388–92.

Gillespie, K.M., Qasim, F.J., Tibbatts, L.M., Thiru, S., Oliveira, D.B.G. and Mathieson, P.W. (1995) Interleukin-4 gene expression in mercury-induced autoimmunity. *Scand. J. Immunol.*, **41**, 268–72.

Gleichmann, E., Pals, S.T., Rolink, A.G., Radaszkiewicz, T. and Gleichmann, H. (1984) Graft-versus-host reactions: clues to the etiopathology of a spectrum of immunological diseases. *Immunol. Today*, **5**, 324–32.

Goldman, M., Druet, P. and Gleichmann, E. (1991) Th2 cells in systemic autoimmunity: insights from allogeneic diseases and chemically-induced autoimmunity. *Immunol. Today*, **12**, 223–7.

Griem, P. and Gleichmann, E. (1995) Metal ion induced autoimmunity. *Curr. Opin. Immunol.*, **7**, 831–8.

Griem, P., Panthel, K., Kalbacher, H. and Gleichmann, E. (1996) Alteration of model antigen by Au(III) leads to T cell sensitization to cryptic peptides. *Eur. J. Immunol.*, **26**, 279–87.

Guery, J.C., Tournade, H., Pelletier, L., Druet, E. and Druet, P. (1990) Rat anti-glomerular basement membrane antibodies in toxin-induced autoimmunity and in chronic GVH reactions share recurrent idiotypes. *Eur. J. Immunol.*, **20**, 101–5.

Hall, C.L., Fothergill, N.J., Blackwell, M.M., Harrison, P.R., MacKenzie, J.C. and MacIver, A.G. (1987) The natural course of gold nephropathy: long term study of 21 patients. *Br. Med. J.*, **295**, 745–8.

Hall, C.L., Jawad, S., Harrison, P.R., MacKenzie, J.C., Bacon, P.A., Klouda, P.T. and MacIver, A.G. 1988) Natural course of penicillamine nephropathy: a long term study of 33 patients. *Br. Med. J.*, **296**, 1083–6.

Hirsch, F., Couderc, J., Sapin, C., Fournie, G. and Druet, P. (1982) Polyclonal effect of $HgCl_2$ in the rat, its possible role in an experimental autoimmune disease. *Eur. J. Immunol.*, **12**, 620–5.

Hua, J., Pelletier, L., Berlin, M. and Druet, P. (1993) Autoimmune glomerulonephritis induced by mercury vapour exposure in the Brown Norway rat. *Toxicology*, **79**, 119–29.

Joshua, D., Liu, Y., Williams, G., Smith, C., Gordon, J. and Maclennan, I. (1989) Mechanism of antigen-driven selection in germinal centres. *Nature*, **342**, 929–31.

Kappler, J.W., Roehm, N. and Marrack, P. (1987) T-cell tolerance by clonal elimination in the thymus. *Cell*, **9**, 273–80.

Kenna, G.J. (1991) The molecular basis of halothane-induced hepatitis. *Biochem. Soc. Transact.*, **19**, 191–5.

Kermarrec, N., Dubay, C., De Gouyon, B., Blanpied, C., Gauguier, D., Gillespie, K., Mathieson, P.W., Druet, P., Lathrop, M. and Hirsch, F. (1996) Serum IgE concentration and other immune manifestations of treatment with gold salts are linked to the MHC and IL-4 regions in the rat. *Genomics*, **31**, 111–14.

Kiely, P.D.W., Thiru, S. and Oliveira, D.B.G. (1995a) Inflammatory polyarthritis induced by mercuric chloride in the Brown Norway rat. *Lab. Invest.*, **73**, 284–93.

Kiely, P.W., Wolfreys, K.J. and Oliveira, D.B.G. (1995b) T-cell-dependent and -independent vasculitis. *Nephrol. Dial. Transplant.*, **10**, 721(Abstract).

Kiely, P.D.W., O'Brien, D. and Oliveira, D.B.G. (1996) Anti-CD8 treatment reduces the severity of inflammatory arthritis, but not vasculitis, in mercuric chloride-induced autoimmunity. *Clin. Exp. Immunol.*, **106**, 280–5.

Kitteringham, N.R., Lambert, C., Maggs, J.L., Colbert, J., and Park, B.K. (1988) A comparative study of the formation of chemically reactive drug metabolites by human liver microsomes. *Br. J. Clin. Pharmacol.*, **826**, 13–21.

Kubicka-Muranyi, M., Kremer, J., Rottman, N., Lubben, B., Albers, R., Bloksma, N., Luhrman, R. and Gleichmann, E. (1996) Murine systemic autoimmune disease induced by mercuric chloride – T helper cells reacting to self proteins. *Int. Arch. Allergy Immunol.*, **109**, 11–20.

Lanzavecchia, T. (1985) Antigen-specific interaction between T and B cells. *Nature*, **314**, 537–9.

Le Gros, G., Ben-Sasson, S.Z., Seder, P., Finkelman, F.D. and Paul, W.E. (1990) Generation of interleukin 4 (IL-4)-producing cells *in vivo* and *in vitro*: IL-2 and IL-4 are required for *in vitro* generation of IL-4-producing cells. *J. Exp. Med.*, **172**, 921–30.

Lehmann, P.V., Forsthuber, T., Miller, A. and Sercarz, E.E. (1992) Spreading of T cell autoimmunity to cryptic determinants of an autoantigen. *Nature*, **358**, 155–7.

Lehmann, P.V., Sercarz, E.E., Forsthuber, T., Dayan, C.M. and Gammon, G. (1993) Determinant spreading and the dynamics of the autoimmune T-cell repertoire. *Immunol. Today*, **14**, 203–8.

Lerner, W., Carusa, R., Faig, D. and Kaspatkin, S. (1985) Drug-dependent and non-drug-dependent antibodies in drug-induced thrombocytopenia purpura. *Blood*, **66**, 306–11.

Levine, B. and Redmond, A. (1967) Immunochemical mechanisms of penicillin-induced Coombs positivity and haemolytic anaemia in man. *Int. Arch. Allergy Appl. Immunol.*, **31**, 594–606.

Lundgren, M., Persson, U., Larsson, P., Magnusson, C., Edvard Smith, C.I., Hammarström, L. and Severinson, E. (1989) Interleukin 4 induces synthesis of IgE and IgG4 in human B cells. *Eur. J. Immunol.*, **19**, 1311–15.

Marriott, J.B., Qasim, F. and Oliveira, D.B.G. (1994) Anti-phospholipid antibodies in the mercuric chloride treated Brown Norway rat. *J. Autoimmun.*, **7**, 457–67.

Martin, S. and Weltzien, H.U. (1994) T cell recognition of haptens, a molecular view. *Int. Arch. Allergy Immunol.*, **104**, 10–16.

Martin, S., Ortmann, B., Pflugfleder, U., Birsber, U. and Weltzien, H.U. (1992) Role of hapten-anchoring peptides in defining hapten epitopes for MHC-restricted cytotoxic T cells. Cross-reactive TNP determinants on different peptides. *J. Immunol.*, **149**, 2569–75.

Martin, S., Ruh, H., Hebbelmann, S., Pflugfelder, U., Rude, B. and Weltzien, H.U. (1995) Carrier-reactive hapten-specific cytotoxic T lymphocyte clones originate from a highly preselected T cell repertoire: implications for chemical-induced self reactivity. *Eur. J. Immunol.*, **25**, 2788–96.

Mathieson, P.W., Stapleton, K.J., Oliveira, D.B.G. and Lockwood, C.M. (1991) Immunoregulation of mercuric chloride-induced autoimmunity in Brown Norway rats: a role for CD8+ T cells revealed by *in vivo* depletion studies. *Eur. J. Immunol.*, **21**, 2105–9.

Mathieson, P.W., Thiru, S., and Oliveira, D.B.G. (1992) Mercuric chloride-treated Brown Norway rats develop widespread tissue injury including necrotizing vasculitis. *Lab. Invest.*, **67**, 121–9.

Mathieson, P.W., Thiru, S. and Oliveira, D.B.G. (1993) Regulatory role of OX22+ T cells in mercury-induced autoimmunity in the Brown Norway rat. *J. Exp. Med.*, **177**, 1309–16.

Meyer, M., Schreck, R. and Baeuerle, P.A. (1993) H_2O_2 and antioxidants have opposite effects on activation of NF-kB and AP-1 in intact cells: AP-1 as secondary antioxidant-responsive factor. *EMBO J.*, **12**, 2005–15.

Mirtcheva, J., Pfeiffer, C., Bruijn, J.A., Jaquesmart, F. and Gleichmann, E. (1989) Immunological alterations inducible by mercury compounds III. H-2A acts as an immune response and H-2E as an immune 'suppression' locus for HgCl2-induced antinucleolar autoantibodies. *Eur. J. Immunol.*, **19**, 2257–61.

Mitchison, N.A. (1971) The carrier effects in the secondary response to hapten-protein conjugates II. Cellular cooperation. *Eur. J. Immunol.*, **1**, 18–27.

Moudgil, K.D. and Sercarz, E.E. (1993) Dominant determinants in hen eggwhite lysozyme correspond to the cryptic determinants within its self-homologue, mouse lysozyme: Implications in shaping of the T cell repertoire and autoimmunity. *J. Exp. Med.*, **178**, 2131–8.

Moudgil, K.D. and Sercarz, E.E. (1994) The T cell repertoire against cryptic self determinants and its involvement in autoimmunity and cancer. *Clin. Immunol. Immunopathol.*, **73**, 283–9.

Nakashima, I., Pu, M.-Y., Nishizaki, A., Rosila, I., Ma, L., Katano, Y., Ohkusu, K., Rahman, S.M.J., Isobe, K.-I., Hamaguchi, M. and Saga, K. (1994) Redox mechanism as alternative to ligand binding for receptor activation delivering disregulated cellular signals. *J. Immunol.*, **152**, 1064–71.

Nagata, N., Hurtenbach, U. and Gleichmann, E. (1986) Specific sensitization of Lyt-$1^{+}2_{-}$ T cells to spleen cells modified by the drug D-penicillamine or a stereoisomer. *J. Immunol.*, **136**, 136–42.

Noble, A., Macary, P.A. and Kemeny, D.M. (1995) IFN-gamma and IL-4 regulate the growth and differentiation of CD8+ T cells into subpopulations with distinct cytokine profiles. *J. Immunol.*, **155**, 2928–37.

Ochel, M., Vohr, H.W., Pfeiffer, C. and Gleichmann, E. (1991) IL4 is required for the IgE and IgG1 increase and IgG1 autoantibody formation in mice treated with mercuric chloride. *J. Immunol.*, **146**, 3006–11.

O'Donnell, C.A. and Coleman, J.W. (1992) A T-cell response to the anti-arthritic drug penicillamine in the mouse: requirements for generation of the drug-derived antigen. *Immunology*, **76**, 604–9.

Oliveira, D.B.G., Foster, G., Savill, J., Syme, P.D. and Taylor, A. (1987) Membranous nephropathy caused by mercury-containing skin lightening cream. *Postgrad. Med. J.*, **63**, 303–4.

Oliveira, D.B.G., Gillespie, K.M., Wolfreys, K., Mathieson, P.W., Qasim, F. and Coleman, J.W. (1995) Compounds that induce autoimmunity in the Brown Norway rat sensitize mast cells for mediator release and IL-4 expression. *Eur. J. Immunol.*, **25**, 2259–64.

Ortmann, B., Martin, S., von Bonin, A., Schiltz, E., Hoschutzky, H. and Weltzien, H.U. (1992) Synthetic peptides anchor T cell-specific TNP epitopes to MHC antigens. *J. Immunol.*, **148**, 1445–50.

Park, B.K. and Kitteringham, N.R. (1990) Drug-protein conjugation and its immunological consequences. *Drug Metab. Rev.*, **22**, 87–144.

Park, B.K., Coleman, J.W. and Kitteringham, N.R. (1987a) Drug disposition and drug hypersensitivity. *Biochem. Pharmacol.*, **36**, 581–90.

Park, B.K., Tingle, M.D., Grabowski, P.S., Coleman, J.W. and Kitteringham, N.R. (1987b) Drug-protein conjugates XI. Disposition and immunogenicity of dinitro fluorobenzene, a model compound for the investigation of drugs as haptens. *Biochem. Pharmacol.*, **36**, 591–9.

Pelletier, L., Pasquier, R., Hirsch, F., Sapin, C. and Druet, P. (1986) Autoreactive T cells in mercury-induced autoimmune disease. *J. Immunol.*, **137**, 2548–54.

Pelletier, L., Pasquier, R., Vial, M., Mandet, C., Moutier, R., Salomon, J.C. and Druet, P. (1987) Mercury-induced autoimmune glomerulonephritis: requirement for T cells. *Nephrol. Dial. Transplant.*, **1**, 211–18.

Pelletier, L., Pasquier, R., Rossert, J., Vial, M., Mandet, C. and Druet, P. (1988) Autoreactive T cells in mercury-induced autoimmunity: ability to induce the autoimmune disease. *J. Immunol.*, **140**, 750–4.

Pelletier, L., Rossert, J., Pasquier, R., Vial, M. and Druet, P. (1990) Role of $CD8^+$ cells in mercury-induced autoimmunity or immunosuppression in the rat. *Scand. J. Immunol.*, **31**, 65–74.

Pelletier, L., Bellon, B., Tournade, H., Dudey, C., Guery, C., Saoudi, A., Hirsch, F. and Druet, P. (1991) Chemical-induced autoimmunity. *Immunology Series*, **55**, 315–53.

Pelletier, L., Castedo, M., Bellon, B. and Druet, P. (1994) Mercury and autoimmunity, in *Immunotoxicology and Autoimmunity* (eds J.H. Dean, M.I. Luster, A.E. Munson and I. Kimber), Raven Press, New York, pp. 539–52.

Petz, L.D. and Branch, D.R. (1973) Drug-induced hemolytic anaemia. *Meth. Haematol.*, **12**, 47–94.

Pfueller, S.L., Bilston, R.A., Logan, D., Gibson, J.M. and Firkin, B.G. (1988) Quinine and quinidine induced thrombocytopenia. *Blood*, **72**, 1155–62.

Pietsch, P., Vohr, H.W., Degitz, K. and Gleichmann, E. (1989) Immunological alterations inducible by mercury compounds II. Mercuric chloride and gold sodium thiomalate enhance serum IgE and IgG in susceptible mouse strains. *Int. Arch. Allergy Appl. Immunol.*, **90**, 47–53.

Powrie, F., and Mason, D. (1989) The MRC $OX\text{-}22^-$ $CD4^+$ T cells that help B cells in secondary immune responses derive from naive precursors with the MRC $OX\text{-}22^+$ $CD4^+$ phenotype. *J. Exp. Med.*, **169**, 653–62.

Prigent, P., Saoudi, A., Pannetier, C., Graber, P., Bonnefoy, J., Druet, P. and Hirsch, F. (1995) Mercuric chloride, a chemical responsible for Th cell (Th) 2-mediated autoimmunity in Brown Norway rats, directly triggers T cells to produce interleukin-4. *J. Clin. Invest.*, **96**, 1484–9.

Prouvost-Danon, A., Abadie, A., Sapin, C., Bazin, H. and Druet, P. (1981) Induction of IgE synthesis and potentiation of anti-ovalbumin IgE antibody response by HgCl2 in the rat. *J. Immunol.*, **126**, 699–702.

Pusey, C.D., Bowman, C., Morgan, A., Weetman, A.P., Hartley, B. and Lockwood, C.M. (1990) Kinetics and pathogenicity of autoantibodies induced by mercuric chloride in the Brown Norway rat. *Clin. Exp. Immunol.*, **81**, 76–82.

Qasim, F.J., Mathieson, P.W., Thiru, S. and Oliveira, D.B.G. (1995) Time course and characterisation of mercuric chloride induced autoimmunity in the Brown Norway rat. *J. Autoimmun.* **8**, 195–208.

Quintans, J., Suzuki, H., Sosman, J.A. and Shah, P.D. (1986) Immunoregulation by T cells. I. Characterisation of the Iek-specific Lbd self reactive T cell clone that helps, suppresses and contrasuppresses B cell responses. *J. Immunol.*, **136**, 1974–81.

Richardson, B. (1986) Effect of an inhibitor of DNA methylation on T cells. II. 5-azacytidine induces self reactivity in antigen-specific T4+ cells. *Hum. Immunol.*, **17**, 456–70.

Robinson, C.J.G., Balazs, T. and Egorov, I.K. (1986) Mercuric chloride-, gold sodium thiomalate- and D-penicillamine-induced antinuclear antibodies in mice. *Toxicol. Appl. Pharmacol.*, **86**, 159–69.

Rolink, A.G. and Gleichmann, E. (1983) Allosuppressor and allohelper T cells in acute and chronic graft v host disease *J. Exp. Med.*, **158**, 546–58.

Rolink, A.G., Pals, S.T. and Gleichmann, E. (1983) Allosuppressor and allohelper T cells in acute and chronic GVH disease. II. F1 recipients carrying mutations at H-2K and/or IA. *J. Exp. Med.*, **157**, 755–71.

Roman-Franco, A.A., Turiello, M., Albini, B., Ossi, E., Milgrom, F. and Andres, G.A. (1978) Anti-basement membrane antibodies and antigen-antibody complexes in rabbits injected with mercuric chloride. *Clin. Immunol. Immunopathol.*, **9**, 464–81.

Romagnoli, P., Labhardt, A.M. and Sinigaglia, F. (1991) Selective interaction of Ni with an MHC-bound peptide. *EMBO J.*, **10**, 1303–6.

Romagnoli, P., Spinas, G. and Sinigaglia, F. (1992) Gold-specific T cells in rheumatoid arthritis patients treated with gold. *J. Clin. Invest.*, **89**, 254–8.

Rossert, J., Pelletier, L., Pasquier, R. and Druet, P. (1988) Autoreactive T cells in mercury-induced autoimmunity. Demonstration by limiting dilution analysis. *Eur. J. Immunol.*, **18**, 1761–6.

Rubin, R.L. (1994) Autoimmune reactions induced by procainamide and hydralazine, in: *Autoimmunity and Toxicology. Immune Disregulation Caused by Drugs and Chemicals* (eds M.E. Kammuller, N. Bloksma and W. Seinen), Elsevier, Amsterdam, pp. 119–50.

Sapin, C., Druet, E. and Druet, P. (1977) Induction of anti-glomerular basement membrane antibodies in the Brown-Norway rat by mercuric chloride. *Clin. Exp. Immunol.*, **28**, 173–9.

Sapin, C., Hirsch, F., Delaporte, J., Bazin, H., and Druet, P. (1984) Polyclonal IgE increase after $HgCl_2$ injections in BN and LEW rats: a genetic analysis. *Immunogenetics*, **20**, 227–36.

Schuhmann, D., Kubicka-Muranyi, M., Mirtschewa, J., Gunther, J., Kind, P. and Gleichmann, E. (1990) Adverse immune reactions to gold. I. Chronic treatment with a Au(I) drug sensitizes mouse T cells not to Au(I) but to Au(III) and induce autoantibody formation. *J. Immunol.*, **145**, 2132–9.

Shearer, G.M. (1974) Cell-mediated cytotoxicity to trinitrophenyl modified syngeneic lymphocytes. *Eur. J. Immunol.*, **4**, 527–33.

Sinigaglia, F., Scheidegger, D., Garotta, G., Sheper, R., Pletscher, M. and Lanzevecchia, A. (1985) Isolation and characterisation of Ni-specific T cell clones from patients with Ni-contact dermatitis. *J. Immunol.*, **135**, 3929–32.

Swain, S.L., Weinberg, A.D., English, M. and Huston, G. (1990) IL-4 directs the development of Th2-like helper effectors. *J. Immunol.*, **145**, 3796–806.

Tournade, H., Pelletier, L., Pasquier, R., Vial, M.-C., Mandet, C. and Druet, P. (1990) D-Penicillamine-induced autoimmunity in Brown-Norway rats. Similarities with $HgCl_2$-induced autoimmunity. *J. Immunol.*, **144**, 2985–91.

Tournade, H., Guery, J., Pasquier, R., Vial, M., Mandet, C., Druet, E., Dansette, P.M., Druet P. and Pelletier, L. (1991) Effect of the thiol group on experimental gold-induced autoimmunity. *Arthritis Rheum.*, **34**, 1594–9.

Uetrecht, J.P. (1989) Idiosyncratic drug reactions: possible role of reactive metabolites generated by leukocytes. *Pharm. Res.*, **6**, 265–73.

van der Meide, P.H., de Labie, M.C.D.C., Botman, C.A.D., van Bennekon, W.P., Olsson, T., Aten, J. and Weening, J.J. (1993) Mercuric chloride down-regulates T cell interferon-γ production in Brown Norway but not in Lewis rats; role of glutathione. *Eur. J. Immunol.*, **23**, 675–81.

van Rappard-van der Veen, F.M., Rolink, A.G. and Gleichmann, E. (1982) Diseases caused by reactions of T lymphocytes towards incompatible structures of the MHC. VI. Autoantibodies characteristic of SLE induced by abnormal T-B cell cooperation across I-E. *J. Exp. Med.* **155**, 1555–60.

Vistentin, G.P., Newman, P.J. and Aster, R.H. (1991) Characteristics of quinine- and quinidine-induced antibodies specific for platelet glycoproteins IIb and IIIa. *Blood* **77**, 2668–75.

Weigle, W.O. (1965) The production of thyroiditis and antibody following injection of thyroglobulin into rabbits previously stimulated with altered thyroglobulin. *J. Exp. Med.*, **122**, 1049–62.

Wolfreys, K. and Oliveira, D.B.G. (1997) Alterations in intracellular reactive oxygen species generation and redox potential modulate mast cell function. *Eur. J. Immunol.*, Abstract in press.

Worlledge, S.M. (1973) Immune drug-induced hemolytic anemias. *Semin. Haematol.*, **10**, 327–44.

Yamashita, U., Takami, T., Hamasaka, T. and Kitigawa, M. (1976) The role of hapten-reactive T-lymphocytes in the induction of autoimmunity in mice. II. Termination of self tolerance to erythrocytes by immunization with hapten-isologous erythrocytes. *Cell. Immunol.*, **25**, 32–40.

Index